TECHNETIUM IN THE ENVIRONMENT

Proceedings of a Seminar on the Behaviour of Technetium in the Environment, organised by the Commission of the European Communities, Radiation Protection Programme, and the Service d'Etudes et de Recherches sur l'Environnement of the Commissariat à l'Energie Atomique, France, in collaboration with the Office of Health and Environmental Research of the Department of Energy, USA, and held at Cadarache (France), 23–26 October 1984.

PROGRAMME COMMITTEE

A. Grauby, *France*
H. Smith, *UK*
R. Kirchmann, *Belgium*
P. Scoppa, *Commission of the European Communities*
C. Myttenaere, *Commission of the European Communities*
A. Cigna, *Italy*
E. Holm, *Sweden*
O. Hoffman, *USA*

TECHNETIUM IN THE ENVIRONMENT

Edited by

G. DESMET and C. MYTTENAERE

Commission of the European Communities, Radiation Protection Programme

ELSEVIER APPLIED SCIENCE PUBLISHERS
LONDON and NEW YORK

ELSEVIER APPLIED SCIENCE PUBLISHERS LTD
Crown House, Linton Road, Barking, Essex IG11 8JU, England

Sole Distributor in the USA and Canada
ELSEVIER SCIENCE PUBLISHING CO., INC.
52 Vanderbilt Avenue, New York, NY 10017, USA

WITH 102 TABLES AND 114 ILLUSTRATIONS

British Library Cataloguing in Publication Data

Technetium in the environment.
1. Technetium—Environmental aspects
I. Desmet, G. II. Myttenaere, C.
363.7'384 TD196.T4

Library of Congress Cataloging-in-Publication Data

Seminar on the Behaviour of Technetium in the Environment (1984: Cadarache, France)
Technetium in the environment.

"Proceedings of a Seminar on the Behaviour of Technetium in the Environment, organised by the Commission of the European Communities, Radiation Protection Programme, and the Service d'études et de recherches sur l'environnement of the Commissariat à l'énergie atomique, France, in collaboration with the Office of Health and Environmental Research of the Department of Energy, USA, and held at Cadarache (France), 23–26 October 1984"—P. ii.
Bibliography: p.
Includes index.
1. Technetium—Environmental aspects—Congresses.
I. Desmet, G. II. Myttenaere, C. III. Commission of the European Communities. Radiation Protection Programme. IV. France. Service d'études et de recherches sur l'environnement. V. United States. Dept. of Energy. Office of Health and Environmental Research. VI. Title.
QH545.T37S46 1984 574.5'222 85-29221

ISBN-13:978-94-010-8361-4 e-ISBN-13:978-94-009-4189-2
DOI: 10.1007/978-94-009-4189-2

Publication arrangements by: Commission of the European Communities, Directorate-General for the Information Market and Innovation, Luxembourg
EUR 10102 EN

PREFACE

Since the discovery of technetium (in 1937, by Perrier and Segre), technetium-99 and technetium-99m have become most familiar in the specialised literature, the behaviour of the former being of concern regarding its health impact, the latter having been used intensively in medical tests for many years. The knowledge of the characteristics and behaviour of this element in nature is currently increasing exponentially. It has become clear, however, that there is still a serious shortage of information needed to assess properly its burden and impact on man and his environment.

Technetium may indeed be released from nuclear installations or as a waste product from medical applications. It is generally considered to be highly mobile as the anion TcO_4^-, but it is also suspected to be readily reduced in soils and metabolised in living beings into lower oxidation states.

The seminar was planned to analyse thoroughly the experimental results obtained up to now and to make recommendations of specific issues requiring further research. This book presents the papers of the different sessions in the order of presentation, and the conclusions drawn by three discussion groups which dealt separately with:

— problems of sources and distribution of man-made technetium and its radiological consequences;
— the chemical behaviour of technetium in different ecosystems;
— biochemical reactions and binding with macromolecules in various life systems.

The editors, together with the Programme Committee of the Seminar, wish to express their gratitude to the chairmen of the different sessions for their summary of the group discussions.

The publication of these proceedings enables the organisers of this seminar to draw attention to this specific element in the table of Mendeleev, with its multifarious and intensive utilisation in medical research, but with its paradoxically relatively unknown basic behaviour in the environment.

THE EDITORS

CONTENTS

1

Source Terms for Technetium-99 from Nuclear Fuel Cycle Facilities

JOHN E. TILL

Radiological Assessments Corporation, Rt 2 Box 122, Neeses, South Carolina 29107, USA

1. INTRODUCTION

1.1 Basis for technetium-99 source term calculations

In the United States, releases of key radionuclides from nuclear fuel cycle facilities are reported in Table S-3 of 10 CFR Part 51 [1]. This table must be included in each environmental impact report when applying for a construction permit or an operating licence and forms an authoritative basis for evaluating the environmental impact of the light-water reactor fuel cycle. As a result, most of the environmental impacts of the uranium fuel cycle have not been subject to litigation in individual reactor licensing proceedings.

The values reported in Table S-3 are based on a 'normalised' light-water reactor, defined as a 1000-MW(e) core which is assumed to operate with an annual capacity factor of 0·8, thus producing 800 MW years of electricity. The front end of the cycle covers the annual supply of fuel for the model reactor and is dictated by the level to which the fuel is burned (normally expressed as megawatt days thermal power per metric ton, MWd(t)/MT (or MWd/MT)), removed from the core, and replaced by fresh fuel. The back end of the fuel cycle includes steps associated with recovery of the uranium and plutonium, processing, and disposal of waste; likewise it is affected by the level to which the fuel is burned. For Table S-3, it is assumed that the fuel is burned to 33 000 MWd/MT.

Releases of technetium-99 (^{99}Tc) to the environment from the nuclear fuel cycle are not included in Table S-3. This omission exists because the Fuel Cycle Hearing Board originally concluded that the conservative assumption of complete release of ^{129}I tended to compensate for no emission of ^{99}Tc. It has since been decided that releases of ^{99}Tc, together with an appraisal of its environmental impacts, could be considered in individual licensing proceedings.

The purpose of this paper is to report on research to develop source terms for ^{99}Tc released to the environment by the nuclear fuel cycle. The basic assumptions used in 10 CFR Part 51 to normalise the throughput of various stages of the fuel cycle will remain in effect. Data reported in this paper summarise the results of a more comprehensive study by Till *et al.* [2] on technetium production and behaviour in the environment.

1.2 Physical properties of technetium

Sixteen isotopes of technetium are known with mass numbers from 92 to 107 along with six isomers. None of these are stable. Among the shorter-lived isotopes, those of greatest practical use are ^{95m}Tc (61 days), ^{97m}Tc (90·5 days), and ^{99m}Tc (6 hours). All are obtained through nuclear interactions with molybdenum. Long-lived isotopes include ^{97}Tc ($2{\cdot}6 \times 10^6$ years), ^{98}Tc ($1{\cdot}5 \times 10^6$ years), and ^{99}Tc ($2{\cdot}1 \times 10^5$ years).

Technetium-99, the only environmentally significant technetium isotope produced in fission reactors, decays by beta emission with a maximum energy of 0·292 MeV and a specific activity of $1{\cdot}7 \times 10^{-2}$ Ci/g. The maximum yield from thermal neutron fission of ^{235}U is reported to be 6·06%, making ^{99}Tc relatively high in abundance among fission products. Several mechanisms exist for production of ^{99}Tc aside from direct fission, including (n, γ) reaction with ^{98}Mo as shown below:

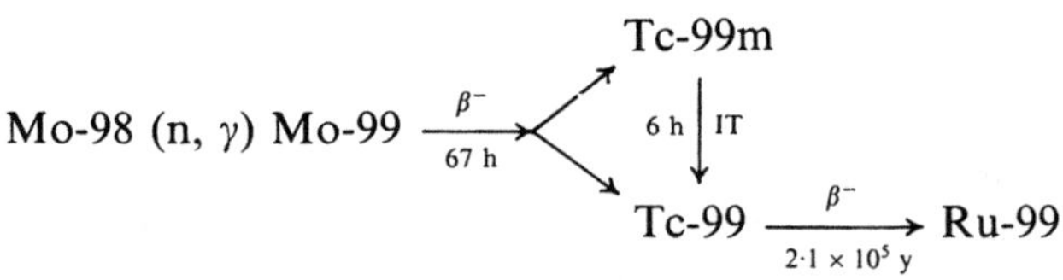

In addition, small amounts are produced through the fission of heavy metals other than ^{235}U.

1.3 Chemical properties of technetium

Several reviews on technetium chemistry are available. The environmental chemistry is reviewed in articles by McFadden [3], Paquette *et al.* [4], and Turcotte [5].

The most stable forms of technetium in aqueous solutions are: pertechnetate ion, TcO_4^-, and the very insoluble hydrated oxide, TcO_2. In water systems such as those that would contain living organisms in a pH range of 3–10, the reported valences of technetium are confined to VII, IV, V, and III. Much of the work concerning technetium movement in the environment can be categorised by the two rough classes: TcO_4^- as an aqueous ion is highly mobile and $TcO_2{\cdot}H_2O$ is highly insoluble and thus immobile in aqueous systems.

The pertechnetate ion (TcO_4^-) is stable in water over a wide pH range in the absence of reducing substances. Strong reducing agents such as borohydride or stannous ions, hydrazine, ascorbic acid, and zinc are used to ensure reduction of the TcO_4^-. Conversely, strong oxidants like bromine water, ceric ion, alkaline peroxide, or acid persulphate, are employed to obtain rapid oxidation of the TcO_2 to the VII valence state.

The presence of substances that may induce other reactions such as the formation of insoluble species or complex ions cannot be underestimated, but little is known of any of the environmental species except those of an insoluble IV hydrous oxide or TcO_4^-. However, it is known that TcO_4^- reduced in acid will produce the technetium (IV) which disappears from the solution as a precipitate above a pH of three or four [6].

The halides of technetium (IV) form complexing ions of the form TcX_6^{-2}. Thus the stability of $TcCl_6^{-2}$ is demonstrated by the reduction of TcO_4^- in greater than 5 M HCl solutions [7]. Complexing substances found in the environment that have given evidence of forming lower valence complexes with technetium include carbonates and phosphates in addition to halides among inorganic substances [4]. Organic compounds including sugars, ions of polybasic acids and polyhydric alcohols such as citrates, succinates, maleates, tartrates, and mannitol also have been reported to form technetium (IV) complexes [4]. Numerous references in the environmental literature suggest reactions with sulphur-containing materials, especially those of organic origin. Cysteine and some of its derivatives form soluble complexes of the IV and V state that have been characterised [8]. Reduction of TcO_4^- with $SnCl_2$ in a citrate buffer at pH 7 forms soluble IV and V citrate complexes that have been characterised [9].

Volatile compounds, technetium hexafluoride, TcF_6 (VI), and technetium oxyfluoride, TcO_3F (VII), have been prepared [10, 11]. These compounds are particularly interesting in the nuclear fuel cycle because technetium is associated with recovered uranium as UF_6. The gas, UF_6, is the chemical form of uranium in gaseous diffusion plants operated to attain isotopic enrichment of ^{235}U. The largest source of production of technetium from recovered uranium has been its separation in the UF_6

gas stream with solid MgF_2. It is eluted from MgF_2 with nitric acid [12]. Conversion steps for the process that produces UF_6 are as follows:

$$UO_3 \xrightarrow{H_2} UO_2 \xrightarrow[\text{HF}]{\text{Anhydrous}} UF_4 \xrightarrow{F_2} UF_6$$

The pure TcF_6 was prepared from the metal and F_2, and the TcO_3F from F_2 and TcO_2. In either case, the products would be expected to hydrolyse in the presence of water. By analogy with ReF_6, TcF_6 also would disproportionate with water to TcO_4^- and TcO_2 or some variety of technetium (IV) hydrous oxide.

To summarise, in non-complexing aqueous solution, the VII valence state is most stable under oxidising conditions. In moderately reducing conditions, the IV is the most stable. Technetium (V) and (VI) generally disproportionate to other valence states. Technetium (III) is known but is readily oxidised to (IV) [6].

2. DESCRIPTION OF THE NUCLEAR FUEL CYCLE AND THE PRODUCTION OF TECHNETIUM IN NUCLEAR REACTIONS

2.1 The model reactor and inventory of technetium

A description of the nuclear fuel cycle begins with the key facility on which 'normalisation' is based, the reactor. The model reactor was first defined in WASH-1248 [13] as a light-water reactor (LWR) designed to produce 1000 MW(e) of energy. The assumed capacity factor is 0·8; thus the model reactor produces 800 MW years of electricity annually. The useful life of the reactor is considered to be 30 years. In WASH-1248 the fuel needed each year to sustain operation at the nominal power level was characterised as the sum of one initial core loading plus 29 annual reloads, averaged over the 30-year life of the plant. This was originally referred to as the 'lifetime average annual fuel requirement' (AFR) but has subsequently been changed to 'reference reactor year' (RRY) due to a conflict of the AFR acronym with the more popular meaning, 'away-from-reactor storage of spent fuel'.

Two types of light-water reactors are in use today in the United States, the pressurised water reactor (PWR) and the boiling water reactor (BWR). In the PWR, the cooling water is kept from boiling by high system pressures. The heated, high-pressure water subsequently transfers heat through metal tubes in a steam generator to a secondary system where steam is produced to turn a turbine generator. In a BWR, the cooling water is converted to steam in the reactor vessel and flows through the system to the turbine generator. The thermal energy is predominantly produced in both reactors by the fission of ^{235}U atoms in the fuel elements. Slight amounts of energy are also released through fission of ^{238}U and two heavy metals created during the reactor operation, ^{239}Pu and ^{241}Pu. The fuel is considered to be 'spent' when it can no longer sustain a chain reaction at economic power levels because of the depletion of fissile isotopes and its content of neutron 'poisons' produced as a by-product of fission. However, at this time the fuel still contains approximately 8 g of ^{235}U per kg of uranium and 98% of the ^{238}U that was originally loaded. Although slight differences exist in the performance and fuel cycle management for the two reactors, little difference is noted between BWRs and PWRs in the amounts of ^{99}Tc produced in spent fuel and therefore available for release to the environment during other steps of the nuclear fuel cycle.

For the purpose of this study, it is assumed that annual feed of new fuel for one RRY is 35 MT of uranium (40 MT of UO_2). The fresh fuel is enriched to contain 3·2 wt% ^{235}U. Once the fuel is burned to 33 000 MWd/MT, the spent fuel contains 0·84 wt% ^{235}U [13]. These parameters are the fundamental bases on which calculations for ^{99}Tc production and release are made.

The position of most technetium isotopes on the yield curve for both fast and thermal neutron fission of heavy metals indicates that it is produced in high abundance relative to other by-products. Specific yield data for thermal neutron fission of ^{235}U as reported by Katcoff [14] is shown in Table I. Since those technetium isotopes having atomic numbers greater than ^{99}Tc have half-lives of less than 18 min, they are not present in appreciable quantities as a fission product. Technetium is also formed in high abundance from thermal neutron fission of ^{233}U (4·8%), ^{239}Pu (5·9%), and fast neutron fission of ^{239}Pu (5·9%), ^{238}U (6·3%), and ^{232}Th (2·7%). In a nuclear reactor, all of these sources may contribute to the accumulated inventory of ^{99}Tc.

Table I

Yield of technetium isotopes for thermal neutron fission of ^{235}U

Isotope	^{99}*Tc*	^{101}*Tc*	^{102}*Tc*	^{103}*Tc*	^{104}*Tc*	^{105}*Tc*	^{107}*Tc*
Yield (%)	6·06	5·6	4·3	3·0	1·8	0·9	0·19

Table II shows the concentration of ^{99}Tc per MT of fuel for six types of reactor systems based on data reported by Kee *et al.* [15]. The referenced burn-up in the second column is the level at which the calculations were made. The right column indicates the ^{99}Tc level in the fuel 'normalised' assuming strict linearity for 33 000 MWd/MT regardless of system.

Table II
Concentration of ^{99}Tc in nuclear fuel for several types of reactor systems

Type of reactor	*Assumed burn-up (MWd/MT)*	*^{99}Tc (Ci/MT)*	*^{99}Tc normalised to 33 000 MWd/MT (Ci/MT)*
PWR-U	33 000	14·4	14·4
PWR-MOX*	33 000	14·5	14·5
BWR-U	27 500	12·1	14·5
BWR-MOX	27 500	12·1	14·5
HTGR	93 972	33·9	11·9
LMFBR	37 137	16·0	14·2

*MOX = mixed U–Pu oxide fuel.

These levels of production provide the foundation for determining releases to the environment from nuclear fuel cycle facilities. As a comparison, it is interesting to note that a PWR having uranium fuel with a burnup of 33 000 MWd/MT contains approximately $7{\cdot}6 \times 10^4$ Ci (0·54 kg/MT) of ^{90}Sr and $1{\cdot}1 \times 10^5$ Ci (1·2 kg/MT) of ^{137}Cs per MT of fuel immediately after discharge from the reactor [15].

2.2 The nuclear fuel cycle

This paper reviews source terms that have been calculated for the nuclear fuel cycle when there is reprocessing of nuclear fuel and uranium not burned in the reactor is returned for re-use. Although this fuel cycle option is not available in the United States and its implementation appears to be uncertain at this time, the 'closed' uranium fuel cycle is still being considered by other countries and ^{99}Tc releases must be carefully considered. Since the nuclear fuel cycle with reprocessing yields higher source terms for ^{99}Tc than the fuel cycle without reprocessing, only source terms for the recycle option will be addressed in this report.

The following sections provide the reader with a brief description of each step of the nuclear fuel cycle with emphasis on data needed to predict the quantity of ^{99}Tc releases to the environment at each stage.

2.2.1. Mining and milling

Since technetium exists in uranium ore only in trace amounts, it is assumed that sources of ^{99}Tc from the mining and milling of uranium ore are negligible. Therefore, no further description of these facilities will be provided.

2.2.2. Uranium hexafluoride production

Prior to enrichment, U_3O_8 is converted to the volatile chemical form of hexafluoride. This process is performed by two methods, both contributing approximately equal amounts to the total UF_6 produced and differing only in the front end of the process. One method, known as the wet process, employs chemical solvent extraction initially to prepare a concentrated uranium feed followed by conversion to uranium trioxide (UO_3), reduction to uranium dioxide (UO_2), and hydrofluorination [16]. The other method, known as the hydrofluor (or dry) process, consists of continuous successive reduction, hydrofluorination and fluorination of the ore followed by fractional distillation. Only 0·1% of the uranium is lost during conversion. The input to the model UF_6 plant is assumed to be 182 MT of natural uranium which in turn is converted to 270 MT of natural UF_6. Of this 270 MT of UF_6, 13·2% is from uranium recycled back to the system (see *Reprocessing* section).

2.2.3. Enrichment

Following conversion to UF_6, the uranium is enriched from approximately 0·7% of the fissile ^{235}U found in natural ores to between 3 and 4% required to sustain a chain reaction in the reactor. Enrichment for the model reactor in this study is 3·2%.

Currently enrichment is by the gaseous diffusion process. It is founded upon the concept that the velocities of gas molecules at a given temperature are based upon their masses, and by passing the UF_6 through a series of porous barriers, the flow of lighter isotopes of uranium is enhanced, thus creating a greater concentration of ^{235}U in the final product than in the feed. The maximum theoretical enrichment in a single stage is a factor of 1·0043; therefore, approximately 1200 stages are required to produce uranium enriched to 4% ^{235}U. The solid residue or depleted by-product from the process is known as 'tails' and still contains some ^{235}U. The level of ^{235}U left in the tails affects the amount of UF_6 required as input to the plant in order to give the 35 MT of 3·2% fuel required per RRY. It is controlled by several factors, including the availability of feed material and the plant capacity. For this study, we have assumed the same ^{235}U assay as in tails used in WASH-1248 [13] of 0·25%.

Of the 270 MT of UF_6 feed into the plant, 218 MT are rejected in the depleted uranium tails. The remaining 52 MT of enriched uranium product is feed material to the fuel fabrication plant. Again 13·2% (6·7 MT) is uranium that has been recycled back into the system.

2.2.4. Fuel fabrication

The UF_6 enriched to 3·2% ^{235}U constitutes the feed material to the model fuel fabrication plant. The process consists of three basic steps: chemical conversion of the UF_6 to UO_2, mechanical conversion to pellets and insertion into metal tubes, and recovery of uranium from scrap and off-specification material.

The 52 MT/RRY of enriched UF_6 fed into the fuel fabrication plant is converted to 40 MT of UO_2 as fuel for the model reactor.

2.2.5. Reprocessing

The accepted method for recovering uranium from spent fuel is a chemical technique known as the Purex process. In this process the uranium is separated from fission products and heavy metals produced during irradiation. Elements are stored for 150 days after discharge from the reactor to permit short-lived radionuclides to decay, and then are sheared into small pieces to allow nitric acid dissolution of the uranium and by-products. The liquid containing the dissolved fuel is then processed through a series of solvent extraction or ion exchange systems, producing a purified uranyl nitrate solution which is then converted to UF_6 [17].

As a basis for this study, the reprocessing plant handles 35 MT (including 34 MT of uranium and 1 MT of by-products) per RRY. This is a significantly smaller amount than the total plant capacity, generally assumed to be 2000 MT per year. If it is assumed that 100% of the uranium is recovered, then 34 MT of uranium having 0·84% ^{235}U is returned to the fuel cycle.

2.2.6. Solidification of high-level wastes and high-level waste disposal

During reprocessing, fission products and other by-products formed during irradiation are separated from the uranium and retained in liquid tanks for solidification prior to disposal as high-level wastes. Two principal solidification processes are available, calcination and glassification [18]. The glassification process is used as the model in this study. Details on the mechanical and chemical techniques can be found in NUREG-0116

[17] and are not repeated here. Effluents from the plant include activation and fission products and transuranium radionuclides. The primary non-radioactive effluent consists of oxides of nitrogen. It is assumed there will be no liquid effluent.

Once solidified, high-level waste including technetium will be transferred to a geologic repository. It is assumed that one RRY creates 2 m^3 of solidified high-level waste [13]. Effluents to the atmosphere from the disposal facility include small amounts of radioactive gases, particulates, and chemicals. A more in-depth discussion of the waste repository can be found in Vol. 3 of NUREG-0002 [19].

2.2.7. Shallow land burial of low-level wastes

Nuclear fuel cycle facilities create solid radioactive waste contaminated with isotopes having relatively short half-lives (usually less than 30 years). The waste is often compacted or solidified prior to final disposal. It is generally buried in shallow trenches and migration of radionuclides beyond the site boundary is restricted provided several precautions are taken. These include: (1) no permanent surface water; (2) good surface drainage with little potential for erosion; (3) the most shallow water table not less than 50 ft below the surface, with small fluctuations; and (4) soil of low permeability. An overburden of soil both provides a barrier for erosion and acts as a radiation shield [19].

2.3. Calculation of source terms for ^{99}Tc

2.3.1. Reprocessing

It is assumed that the technetium produced during fission is in its elemental state and that it is converted to pertechnetic acid ($HTcO_4$) during the dissolution of the fuel elements. Part of the technetium follows the uranyl nitrate as $HTcO_4$ through the extraction process and is converted to heptaoxide (Tc_2O_7) during the denitration or calcining step.

Pruett [20] examined solvent extraction of heptavalent technetium from aqueous nitric acid or hydrochloric acid by tributyl phosphate (TBP) over a wide range of concentrations and temperatures. He determined distribution coefficients as a function of temperature and the concentrations of TBP, HNO_3, HCl, and $UO_2(NO_3)_2$. The purpose of this work was to investigate the possible extraction of ^{99}Tc from the waste stream during reprocessing in order to concentrate and isolate this isotope to facilitate permanent disposal. These data and a subsequent report by Pruett [21] reflect the most comprehensive studies performed to date on the behaviour of technetium during fuel reprocessing and can be used to calculate the fraction of technetium that is carried through each stream when specific operating conditions of the process are known.

In the normal Purex process, 20% of the ^{99}Tc produced in the fuel elements remains in the separated uranium fraction and the remaining 80% follows the fission product waste stream or the plutonium fraction [22, 23]. If it is assumed that the irradiated fuel from the model reactor contains 14·5 Ci (0·84 kg) of ^{99}Tc per MT of fuel (see Table II) after its initial discharge and that there are 35 MT of fuel per RRY, then the reprocessing plant handles approximately 510 Ci (30 kg) of ^{99}Tc per RRY of fuel processed. Assuming 20% is returned to the fuel cycle, the 34 MT of uranium returned for re-enrichment contains about 100 Ci (6 kg) or about 176 ppm of ^{99}Tc.

It is important to consider the limitations of handling the uranium that is returned to the fuel cycle containing nearly 220 ppm of ^{99}Tc. The specifications for feed UF_6 described in the Federal Register [24] state that the maximum beta activity due to fission products shall be no more than 10% of the beta activity of aged natural uranium. Assuming a specific activity of aged natural uranium of $6{\cdot}8 \times 10^{-7}$ Ci/g and that for ^{99}Tc of $1{\cdot}7 \times 10^{-2}$ Ci/g, a ratio of specific activities of $2{\cdot}5 \times 10^{4}$ implies that it would take a concentration of about 40 ppm of ^{99}Tc to contribute an activity equal to that for the uranium daughters. Ten percent of this value would imply a maximum ^{99}Tc concentration of 4 ppm of ^{99}Tc in the recycled uranium if this material were returned for enrichment. Levin [25] also calculates a value of 4 ppm of ^{99}Tc in recycled uranium as maximum for compliance with the beta activity standard. However, he suggests reducing this by a factor of 10 to 0·4 ppm as a new specification. Since this lower value has not yet been adopted, 4 ppm ^{99}Tc is used as the basis of source terms in this study. The reduction in ^{99}Tc concentration could be accomplished by solvent extraction techniques of the uranyl nitrate prior to conversion to UF_6 [21], by chemical separation after conversion to UF_6 [12] or by diluting the recovered uranium containing ^{99}Tc with the fresh uranium coming into the cycle.

Because of this restriction on impurities of recycled uranium, it is assumed that the ^{99}Tc is reduced to a concentration of 4 ppm before it re-enters the fuel cycle. This assumption implies that of 34 MT of uranium recovered, 2·3 Ci (0·14 kg) of ^{99}Tc returns to the fuel for any year after recycle begins. The approximately 508 Ci/RRY remaining enters the high-level waste.

It is expedient at this time to determine the release of ^{99}Tc from the reprocessing site. As mentioned earlier, there will be no liquid effluent.

Release of technetium to the atmosphere can be estimated by using a 'confinement factor' which, when divided into the quantity of ^{99}Tc reprocessed in fuel for one RRY, gives the amount that escapes the plant to the atmosphere. This confinement factor can be derived through an analysis of technetium chemistry and comparison with other chemically similar radionuclides.

If ^{99}Tc is released during the dissolution step, volatile chemical forms would likely be Tc_2O_7 or $HTcO_4$ since the formation of stable alkyl reaction products as occurs with iodine has not been demonstrated and appears improbable [26]. The gas scrubbers on the stack would be expected to strip the majority of the ^{99}Tc from the effluent, as the reaction of technetium compounds with water in the presence of an oxidiser would result in non-volatile pertechnetate. Therefore, it is assumed that most of the ^{99}Tc that is cycled through the plant remains in process streams and does not escape. For purposes of this study, a confinement factor for ^{99}Tc of 1×10^8 is assumed. This value was selected due to similarities in the chemical properties and volatility of technetium with ruthenium and an accepted confinement factor for ^{106}Ru for reprocessing [17]. Applying this factor to 510 Ci of ^{99}Tc per RRY for reprocessing yields a source term of $5{\cdot}1 \times 10^{-6}$ Ci/RRY of ^{99}Tc released to the atmosphere.

2.3.2. Uranium hexafluoride production

As previously described, the output of the model conversion plant is 270 MT/RRY of UF_6. It was determined earlier that 2·3 Ci of ^{99}Tc per RRY is returned to the system with the recycled fuel. It is assumed that the ^{99}Tc entering the plant for conversion reacts with fluorine to give TcF_6 and TcO_3F, but predominantly exists as TcF_6. An estimate of the source term for ^{99}Tc released to the atmosphere and water can be made by assuming that all of the ^{99}Tc exists as TcF_6 within the plant and that it is released to each pathway in the same fraction of F^- appearing in the effluent to the total F^- used in the process.

The total amount of F^- used in the process can be calculated by summing the quantity being transferred to the enrichment plant as UF_6 (87 MT of F^-) plus losses from the facility in liquid or gaseous effluent and solid residue. Liquid releases account for 0·22 MT/RRY of F^- and gaseous effluent contains 0·11 MT/RRY of F^-. Since no breakdown of F^- for solid waste was reported in WASH-1248 [13], this loss is neglected. Therefore, the total amount of F^- required for UF_6 conversion is assumed to be approximately 87 MT/RRY. The fraction of F^- release to liquids is 0·22/87 and that for atmospheric releases is about 0·11/RRY. Assuming that all 2·3 Ci of ^{99}Tc is TcF_6, a source term of $1{\cdot}0 \times 10^{-3}$ Ci/RRY is the source term to liquids and $5{\cdot}1 \times 10^{-4}$ Ci/RRY is the source term to atmosphere. It is emphasised that these source terms are somewhat speculative for ^{99}Tc released at the site of UF_6 conversion. Monitoring data for ^{99}Tc at existing UF_6 conversion plants handling recycled fuel that could provide a more realistic basis for establishing a confinement factor are not available.

2.3.3. Enrichment

The behaviour of ^{99}Tc during the enrichment process is still being investigated. It is known that significant amounts of ^{99}Tc can accumulate in the cascades and a portion of this is released to the environment. The airborne emissions result from its presence in the purge cascade's effluent, while the major source to liquids is equipment decontamination operation. Several important assumptions can be made regarding the characteristics of ^{99}Tc at gaseous diffusion plants [27]:

1. ^{99}Tc moves toward the product end (top) of the diffusion cascade,
2. ^{99}Tc can be selectively removed from the UF_6 stream by use of magnesium fluoride traps; however, removal efficiency varies considerably according to technetium concentration, and
3. since most technetium compounds are highly soluble in water, they can be removed from decontamination solutions by either reduction–precipitation or ion exchange techniques.

The key to quantifying the release of ^{99}Tc to the environment during the enrichment process lies in establishing a confinement factor to determine the fraction of ^{99}Tc that passes through the plant cascades which ultimately escapes. Unfortunately, data on which this factor could be based are not well documented; therefore, assumptions must be made to derive a realistic estimate.

Monitoring data are available that report the release of ^{99}Tc to surface streams and the atmosphere from each of the gaseous diffusion plants in operation today.These data are summarised in Table III. The values in Table III suggest that releases of ^{99}Tc to both water and air at the Oak Ridge site have decreased over recent years; however, this decrease is likely due to the fact that incoming feed UF_6 no longer contains recycled uranium and what is present in the source terms results from ^{99}Tc that had accumulated on the cascades [27]. The difficulty in applying these data is that the amount of ^{99}Tc coming into the plant is not reported.

Therefore, it is impossible to use these release rates to calculate a confinement factor.

Table III also lists the ratio of the source term for water to that for air. It is obvious that releases to water significantly exceed those to air. If the 1977 Oak Ridge data are neglected (no explanation could be found as to why releases that year were exceptionally small), then the mean source term ratio of water/air is approximately 40. Thus approximately 97·5% of the ^{99}Tc that escapes the plant enters the liquid pathway and 2·5% enters the gaseous effluent. This information can be used to assist in predicting a source term for the model enrichment plant for one RRY of uranium containing 2·3 Ci (0·14 kg) of ^{99}Tc.

Table III

Release of ^{99}Tc to water and air during gaseous diffusion enrichment of uranium

Plant location	*Year*	*Annual release (Ci)*		*Ratio (water/air)*	*Reference*
		Water	*Air*		
Oak Ridge	1975	6·4	0·3	21	28
Paducah	1975	6·4	0·8	8	29
Oak Ridge	1976	24	6·8	3·5	30
Paducah	1976	16	0·1	161	31
Oak Ridge	1977	15	2×10^{-6}	$7{\cdot}5 \times 10^{6}$	32
Paducah	1977	10	0·1	103	33
Portsmouth	1977	31	4·5	6·9	34
Oak Ridge	1978	4	0·3	13	35
Paducah	1978	9·2	0·6	15	36
Oak Ridge	1980	5·1	0·9	5·7	37
Oak Ridge	1981	3·5	0·04	87	38
Oak Ridge	1982	1·7	0·03	57	39

Unfortunately, there are no studies available in the literature that carefully analyse the fraction of ^{99}Tc that enters each waste stream or is carried through to the final product during enrichment at an equilibrium recycle situation. It is accepted that a significant build-up will occur in the cascade at the start of recycle, and that eventually a steady-state situation will develop with ^{99}Tc either being released to the environment, remaining with the enriched UF_6, or being selectively removed with the use of traps [40] and disposed of as a waste product.

Because quantitative data are not readily available from which a confinement factor for ^{99}Tc can be calculated, several assumptions are now made. First, it is assumed that 50% of the ^{99}Tc entering the plant remains with the UF_6 product (about 1·15 Ci or 0·071 kg). This assumption accounts for the fact that the lighter TcF_6 moves to the top (product end) of the cascade rather than to the bottom (tails end). Second, it is assumed that 40% (0·92 Ci or 0·054 kg) of the ^{99}Tc remains with tails or is selectively removed. Third, it is assumed that 10% (0·23 Ci or 0·013 kg) of the ^{99}Tc escapes to the environment, employing an overall confinement factor of ten. This assumption may be high; however, it accounts for reasonable precautions being taken to minimise ^{99}Tc releases, yet it gives sufficient weight to the source term to make it noteworthy. Of that released to the environment, 97·5% or 0·22 Ci/RRY enters the liquid pathway and 0·0058 Ci/RRY enters the atmosphere. It is emphasised that there are no experimental data in the open literature on which to base these assumptions.

2.3.4. Fuel fabrication

The incoming enriched UF_6 still contains approximately 1·15 Ci (0·07 kg) of ^{99}Tc. Conversion to UO_2 begins with the evaporation of UF_6 and precipitate formation. The precipitate is calcined to UO_3 and then reduced to UO_2 powder. Because of its high vapour pressure, the Tc_2O_7 may vaporise at this point and be carried through the conversion process.

Data do not exist that permit calculation of a confinement factor for fuel fabrication plant handling fuel at equilibrium recycle based upon operational experiences. An estimate of source terms can be made by assuming the TcF_6 is released to water or air in the same fraction as fluorides. In WASH-1248 [13] it is shown that for a plant handling 900 MT of uranium each year, 0·1 MT of fluorides escape to the atmosphere and 106 MT of fluorides go to surface water. In order to calculate an upper bound for ^{99}Tc release, several assumptions are made. First, again assume that 50% of the ^{99}Tc is carried through the process and remains with the UO_2. Second, assume that of the remaining 50% (0·58 Ci or 0·03 kg) the fraction $9{\cdot}4 \times 10^{-4}$ enters the atmospheric waste stream and the remainder (99·9%) enters the liquid waste stream. Furthermore, assume that the atmospheric waste is not treated, thus yielding a source term of $5{\cdot}4 \times 10^{-4}$ Ci/RRY. Storton [41] investigated the removal of ^{99}Tc from the aqueous discharge of a fuel fabrication facility. He based his experiments on the assumption that the concentration in liquid effluent should be reduced to less than 200 pCi/ml of ^{99}Tc using a strong-base ion-exchange resin. A confinement factor of 200 for this process was determined. If we assume that ion exchange treatment is provided for the pathway and that a confinement factor of 200 is achievable at reasonable cost, then a source term of $2{\cdot}9 \times 10^{-3}$ Ci/RRY results.

2.3.5. Solidification of high-level waste and high-level waste disposal
Descriptions of the environmental impact of solidification of high-level wastes have been reported [42] and one study has been completed in which the behaviour of ^{99}Tc during calcining was evaluated [43]

It is assumed that there are no liquid radioactive wastes associated with the process and that the confinement factors for radionuclides are identical to those for reprocessing [17]. Thus the confinement factor of 1×10^8 is again used. Assuming that approximately 510 Ci of ^{99}Tc is included in the high-level waste residue, a source term to atmosphere of $5{\cdot}1 \times 10^{-6}$ Ci/RRY is calculated.

The amount of ^{99}Tc entering the waste repository per RRY is approximately the same for each fuel cycle option. The only difference is the physical form in which the waste exists inside the storage canisters: the form for no recycle is spent fuel elements and that for recycle is solidified glass pellets. In NUREG-0002 [19], it is assumed that no breach of containment occurs during the storage of the waste and that the only releases of radionuclides are those occurring during the operation of the facility. Based upon data on the content of radionuclides per MT of heavy metal in the waste, and the source terms for the model facility, an overall confinement factor for ^{106}Ru of 6×10^7 can be derived. Assuming that the fuel contains 14·5 Ci/MTHM of ^{99}Tc, then the waste disposal facility would release $2{\cdot}4 \times 10^{-7}$ Ci of ^{99}Tc per year of operation. Since this is for operation of the entire facility at 2060 m^3 per year and the high-level waste from one RRY is approximately 2 m^3, then the source term of $2{\cdot}4 \times 10^{-10}$ Ci/RRY to the air for high-level waste disposal is obtained. This release rate of ^{99}Tc is expected to become zero when the facility is closed after 100 years of operation.

2.3.6. Shallow-land burial of low-level wastes
Although surveys of the volume and quantity of waste from nuclear power plants do not indicate that any ^{99}Tc has been buried due to nuclear fuel cycle operations [45], it is likely that when equilibrium recycle is reached, some ^{99}Tc will be present in low-level waste. Most of the ^{99}Tc comes from UF_6 conversion, re-enrichment of fuel, and fuel fabrication facilities; however, it would be limited to cleaning materials or contaminated equipment.

Of the original 510 Ci produced per RRY, all but 2·3 Ci is expected to remain at the reprocessing site (where it is prepared for disposal as high-level waste). Of the 2·3 Ci/RRY returned with the uranium, it is unlikely that more than 0·1% or 0·002 Ci would be received at a low-level waste site for permanent disposal per RRY.

The length of time that the ^{99}Tc remains confined at the low-level site depends upon the integrity of the containers and the characteristics of the site. Once containment is broken, the release rate is still a function of time and only a fraction escapes to air or water each year. Obviously, this amount is very small and its quantification is necessarily complex, requiring the use of hydrological models. An approximation of the source term can be made using a comprehensive assessment model currently being developed by the U.S. Environmental Protection Agency [45]. This model predicts that for a site such as Barnwell, South Carolina, for 100 years of burial of one Ci of ^{99}Tc, $1{\cdot}6 \times 10^{-3}$ Ci ends up in a well near the site boundary. This implies that a confinement factor of 625 exists at 100 years after initial disposal. Releases to the atmosphere at the 100-year point are negligible.

Applying this confinement factor to our 0·002 Ci/RRY yields a source term to well water of $3{\cdot}2 \times 10^{-6}$ Ci/RRY. As stated, the source term is assumed to be zero.

3. CONCLUSIONS

The purpose of this paper has been to review source terms calculated for ^{99}Tc releases to the environment from nuclear fuel cycle facilities when there is reprocessing of spent fuel. These source terms are summarised in Table IV. It is evident from the study that ^{99}Tc may be released to the environment during reprocessing, UF_6 production, enrichment, fuel fabrication, high-level waste solidification, high-level waste disposal, and low-level waste disposal. The amounts expected to be released were calculated and tabulated on the basis of one reference reactor year (RRY). Among these sources, the largest anticipated release of ^{99}Tc is via the liquid effluent at uranium enrichment facilities. The estimated amount of this source term (0·22 Ci/RRY) exceeds atmospheric releases at this site by an order of magnitude. This source term is about two orders of magnitude greater than releases estimated for fuel fabrication plants and UF_6 conversion facilities.

It is emphasised that the calculations presented here have been made in the absence of published data quantifying the fate of ^{99}Tc in the

Table IV
A summary of ^{99}Tc source terms for nuclear fuel cycle facilities

Fuel cycle facility	*Release (Ci/RRY)*	
	Air	*Liquid*
Mining/milling	0	0
Reprocessing	$5{\cdot}1 \times 10^{-6}$	0
UF_6 conversion	$5{\cdot}1 \times 10^{-4}$	$1{\cdot}0 \times 10^{-3}$
Enrichment	$5{\cdot}8 \times 10^{-3}$	$2{\cdot}2 \times 10^{-1}$
U-fuel fabrication	$5{\cdot}4 \times 10^{-4}$	$2{\cdot}9 \times 10^{-4}$
High-level waste solidification	$5{\cdot}1 \times 10^{-6}$	0
High-level waste disposal	$2{\cdot}4 \times 10^{-10}$	0
Low-level waste disposal	0	$3{\cdot}2 \times 10^{-6}$

uranium enrichment facilities. Improvements in these estimates will require specific information on the fraction of the ^{99}Tc inventory that enters these plants and ends up in depleted uranium, in the enriched product, and in the liquid and gaseous effluent. In addition, one may still question the degree to which the technetium will build up on the cascades during enrichment when at equilibrium recycle. Further consideration should also be given to improving techniques for removing technetium from uranyl nitrate prior to conversion to UF_6. This may be required as a special step at reprocessing plants prior to conversion if ^{99}Tc releases to the environment are to remain acceptably low.

It is evident that because of its unique chemistry and environmental mobility, ^{99}Tc should be given special consideration in nuclear fuel cycle assessments when there is recycling of uranium fuel. In general, published data related to its production, concentration in recycled fuel, and ultimate release to the environment are not readily available. Future research should be directed toward filling this gap in our understanding of ^{99}Tc.

REFERENCES

1. CODE OF FEDERAL REGULATIONS (1980). Licensing and regulatory policy and procedures for environmental protection. 10 CFR, Part 51.
2. TILL, J. E., SHOR, R. W. and HOFFMAN, F. O. (1984). Environmental effects of the uranium fuel cycle. ORNL/TM-9150 (NUREG-3738).
3. MCFADDEN, K. M. (1980). The chemistry of technetium in the environment. PNL-2579, Pacific Northwest Laboratory, Richland, Washington. Available for purchase from National Technical Information Service, Springfield, Virginia 22161.

4. PAQUETTE, J., REID, J. A. K. and ROSINGER, E. L. J. (1980). Review of technetium behavior in relation to nuclear waste disposal. TR-25, Atomic Energy of Canada Ltd., Chalk River, Ontario, Canada.
5. TURCOTTE, M. D. S. (1982). Environmental behavior of technetium-99. DP-1644, E. I. Du Pont de Nemours and Co., Savannah River Plant and Laboratory, Aiken, South Carolina.
6. RARD, J. A. (1983). Critical review of the chemistry and thermodynamics of technetium and some of its inorganic compounds and aqueous species. UCRL-53440, Lawrence Livermore Laboratory, Livermore, California.
7. BOYD, G. E. (1959). Technetium and prometheum. *J. Chem. Educ.*, **36**(1), 3–14.
8. JOHANSEN, B., SYHRE, R., SPIES, H. and MUNZE, R. (1978). Chemical and biological characterization of different Tc complexes of cysteine and cysteine derivatives. *J. Nucl. Med.*, **19**, 816–24.
9. MUNZE, R. (1977). The formation of technetium citrate complexes. *Radiochem. Radioanal. Letters*, **30**, 61–4.
10. SELIG, H., CHERNICK, C. L. and MALM, J. G. (1961). The preparation and properties of TcF_6. *J. Inorg. Nucl. Chem.*, **19**, 377.
11. SELIG, H. and MALM, J. G. (1963). The preparation and properties of pertechnetyl fluoride, TcO_3F, *J. Inorg. Nucl. Chem.*, **250**, 349.
12. GOLLIHER, W. R., LE DOUX, R. A., BERNSTEIN, S. and SMITH, V. A. (1963). Separation of ^{99}Tc from uranium hexafluoride. TIC-18290, USAEC report, Paducah Gaseous Diffusion Plant, Paducah, Kentucky (available NTIS).
13. US ATOMIC ENERGY COMMISSION (1974). Environmental survey of the uranium fuel cycle. WASH-1248.
14. KATCOFF, S. (1958). *Nucleonics*, **16**(78).
15. KEE, C. W., CROFF, A. G. and BLOMEKE, J. O. (1976). Updated projections of radioactive wastes to be generated by the U.S. nuclear power industry, Oak Ridge National Laboratory report ORNL/TM-5427.
16. PAGE, H. (1980). United Kingdom experience of production of uranium fluorides. In: *Production of Yellow Cake and Uranium Fluorides*, IAEA, Vienna.
17. US NUCLEAR REGULATORY COMMISSION (1976). Environmental survey of the reprocessing and waste management portions of the LWR fuel cycle. NUREG-0116.
18. US ENERGY RESEARCH AND DEVELOPMENT ADMINISTRATION (1976). Alternatives for managing wastes from reactors and post fission operations in the LWR fuel cycle, Vol. 4. ERDA-76-43.
19. US NUCLEAR REGULATORY COMMISSION (1976). Final generic environmental statement on the use of recycle plutonium in mixed oxide fuel in light water cooled reactors. NUREG-0002.
20. PRUETT, D. J. (1981). The solvent extraction of heptavalent technetium by tributyl phosphate. *Separation Science and Technology*, **16**(19), 1157–79.
21. PRUETT, D. J. (1984). The solvent extraction of heptavalent technetium and rhenium by tributyl phosphate. Oak Ridge National Laboratory report ORNL/TM-8668, Oak Ridge, Tennessee.
22. ROBERTS, F. P. (1971). Summary of research on Tc, Rh, and Pd by Battelle

Northwest. BNWL-B-49, Battelle Northwest Laboratory report. Available for purchase from National Technical Information Service, Springfield, Virginia 22161.
23. CAMPBELL, M. H. (1961). Extraction of pertechnetate in a purex process. HW-69712, USAEC report. Available from Technical Information Center, Oak Ridge, Tennessee 37830.
24. FEDERAL REGISTER (1966). Uranium hexafluoride charges and specifications. 31 F.R. 16584.
25. LEVIN, R. W. (compiler) (1981). UF_6 specifications for feed for gaseous diffusion plants. K/TL-1092 Rev. 1, Union Carbide Corporation, Nuclear Division, Oak Ridge, Tennessee.
26. WILDUNG, R. E., MCFADDEN, K. M. and GARLAND, T. R. (1979). Technetium sources and behavior in the environment. *J. Environ. Qual.*, **8**, 156–61.
27. U.S. DEPARTMENT OF ENERGY (1979). Environmental assessment of the Oak Ridge gaseous diffusion plant site. DOE/EA-0106.
28. UNION CARBIDE CORPORATION (1976). Environmental monitoring report. Oak Ridge Facilities for Calendar Year 1975, Y/UB-4.
29. UNION CARBIDE CORPORATION (1976). Environmental monitoring report. Paducah Gaseous Diffusion Plant for Calendar Year 1975, Y/UB-5.
30. UNION CARBIDE CORPORATION (1977). Environmental monitoring report. Oak Ridge Facilities for Calendar Year 1976, Y/UB-6.
31. UNION CARBIDE CORPORATION (1977). Environmental monitoring report. Paducah Gaseous Diffusion Plant for Calendar Year 1976, Y/UB-7.
32. UNION CARBIDE CORPORATION (1978). Environmental monitoring report. Oak Ridge Facilities for Calendar Year 1977, Y/UB-8.
33. UNION CARBIDE CORPORATION (1978). Environmental monitoring report. Paducah Gaseous Diffusion Plant for Calendar Year 1977, Y/UB-9.
34. GOODYEAR ATOMIC CORPORATION (1978). Portsmouth Gaseous Diffusion Plant Environmental Monitoring Report for Calendar Year 1977, GAT-955.
35. UNION CARBIDE CORPORATION (1979). Environmental monitoring report. Oak Ridge Facilities for Calendar Year 1978, Y/UB-10.
36. UNION CARBIDE CORPORATION (1979). Environmental monitoring report. Paducah Gaseous Diffusion Plant for Calendar Year 1978, Y/UB-11.
37. UNION CARBIDE CORPORATION (1981). Environmental monitoring report. Oak Ridge Facilities for Calendar Year 1980, Y/UB-15.
38. UNION CARBIDE CORPORATION (1982). Environmental monitoring report. Oak Ridge Facilities for Calendar Year 1981, Y/UB-16.
39. UNION CARBIDE CORPORATION (1983). Environmental monitoring report. Oak Ridge Facilities for Calendar Year 1982, Y/UB-18.
40. SARACENO, A. J. (1980). Technetium removal from plant vent streams. *Proc. Waste Management Seminar*, CONF-800416.
41. STORTON, J. M. (1977). Ion exchange as a technique to remove technetium. In: *Management of Low-Level Radioactive Waste, Vol. 1* (Eds. M. W. Carter, A. A. Moghissi and B. Kahn), New York, Pergamon Press, pp. 413–18.
42. U.S. DEPARTMENT OF ENERGY (1982). Defense waste processing facility, final environmental impact statement. DOE/EIS-0082.

43. KNOX, C. A. and FARNSWORTH, R. K. (1981). Behavior of radioactive iodine and technetium in the spray calcination of high-level waste. PNL-3741, Pacific Northwest Laboratory, Richland, Washington.
44. NUS CORPORATION (1980). Preliminary state-by-state assessment of low-level radioactive wastes shipped to commercial burial grounds. Report NUS-3440, NUS Corporation, Rockville, Maryland.
45. US ENVIRONMENTAL PROTECTION AGENCY (1985). PRESTO-EPA-POP: A low-level radioactive waste environmental transport and risk assessment code—Methodology manual, EPA 520/1-85-001.

2

Technetium Discharges into the Environment

F. LUYKX

Commission of the European Communities,
Bâtiment Jean Monnet, Plateau du Kirchberg,
BP 1907, Luxembourg

1. TECHNETIUM ISOTOPES

Technetium has some 15 known isotopes, ranging from Tc-92 to Tc-107; none of these isotopes are stable but three of them are very long-lived, viz. Tc-97 ($T_{1/2}$: $2{\cdot}6 \times 10^6$y), Tc-98 ($T_{1/2}$: $1{\cdot}5 \times 10^6$y) and Tc-99 ($T_{1/2}$: $2{\cdot}1 \times 10^5$y). Although produced in nature by spontaneous fission of uranium-238, the major source of the long-lived isotopes stems from fission in nuclear weapons and in reactors; the use of technetium in medicine makes a relatively small contribution.

2. TECHNETIUM INVENTORY IN IRRADIATED NUCLEAR FUEL

The calculated inventories of the long-lived technetium isotopes in typical PWR and AGR fuel are given in Table I [1]. They are respectively based

Table I

Isotope	*Fuel inventory*			
	PWR		*AGR*	
	Bq/GW(th)y	*Bq/t*	*Bq/GW(th)y*	*Bq/t*
Tc-97	18·5	1·7	48·1	2·4
Tc-98	$8{\cdot}9 \times 10^5$	$8{\cdot}1 \times 10^4$	$1{\cdot}4 \times 10^6$	$7{\cdot}0 \times 10^4$
Tc-99	$5{\cdot}8 \times 10^{12}$	5.2×10^{11}	$5{\cdot}9 \times 10^{12}$	$3{\cdot}0 \times 10^{11}$

on burn-ups of 33 and 18 GWd/t and an initial enrichment of 3·2 and 2·16% U-235, respectively. From this table it can be seen that technetium-99 is by far the dominant long-lived technetium isotope produced in the nuclear fuel cycle.

3. TECHNETIUM RELEASES FROM THE NUCLEAR FUEL CYCLE

Technetium discharges occur in the following steps of the nuclear fuel cycle:

—reactor operation,
—nuclear fuel reprocessing,
—conversion of recycled uranium into UF_6,
—enrichment of recycled uranium,
—fuel fabrication,
—waste treatment.

3.1. Nuclear power stations

3.1.1. Theoretical estimates

Holtschmidt and Bonka [2] have evaluated technetium isotope releases from LWR fuel; they obtained the release rates and (primary circuit) coolant concentrations of technetium-99m and technetium-99 in 1300 MWe PWRs and BWRs shown in Table II, assuming the presence of 10 and 25 defective fuel elements respectively on average. This evaluation assumes the same fractional release from fuel as observed for strontium.

Table II

Isotope	*Release rate from fuel (Bq/s)*		*Primary coolant concentration (Bq/t)*	
	PWR	*BWR*	*PWR*	*BWR*
Tc-99m	$6·0 \times 10^{6}$	$3·3 \times 10^{6}$	$3·2 \times 10^{8}$	$1·3 \times 10^{8}$
Tc-99	$9·2 \times 10^{-1}$	$4·8 \times 10^{-1}$	$1·1 \times 10^{2}$	$4·2 \times 10^{1}$

On the basis of caesium-137 discharged to atmosphere relative to its concentration in the primary coolant, as measured in German LWRs, the technetium discharge rates to atmosphere given in Table III have been

Table III

Isotope	*Release rate to atmosphere (Bq/y)*	
	PWR	*BWR*
Tc-99m	$3{\cdot}8 \times 10^7$	$8{\cdot}3 \times 10^7$
Tc-99	$3{\cdot}7 \times 10^1$	$1{\cdot}9 \times 10^2$

deduced [2]. Technetium discharges in liquid effluent can be estimated similarly. Assuming an annual caesium-137 release in liquid effluents of $3{\cdot}7 \times 10^9$ Bq for a 1300 MWe PWR and $1{\cdot}9 \times 10^9$ Bq for a 1300 MWe BWR [3], yields the values for technetium given in Table IV. However, the technetium-99m release rates thus obtained are uncorrected for decay subsequent to leaving the primary circuit. Because of the short half-life (6 h) of this isotope, its concentration in the liquors decreases rapidly with reactor shutdown time and storage or treatment time prior to release.

Table IV

Isotope	*Release rate in liquid effluent (Bq/y)*	
	PWR	*BWR*
Tc-99m	$4{\cdot}7 \times 10^9$	$2{\cdot}4 \times 10^8$
Tc-99	$1{\cdot}6 \times 10^3$	$8{\cdot}1 \times 10^2$

Although the decay of technetium-99m leads to the formation of technetium-99 the in-growth of the latter in effluents will make a negligible contribution since the atomic specific activity is a factor of 3×10^8 lower corresponding to its much longer half-life.

3.1.2. Measured releases

For discharges to atmosphere the available data are very limited. Luxenburger *et al.* [4] attempted to measure technetium-99 releases from some German power stations but found that the levels rarely exceeded the limit of detection (*c.* 10^3 Bq/h). The results for technetium-99 m releases from two US power stations in 1977 have been published as $5{\cdot}4 \times 10^8$ Bq and $8{\cdot}0 \times 10^9$ Bq respectively [5].

For liquid effluents Luxenburger found that the concentrations of technetium-99 never exceeded 20 Bq/m^3, this being the limit of detection

Table V

Site	Year	Technetium-99m discharges in liquid effluent (Bq/y)
Tihange 1	1982	9×10^7
Chooz	1978	$1{\cdot}5 \times 10^8$
Bugey 2 + 3	1978	$2{\cdot}0 \times 10^8$
Fessenheim	1978	$2{\cdot}0 \times 10^7$

[4]. However, technetium-99m discharge data have been published for a number of European and US power stations. The European data are for PWRs and are summarised in Table V [6].

The US data [7] show that technetium-99m is detected in liquid effluents from most light water reactors. In 1979 the annual discharges from PWRs ranged from 4×10^4 to $1{\cdot}2 \times 10^9$ Bq. In BWRs the range was from 1×10^5 to $3{\cdot}7 \times 10^9$ Bq.

These large ranges probably illustrate the strong dependency on the delay prior to discharge.

3.2. Nuclear fuel reprocessing plants

In the nuclear fuel cycle the major discharges of technetium-99 to the environment arise from reprocessing nuclear fuel.

Table VI shows the discharges with liquid effluents over the last six years from the Sellafield plant at Seascale, UK [8]. The values show an important decrease since 1981, the reason being that the previous years reflect delayed discharges from storage tanks [9].

Assuming that for the years 1981 to 1983 the technetium-99 discharges correspond to the fuel throughput in the plant, the normalised discharge rate can be calculated as 0·44 TBq/GW(th)y. When compared with the

Table VI

Year	Technetium-99 discharges in liquid effluents from Sellafield (TBq/y)
1978	178
1979	43·5
1980	56·8
1981	5·8
1982	3·6
1983	4·4

theoretical value of about 6 TBq/GW(th)y, this implies that about 7% of the technetium-99 produced in the fuel is discharged with liquid effluents.

The French reprocessing plant at Cap de la Hague released 11·7 TBq to sea in 1983 [10].

No data on technetium discharges to atmosphere are available for either of these reprocessing plants.

At Karlsruhe (FRG), technetium has been measured in both gaseous and liquid effluents [4]. Over a six months sampling period (September 1983 to March 1984) technetium-99 in the distillate discharged from the LAW-evaporator (which receives the liquid effluents from the WAK reprocessing plant) amounted to $2{\cdot}5 \times 10^5$ Bq. In the WAK off-gases concentrations of up to 0·3 Bq/m^3 have been detected.

3.3. Other steps of the nuclear fuel cycle

During fuel reprocessing technetium is partly carried over with uranium. When the recycled uranium is transformed into UF_6, technetium reacts with the fluoride to form volatile compounds. Small amounts are released to the environment during UF_6 conversion, U-235 enrichment and fuel fabrication.

The only data available are those from the Capenhurst enrichment plant in the UK as given in Table VII. No data are available on technetium-99 discharges associated with waste treatment (vitrification, bitumenisation, etc.).

Table VII

Year	*Technetium-99 discharges in liquid effluent from Capenhurst plant (GBq/y)*
1978	2·6
1979	12.2
1980	11·5
1981	6.3
1982	20·4
1983	3·4

3.4 Total technetium discharged into the environment by the nuclear fuel cycle

At the end of 1983 the total electricity generated by nuclear power stations over the world was 762 GW(e)y [11] or about 2500 GW(th)y. At a production rate of 6 TBq/GW(th)y [1] the corresponding amount of technetium-99 generated would be 15 000 TBq.

If it is assumed that some 10% of the fuel from which this electricity has been produced is reprocessed [12], and that before the end of 1980 all technetium was discharged and thereafter only 10%, the total discharge would be of the order of 1000 TBq (total nuclear electricity production at the end of 1980 was 419 GW(e)y).

4. TECHNETIUM RELEASED INTO THE ENVIRONMENT FROM NUCLEAR DETONATIONS

The quantity of technetium-99 produced by nuclear weapon explosions in the atmosphere to date can be estimated to be some 140 TBq [13].

5. TECHNETIUM USED IN MEDICINE

While technetium-99 is not used per se, technetium-99m is widely applied in checking thyroid up-take and scintigraphy of many organs and tissues including liver, spleen, lung, kidney and bone marrow. It is obtained from a generator source containing molybdenum-99 in activities up to 2×10^{10} Bq [14].

Again, on the basis of relative half-lives, the resulting technetium-99 in the environment will be negligible.

6. CONCLUSION

Technetium-99 is the most important technetium isotope released to the environment because of its long life and its relatively high fission yield. Its release to date, mainly as a result of nuclear fuel reprocessing, is estimated to be of the order of 1000 TBq. The quantity from nuclear weapons testing would only be some 10–15% of this value.

REFERENCES

1. Data from a study made by Associated Nuclear Services, Epsom, Surrey, UK, for the Commission of the European Communities, September 1979.
2. Holtschmidt, H. and Bonka, H. Strahlenexposition durch aerosolgebundene Radionuklide mit Halbwertzeiten kleiner als 8 Tagen. (Report in preparation, RWTH, Aachen, 1984.)

3. Bonka, M. (1982). Strahlenexposition durch radioaktive Emissionen aus kern-technischen Anlagen in Normalbetrieb. Verlag TÜV, Rheinland.
4. Luxenburger, H. J., Schüttelkopf, H. and Bohn, B. (1984). Die Emission von Tc-99 aus kerntechnischen Anlagen, KfK 3795.
5. Decker, T. R. (1977). Radioactive materials released from nuclear power plants. Annual Report, NUREG-0521, USNRC.
6. Luykx, F. and Fraser, G. Radioactive effluents from nuclear power stations and nuclear fuel reprocessing plants in the European Community; Discharge data 1978–1982; radiological aspects. C.E.C. report, to be published.
7. UNSCEAR (1982). *Ionising Radiation: Sources and Biological Effects*, New York, United Nations.
8. BRITISH NUCLEAR FUELS LIMITED. Annual report on radioactive discharges and monitoring of the environment, 1978, 1979, 1980, 1981, 1982, 1983.
9. Atherton, R. Personal communication, BNFL.
10. Patti, F., Masson, M., Vergnaud, G. and Jeannaire, L. Data on released activities, dispersion and transfer of Tc-99 in the Channel. Scientific Seminar of the Behaviour of Technetium in the Environment, Oct. 1984, Cadarache, France.
11. Nuclear power reactors in the world. Reference Data series, Vienna, IAEA, April 1984.
12. Luykx, F. and Fraser, G. (1982). The environmental tritium inventory. European Seminar on the Risks from Tritium Exposure, Mol, Report EUR-9065.
13. Aarkrog, A. *et al.* Time trend of Tc-99 in seaweed from Greenland waters (this volume, pp. 69–78).
14. The Radiochemical Centre, Amersham, Medical Products, 1976/7.

3

Radiochemical Determination of Tc-99 and Release of Tc-99 from Nuclear Facilities

B. BOHN, H. J. LUXENBURGER and H. SCHÜTTELKOPF
Kernforschungszentrum Karlsruhe GmbH, Hauptabteilung Sicherheit, Postfach 36 40, D-7500 Karlsruhe, Federal Republic of Germany

1. INTRODUCTION

The experimental determination of the factors of soil to plant and water to marine biota transfer which constitutes the basis for calculating the dose-exposure of the population yielded unexpectedly high values as compared to the iodine concept previously assumed (Tables I and II) [1]. As measured data are not available about the actual emission of Tc-99 from nuclear facilities, an analytical method was developed to determine the emission of Tc-99 with the gaseous and liquid effluents [2].

2. THE RADIOCHEMICAL PROCEDURE

Liquid effluent samples are collected in polyethylene bottles. Sampling in gaseous effluents is performed by conventional NaOH scrubbers. No noticeable adsorption takes place on the walls of plastic bottles from HNO_3/H_2O_2 and NaOH (Table III). A known Tc-95m activity is added before analysis begins in order to evaluate the chemical yield of analysis.

To oxidise Tc to pertechnetate and to make pertechnetate resistant to reduction by the ion exchanger resin, H_2O_2 is added to the samples. Tc-99 is adsorbed on an anion exchanger using a batch procedure. After the exchanger has been transferred into a column, elution takes place with perchlorate/sulphide. By this first step of analysis a very good decontamination from cationic fission products, except for Zr-95/Nb-95, Ru-106/Rh-106, Sb-125, Se-79 and Te-127m, is achieved, on the one

Table I
Transfer factors soil/plant (in Ci/kg dry plant material per Ci/kg dry soil)

Plants	*Species*	*Transfer factor*	*Reference*
Soybeans	*Glycine max*	160	3
stem		46·9	
leaves		115	
buds		12·2	
cotyledon		528	4
		48	
Wheat	*Tricitum aestivum*		3
leaves		389	
blade		2·1	
ear		0·6	
total plant		210	
		162	4
		129	4
Barley	*Hordeum vulgare*	156	4
Oats	*Avena sativa*	144	4
Radish	*Raphanus sativus*		
root		54·7	4
shoot		284	4
Corn	*Zea mays*	76·7	4
Goose foot plant	*Salsola kali*	214	5
		223	
Sweet grass	*Bromus tectorum*	161	5
		252	6
Grasses		0·4–9·5	7
Festuca arundinacea		1·7	
Lespedeza cuneata		5·4	
Panicum sp.		8·2	
Sorghum sp.		0·4	
Rubus sp.		1·5	
various species		9·5	

hand, and the matrix of the sample is eliminated, on the other hand. The perchlorate is separated from the pertechnetate by reduction of technetium and coprecipitation with Fe(II)hydroxide. After dissolution of the precipitate in sulphuric acid and oxidation of technetium with H_2O_2, pertechnetate is extracted with TBP/alkane. To improve decontamination from Zr-95/Nb-95, Ru-106/Rh-106 and Sb-125 re-extraction takes place with NaOH and after acidification with H_2SO_4 a second extraction step is carried out with TBP/alkane. The alkane is stripped with a rotation evaporator, the remaining TBP is dissolved directly in the scintillator (Fig. 1).

Table II
Transfer factors water/aqueous animals and plants (in Ci/kg fresh weight/Ci/kg water)

Plant/animal	*Species*	*Transfer factor*	*Reference*
Algae	*Fucus vesiculosus*	70·000	8
Worms	*Nereis diversicolor*	Tc(VII): 178–343	9
		Tc(IV): 57–131	9
Isopods	*Anilocra meditteranea*	35	9
Shrimp	*Paleamon elegans*	8	9
	Crangon vulgaris	14	
Lobster[a]	*Hommarus gammarus*	1 160	10
Snails	*Haliotis rufescens*	135–205	11
		100	12
	Helisoma	121	13
Mussel	*Mytilus gallo provincialis*	1·3–1·7	9
	Mytilus californias	1·5	12
	Crasso strea gigas	1.4	12
Plaice	*Pleuronectes platerna*	6	10
		8·7	12
Mosquito-fish	*Gambusia affinis*	75	13
Carp	*Caprinus carpio*	11	13

[a] Dryweight basis.

Table III
Adsorption of TcO_4^- on glass and plastic bottles using different solutions

Contact time in d	*Adsorption in %*				
	HNO_3/H_2O_2[a]	HNO_3/H_2O_2[b]	HCl/H_2O_2[a]	H_2SO_4/H_2O_2[a]	$NaOH$[a]
	Glass bottles/plastic bottles				
5	0/0	–/0	0/0	0/0	0/0
10	0/0	–/0	0/0	8/4	0/0
15	0/0	–/0	0/0	6/4	0/0
20	0/0	–/0	0/0	8/4	0/0
25	1/0	–/0	0/2	6/4	0/0
30	3/0	–/0	4/2	5/4	0/0
35	4/0	–/0	2/2	5/4	0/0
40	4/0	–/0	1/2	5/10	0/0

[a] $c(HNO_3) = c(HCl) = c(H_2SO_4) = c(NaOH) = 0{\cdot}1$ mol/litre; $c(H_2O_2) = 0{\cdot}01$ mol/litre.
[b] $c(HNO_3) = 1$ mol/litre.

Oxidation: H_2O_2/HNO_3
Anion exchange: Dowex 1 × 8, 50–100 mesh

↓

Elution: $NaClO_4/SO_3^{2-}$

↓

Reduction: Fe^{2+}
Precipitation: $Fe(OH)_2$

↓

Dissolution: $H_2SO_4 + H_2O_2$
Extraction: TBP/Alcane

↓

Re-extraction: NaOH
Acidification: H_2SO_4

↓

Extraction: TBP/Alcane
Evaporation of alcane

↓

Counting of TBP solution by LSC

FIG. 1. Schematic description of the radiochemical procedure for Tc-99.

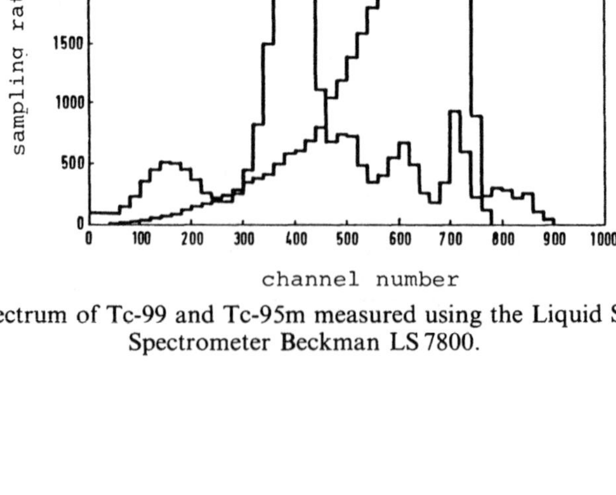

FIG. 2. Spectrum of Tc-99 and Tc-95m measured using the Liquid Scintillation Spectrometer Beckman LS 7800.

The chemical yield of the method of analysis ranges from 50% to 90%. For Zr/Nb, Ru/Rh, Sb, I, Se and Te the method yields decontamination factors $> 10^4$. At present 15 analyses are performed per week.

3. THE COUNTING PROCEDURE

Liquid scintillation spectrometry is used to determine the Tc-99 activity and the chemical yield with Tc-95m. The amount of Tc-95m to be added and the conditions for low-level measurements are optimised with the help of the theory of errors [2]. With a chemical yield of 75%, 30 dpm Tc-95m added, and 600 min of duration of the measurement, a detection limit of 1 pCi per sample is attained. The β-spectrum of Tc-99 and Tc-95m is presented in Fig. 2.

4. APPLICATIONS OF THE PROCEDURE

The investigations are restricted to the emission of Tc-99 from the facilities of the Karlsruhe Nuclear Research Centre and the Karlsruhe Reprocessing Plant. Until now the studies have related to the emission of Tc with the gaseous effluents from the Karlsruhe Reprocessing Plant, the incineration plant, and from the medium active waste evaporator operated by the Decontamination Services. The Tc released by all facilities of the Karlsruhe Nuclear Research Centre with the liquid effluents from the final basins

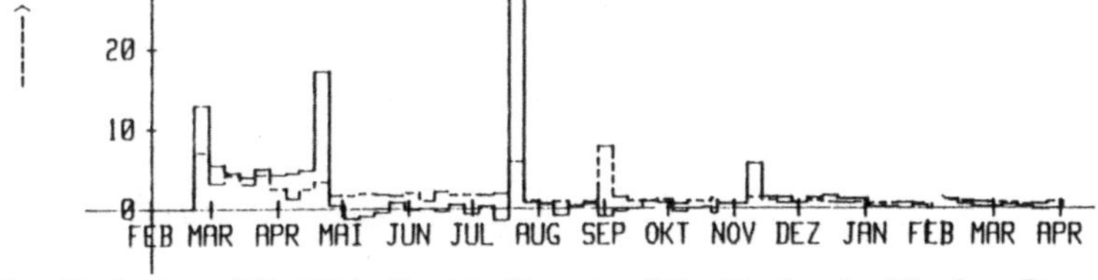

Fig. 3. Emission of Tc-99 in liquid effluents of the Karlsruhe Nuclear Research Centre. —— emission of Tc-99; ---- detection limit for the Tc-99 determination.

Table IV
Emission of Tc-99 in the gaseous effluents of the Neckarwestheim Power Plant

Sampling period	*Activity concentration* (nCi/m^3)	*Emission* (nCi/h)
25.11.83–02.12.83	0·2 ± 0·1	40
02.12.83–09.12.83	0·1 ± 0·2	<35
09.12.83–16.12.83	−0·1 ± 0·3	<25
16.12.83–23.12.83	−0·1 ± 0·2	<25
23.12.83–30.12.83	−0·2 ± 0·1	<10
30.12.83–06.01.84	−0·1 ± 0·2	<20
06.01.84–13.01.84	0·0 ± 0·2	<20
13.01.84–20.01.84	−0·1 ± 0·1	<15
27.01.84–03.02.84	0·0 ± 0·2	<25
03.02.84–10.02.84	0·1 ± 0·2	<25

of the sewage clarification plant was measured, too. Since the main Tc activity is in the liquid effluents waste solutions discharged by the Karlsruhe Reprocessing Plant to the Decontamination Services, the Tc content in the distillate of the low-level waste evaporator is measured in addition prior to its discharge into the final basin.

The activity of Tc in samples collected in final basins (Fig. 3) and in the gaseous effluents from incineration plant and nuclear power plants (Table IV) exceeds the detection limit of 1–2 pCi per sample in rare cases

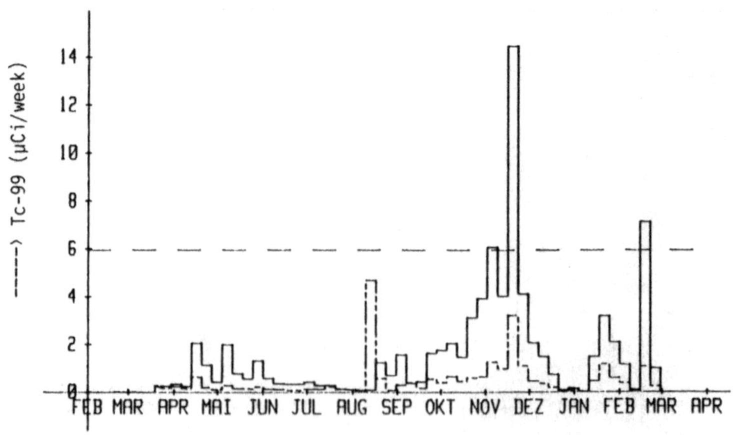

FIG. 4. Emission of Tc-99 in the gaseous effluents of the Karlsruhe Reprocessing Plant. —— emission of Tc-99; ---- detection limit for the Tc-99 determination.

Table V
Tc-99 content of U produced by the Karlsruhe Reprocessing Plant

Campaign	*Burn-up (MWd/t)*	*Tc content (ppb/g uranium)*
HDR	300	0·7
KWO II	24 500	5·9
KRB II	17 600	27
KWO V	29 200	80
KWO V/VI	28 900	98
3. MZFR	6 900	150
4. MZFR	10 500	360
2. FR-2	13 950	350
3. VAK	17 100	230
5. MZFR	12 000	230
3. FR-2	12 000	500
KWO VII	28 000	315
OH(UPII)	29 100	40

only and, if so, by a small amount only. This corresponds to an activity concentration of 0·3 nCi/m^3 liquid effluent and 0·3 pCi/m^3 gaseous effluent. In samples taken from the exhaust air of the medium active waste evaporator, maximum activities of 1·4 pCi/m^3 gaseous effluents can be detected provided that the facility was operating. The detection limits are exceeded most of the time in the gaseous effluent samples of the Karlsruhe Reprocessing Plant (Fig. 4) and in the distillate of the low-level waste evaporator. The highest values of activity concentrations were 7 pCi/m^3 exhaust air from the Karlsruhe Reprocessing Plant and 2 nCi/m^3 distillate in the low-level waste evaporator.

First Tc-99 concentrations have been measured in process solutions and in the uranium product of the Karlsruhe Reprocessing Plant (Table V).

REFERENCES

1. Luxenburger, H. J. and Schüttelkopf, H. (1984). KfK 3418.
2. Luxenburger, H. J., Schüttelkopf, H. and Bohn, B. (1984). KfK 3418.
3. Wildung, R. E., Garland, T. R. and Cataldo, D. A. (1977). *Health Physics*, **32**, 315.
4. Gast, R. G., Landa, E. R., Thorvig, L. J., Grigal, D. F. and Balogh, J. C. (1979). COO-2447-6.

5. ROUTSON, R. C. and CATALDO, D. A. (1978). *Health Physics*, **34**, 685.
6. CATALDO, D. A. (1979). PNL-2740.
7. HOFFMAN, F. O. (1982). ORNL-5856. *Environmental Science and Technology*, **16**, 214.
8. GROMOV, V. V. (1976). *Mar. Sci. Commun.*, **2**, 227.
9. FOWLER, S. W., BENEYOUN, G., PARSI, P. and EASSA, M. W. A. (1980). Impacts of radionuclide releases into marine environment, Vienna.
10. PENTREATH, P. J. (1980). *Ibid.*
11. SPIES, R. B. (1974). *Health Physics*, **29**, 695.
12. BEASLEY, I. M., LORZ, H. V. and GONOR, J. J. (1981). DOE/EV/10251-4.
13. BLAYLOCK, B. G., FRANK, M. L. and DEANGELIS, D. L. (1982). *Health Physics*, **42**, 257.

4

Activités du Technétium 99 Mesurées dans les Eaux Résiduaires, l'Eau de Mer et Deux Bioindicateurs (Littoral de la Manche, 1983)

F. PATTI,[a] M. MASSON,[b] G. VERGNAUD[c] et L. JEANMAIRE[a]

[a] *Service d'Etudes Appliquées de Protection Sanitaire, Département de Protection Sanitaire—IPSN, BP 6, 92260 Fontenay aux Roses, France*

[b] *Service d'Etudes et de Recherches sur l'Environnement, Département d'Etudes et de Recherches en Sécurité—IPSN, Centre de La Hague, BP 270, 50107 Cherbourg, France*

[c] *Service de Prévention et de Radioprotection, Cogema, La Hague, BP 270, 50107 Cherbourg, France*

1. INTRODUCTION

Le technétium 99 ne se trouve pas en quantité appréciable à l'état naturel; il a été introduit dans la biosphère par les retombées atmosphériques consécutives aux explosions nucléaires; actuellement le technétium 99 provient principalement des rejets liquides des usines de retraitement de combustibles irradiés: La Hague en France et Windscale en Grande-Bretagne. La présence de ce radionucléide dans les échantillons biologiques marins, prélevés en Mer d'Irlande [1] et sur les côtes de la Manche [2], implique également sa présence dans l'eau de mer et dans les eaux résiduaires radioactives rejetées; or, très peu de valeurs ont été publiées sur ce sujet, hormis quelques mesures relatives aux rejets de Windscale en 1978, 1979, 1980 [3, 4].

L'objet de ce travail est de présenter les premiers résultats obtenus sur les teneurs en ^{99}Tc:

(1) dans les eaux résiduaires radioactives rejetées par l'Usine de La Hague, au cours de l'année 1983;

(2) dans l'eau de mer où ont été étudiées:
les fluctuations des teneurs en une station de référence: Goury, situé à 5 km au Nord du point de rejet,
les activités mesurées le long de la côte de la Manche, à différentes distances du point de rejet;

(3) dans des échantillons de *Fucus* sp. et *Patella* sp. prélevés concomitamment afin de tenter une estimation des facteurs de transfert 'in situ';

(4) dans des échantillons de Fucus prélevés à La Rochelle et à Damgan (Côte Atlantique)—zones non soumises aux rejets de La Hague—afin de déterminer des valeurs de ^{99}Tc dues aux seules retombées atmosphériques.

2. ECHANTILLONS

2.1. Eaux résiduaires radioactives

Les analyses ont porté sur des échantillons mensuels représentant l'activité volumique moyenne du rejet mensuel; le volume prélevé à chaque rejet a été pondéré en fonction du volume rejeté. Douze échantillons ont été analysés pour l'année 1983.

2.2. Eau de mer

Seize prélèvements ont été effectués à Goury, au cours de l'année 1983, les volumes prélevés ont été de 20, 60 et 100 litres (Tab. II).

La dispersion des teneurs a été déterminée sur 12 stations du littoral, de Cancale à Wimereux (Fig. 1); le volume analysé a été de 60 litres pour Herquemoulin, Siouville et Goury, de 100 litres pour les autres stations plus éloignées du point de rejet.

Au large, au voisinage de l'Ile de Jersey, quatre prélèvements de 100 litres ont été effectués lors de la campagne en bateau 'Pluteus', organisée par le Centre National de la Recherche Scientifique, au printemps 1983.

2.3. *Fucus* sp. et *Patella* sp.

L'échantillonnage est effectué selon une radiale dans la zone à Fucus à raison de 5 kg par point, ce qui correspond à environ 1 kg de produit sec.

La mesure du technétium sur le littoral a été faite sur des prélèvements effectués en avril et mai provenant de 19 stations de Graveline à Roscoff

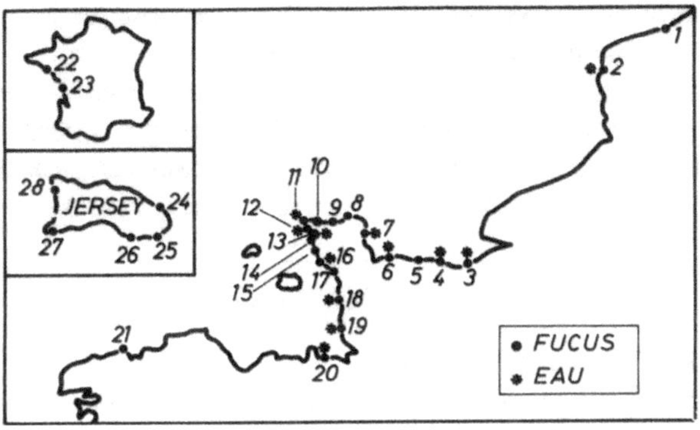

FIG. 1. Repartition géographique des stations de prélèvement.

	km		*km*
1. Graveline	330	12. Herquemoulin	5
2. Wimereux	300	13. Siouville	10
3. Villerville	140	14. Dielette	15
4. Luc sur mer	130	15. Sciotot	20
5. Arromanche	115	16. Le Rozel	25
6. Grancamp	100	17. Carteret	35
7. St Vaast	70	18. Blainville	60
8. Fermanville	40	19. Granville	90
9. La Saline	30	20. Cancale	110
10. St Martin	10	21. Roscoff	180
11. Goury	5		
Point de rejet	0		

(Ouest Bretagne). Par ailleurs cinq prélèvements du littoral de l'Ile de Jersey ont été effectués le 30 mai 1983 et analysés.

Les patelles sont ramassées selon une radiale à la côte, à raison de 10 kg de coquillage correspondant à environ 5 kg de chair soit de l'ordre de 1 kg de produit sec.

3. PRÉPARATION DES ÉCHANTILLONS ET TECHNIQUES RADIOCHIMIQUES

Tous les échantillons d'eau de mer ont été filtrés sur filtre millipore de 0·45 μm de porosité. En ce qui concerne les fucus et patelles, l'analyse a porté sur le produit sec; la préparation des échantillons a été effectuée au Laboratoire de Radioécologie Marine du Centre de La Hague, selon le protocole habituel [5].

Les techniques analytiques utilisées pour le dosage de ^{99}Tc dans ces trois types d'échantillons ont été publiées [6–8].

Lorsque l'activité de l'échantillon le permettait, le rendement chimique a été déterminé par un ajout de ^{95m}Tc (environ 200 pCi); pour les échantillons de faible activité: eau de mer, fucus de La Rochelle, de Roscoff, un rendement chimique moyen a été utilisé pour le calcul de l'activité.

4. RÉSULTATS

4.1. Eaux résiduaires radioactives

Les concentrations moyennes par mètre cube du volume mensuel rejeté sont représentées Tableau I; elles varient de 30 MBq m^{-3} à 322 MBq m^{-3} (0·8 à 8·7 mCi m^{-3}) soit, pour un volume de 92 386 m^3 rejeté au cours de l'année 1983, une activité de 11·7 TBq (316 Ci).

4.2. Activité de l'eau de mer

La Figure 2 représente la variation des concentrations de ^{99}Tc, le long du littoral de la Manche; elles sont de l'ordre de 20 mBq litre^{-1} à Herquemoulin et Siouville, situées respectivement à 5 et 10 km au Sud de point de rejet, et sensiblement du même ordre à Goury (18 mBq litre^{-1}) située 5 km au Nord à 100 km de l'émissaire, les niveaux sont sensiblement supérieurs sur la côte Est: 3 mBq litre^{-1} à Grancamp, 1·5 mBq litre^{-1} à Granville (côte Ouest).

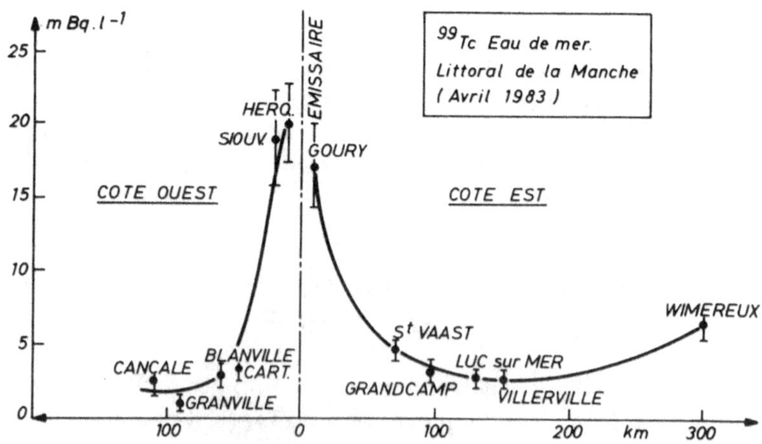

FIG. 2. Variation des concentrations de ^{99}Tc dans l'eau de mer le long du littoral de la Manche (incertitude sur teneurs tenant compte des erreurs sur rendement chimique et comptage, $p = 0{\cdot}95$ pour une mesure).

Tableau I

Rejets: eaux résiduaires radioactives—La Hague 1983

Echantillons mensuels	*Volume rejeté (m^3)*	*Technétium 99*	
		Concentration ($MBq\,m^{-3}$)	*Activité rejetée (GBq)*
Janvier	10 277	44	452
Fevrier	7 473	189	1 412
Mars	9 307	104	968
Avril	8 133	130	1 057
Mai	7 589	322	2 444
Juin	7 032	244	1 716
Juillet	6 326	96	607
Août	5 233	74	387
Septembre	6 667	30	200
Octobre	7 384	104	768
Novembre	7 620	96	732
Decembre	9 345	104	972
Total	92 386		11 715 (316 Ci)

Tableau II

Activités de ^{99}Tc dans l'eau de mer a Goury: prélèvements effectuée en 1983

Mois	*Volume analysé (litre)*	*Activité ($mBq\,litre^{-1}$)*	*Activité moyenne*
Janvier	100	5 ± 1*	—
Fevrier	20	14 ± 3	15
	100	15 ± 2	
Mars	20	17 ± 4	16
	100	15 ± 2	
Avril	20	15 ± 3	18
	100	21 ± 3	
Mai	—	—	—
Juin	100	14 ± 2	—
Juillet	20	19 ± 3	19
	100	19 ± 3	
Août	20	8 ± 2	9
	100	9 ± 2	
Septembre	60	12 ± 2	
October	60	12 ± 2	
Novembre	60	20 ± 3	
Decembre	60	9 ± 2	

* Incertitude sur les teneurs tenant compte de l'erreur sur le rendement chimique ($85 \pm 15\%$) et de l'erreur de comptage ($p = 0{\cdot}95$ pour une mesure).

Cette décroissance des activités de part et d'autre du point de rejet est comparable à celle observée sur les prélèvements de fucus [9] comme le montre la Figure 5 avec cependant une dérive de l'activité plus importante vers l'Est.

Les résultats concernant les prélèvements mensuels effectués à Goury figurent Tableau II. On remarque d'importantes fluctuations mensuelles: 5 mBq $litre^{-1}$ en janvier, 20 mBq $litre^{-1}$ en novembre, soit un facteur quatre entre les deux valeurs extrêmes.

Les prélèvements effectués au large—aux alentours de l'Ile de Jersey—présentent des activités variant de 1 à 4 mBq $litre^{-1}$ comme le montre la Figure 3.

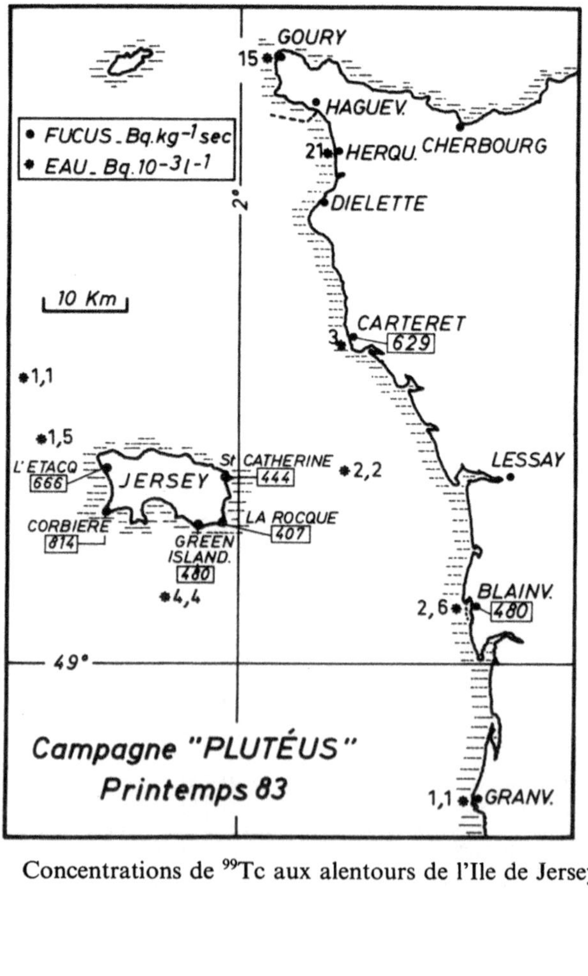

FIG. 3. Concentrations de ^{99}Tc aux alentours de l'Ile de Jersey.

4.3. Activités observées chez *Fucus* sp.

Les Fucus serratus et vesiculosus sont de bons bioindicateurs qui fixent particulièrement le ^{106}Ru-Rh et le ^{99}Tc; au Tableau III, sont représentées les activités de ces radionucléides, estimées sur douze prélèvements mensuels effectués à Goury. Il apparait que l'activité dominante est celle du technétium avec une valeur moyenne annuelle de 2600 Bq kg^{-1} sec, soit approximativement le double de celle de ^{106}Ru-Rh: 1300 Bq kg^{-1} sec, bien que l'activité rejetée en ^{106}Ru-Rh soit très supérieure à celle du technétium: 637 TBq (17 200 Ci) contre 11·7 TBq (316 Ci) pour l'année 1983. Ces chiffres mettent en évidence une affinité plus grande de *Fucus* sp. pour le technétium.

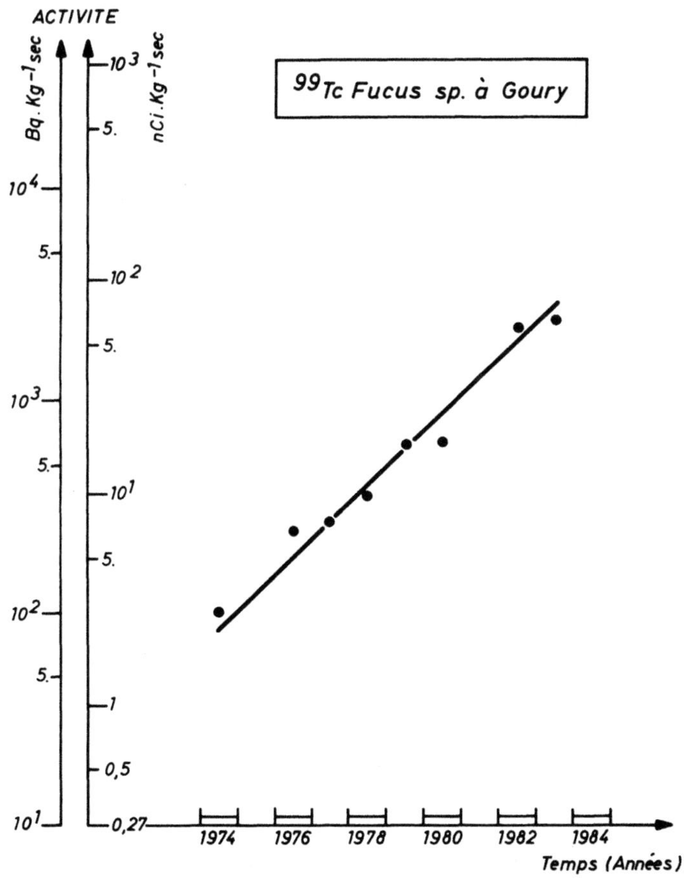

FIG. 4. Evolution dans le temps de l'activité en ^{99}Tc, à Goury, de 1974 à 1983.

Ce tableau montre également que l'activité en potassium 40, isotope radioactif naturel, est du même ordre que celle du ^{106}Ru-Rh: 1200 Bq kg^{-1} sec.

Par ailleurs, les fluctuations mensuelles des teneurs en ^{99}Tc sont assez peu marquées, elles varient entre 3330 Bq kg^{-1} sec (maximum) et 2000 Bq kg^{-1} (minimum) soit un facteur 1·7.

4.3.1. Evolution dans le temps de l'activité en ^{99}Tc à Goury
La courbe représentée Figure 4 montre l'évolution de l'activité entre 1974 et 1983. Elle a varié de 100 à 2600 Bq kg^{-1} sec (3·7 à 96 Bq kg^{-1} frais), soit une croissance d'un facteur 26 en 9 ans—croissance exponentielle—correspondant à une période d'environ 1·8 an, jusqu'à l'obtention d'un état d'équilibre.

4.3.2. Variation des teneurs le long du littoral de la Manche
Le technétium 99 est détectable de Graveline: 300 Bq kg^{-1} sec, à Roscoff: 7 Bq kg^{-1} sec (extrême Ouest de la Bretagne); la variation est représentée Figure 5; elle est comparable à celle obtenue pour l'eau de mer (Fig. 2), avec une teneur sensiblement plus élevée à Herquemoulin qu'à Goury, respectivement 3100 et 2600 Bq kg^{-1} sec, soit environ 15%; les intervalles

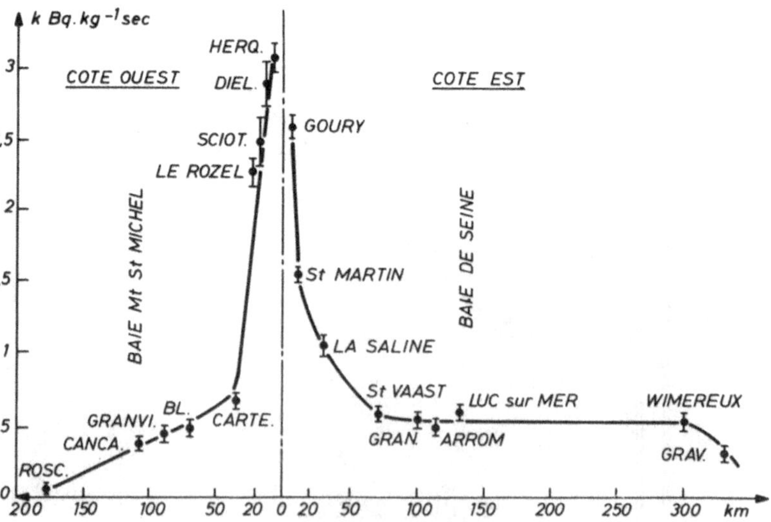

FIG. 5. Variation des concentrations de ^{99}Tc chez *Fucus* sp. le long du littoral de la Manche (incertitude sur teneurs tenant compte des erreurs sur rendement chimique et comptage, $p = 0{\cdot}95$ pour une mesure).

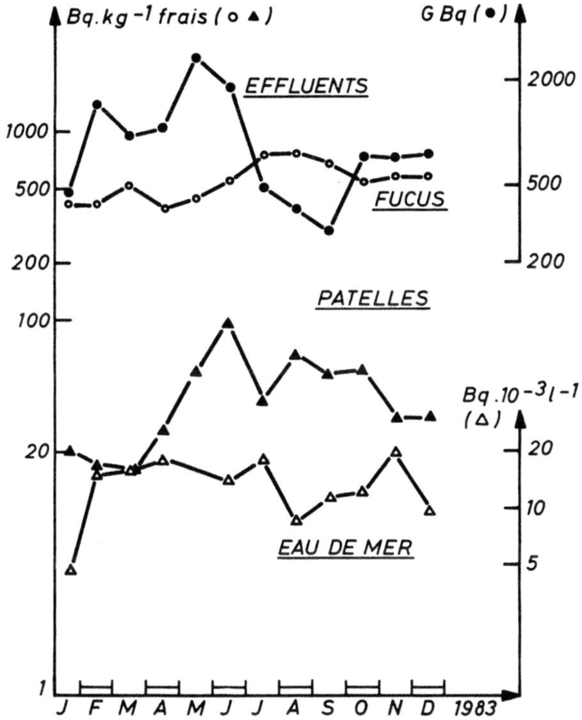

FIG. 6. Variation annuelle de l'activité en ^{99}Tc dans les eaux résiduaires, l'eau de mer, *Fucus* sp. et *Patella* sp. (eau de mer, fucus et patelles ont été prélevés à Goury au même moment).

de confiance portés sur les valeurs de la Figure 5 représentent l'incertitude inhérente à l'analyse (erreur sur le rendement chimique et le comptage) qui ne tient pas compte des fluctuations biologiques; de ce fait il est difficile d'affirmer que la différence est bien significative: un échantillonnage plus important s'avèrerait nécessaire.

Les valeurs du Tableau III—reproduites graphiquement Figure 6—représentant les teneurs en ^{99}Tc des échantillons mensuels prélevés à Goury montrent les fluctuations, selon le mois de prélèvement, qui varient d'un facteur 1·7 chez *Fucus* sp.

Les valeurs concernant les prélèvements provenant du littoral de l'Ile de Jersey sont comprises entre 400 et 600 Bq kg^{-1} sec et comparables à celles observées à Blainville et Carteret (480 Bq kg^{-1} sec et 630 Bq kg^{-1} sec), stations situées sur le littoral de la Manche (Fig. 3).

Tableau III

Activités chez *Fucus* sp. et *Patella* sp. Goury: prélèvements effectués en 1983

Date de prélèvement	Fucus *sp.*				Patella *sp.*			
	$P^{\%}_{sec}$	*kBq kg*$^{-1}$ *sec*			$P^{\%}_{sec}$	*kBq kg*$^{-1}$ *sec*		
	P_{frais}	^{106}Ru-Rh	^{99}Tc	^{40}K	P_{frais}	^{106}Ru-Rh	^{99}Tc	^{40}K
31/1	20	2·0	2·1	1·4	14	1·4	0·15	0·37
28/2	20	1·9	2·1	1·3	17	2·1	0·10	0·39
29/3	20	1·7	2·6	1·4	18	3·0	0·90	0·43
27/4	19	1·3	2·0	1·2	18	1·9	0·14	0·36
26/5	21	1·0	2·1	1·4	21	1·4	0·25	0·31
13/6	18	1·1	3·0	1·2	21	1·1	0·46	0·35
11/7	25	1·1	3·1	1·0	22	1·0	0·17	0·31
10/8	24	1·2	3·3	0·9	21	1·7	0·32	0·38
8/9	21	1·2	3·3	1·1	20	1·1	0·26	0·33
6/10	23	1·2	2·3	1·1	21	0·8	0·26	0·32
3/11	23	1·1	2·6	1·1	18	0·7	0·17	0·34
5/12	23	1·1	2·6	1·2	15	1·0	0·20	0·34
Moyenne	21	1·3	2·6	1·2	19	1·4	0·21	0·35

Mesures de ^{106}Ru-^{106}Rh et de ^{40}K effectuées par le Groupe d'Etudes Atomiques—Marine Nationale—Cherbourg.

Tableau IV
Rapport des teneurs en ^{99}Tc: *Fucus* sp. frais/eau de mer Goury: prélèvements effectués en 1983

Mois	*Eau de mer* (*mBq litre^{-1}*)	*Fucus* (*Bq kg^{-1} frais*)	*Rapport* Fucus *sp. frais* / *Eau*
Janvier	5	430 ± 10*	86 000
Fevrier	15	422 ± 15	28 133
Mars	16	518 ± 20	32 375
Avril	18	380 ± 23	21 100
Mai	—	443 ± 9	—
Juin	14	540 ± 13	38 570
Juillet	19	787 ± 39	41 420
Août	9	799 ± 40	88 770
Septembre	12	699 ± 28	58 250
Octobre	12	534 ± 27	44 500
Novembre	20	596 ± 12	29 800
Decembre	9	593 ± 30	65 890
Moyenne			48 619

* Incertitude sur le rendement chimique et le comptage ($p = 0{\cdot}95$ pour une mesure).

En ce qui concerne les prélèvements effectués dans des zones non soumises aux rejets: Damgan, La Rochelle (prélèvements provenant du réseau d'observation de la radioactivité en milieu marin du Service d'Etudes Appliquées de Protection Sanitaire, DPS (Centre d'Etudes Nucléaires de Fontenay aux Roses)), les concentrations sont respectivement de 0·5 et 0·9 Bq kg^{-1} sec.

4.3.3. Transfert eau—Fucus

La capacité de fixation du technétium 99 par cette algue peut être évaluée par le rapport de l'activité Fucus frais en Bq kg^{-1} sur l'activité de l'eau (Bq litre^{-1}). Ce rapport représente le facteur de concentration ou de transfert. Il a été calculé pour les prélèvements mensuels effectués à Goury en 1983; le Tableau IV indique les valeurs obtenues. Il apparait une importante variation du rapport, de 21 000 à 89 000, qui est liée aux fluctuations des teneurs de ^{99}Tc dans l'eau de mer et dans l'échantillon biologique lui-même comme le montre la Figure 6.

Pour les stations plus éloignées du point de rejet, les valeurs de ce rapport sont données Tableau V; elles varient entre 14 000 et 50 000.

Tableau V

Rapport des teneurs en ^{99}Tc: Fucus frais/eau de mer dans des zones eloignées du point de rejet: prélèvements effectués en 1983

Lieu de prélèvement	*Distance (km)*	*Eau de mer (mBq litre^{-1})*	*Fucus (Bq kg^{-1} frais)*	*Rapport Fucus frais / Eau*
Wimereux	300	7 ± 1*	99 ± 5*	14 140
Luc sur mer	130	$3 \pm 0{\cdot}5$	94 ± 5	31 330
Grancamp	100	$3 \pm 0{\cdot}5$	83 ± 3	27 660
Rejet	0			
Blainville	60	$3 \pm 0{\cdot}5$	$52 + 2$	17 330
Granville	90	$1{\cdot}5 \pm 0{\cdot}3$	76 ± 3	50 660
Cancale	110	$3 \pm 0{\cdot}5$	87 ± 9	29 000
Moyenne				28 550

* Incertitude sur le rendement chimique et le comptage ($p = 0{\cdot}95$ pour une mesure).

4.4. Activités observées chez *Patella* sp.

Ce mollusque est considéré comme un bon bioindicateur. Le Tableau III représente les activités de ^{106}Ru-Rh et ^{99}Tc déterminées sur douze prélèvements mensuels effectués à Goury; l'activité est exprimée en Bq kg^{-1} de chair sèche.

Pour ce type de prélèvement, l'activité en ^{106}Ru-Rh est supérieure à celle du technétium respectivement, en moyenne 1400 Bq kg^{-1} sec, et 210 Bq kg^{-1} sec, soit une activité en ^{99}Tc sept fois inférieure.

Le Tableau III montre également que les concentrations en ^{99}Tc fluctuent de manière plus importante que chez *Fucus* sp. (minimum: 90 Bq kg^{-1} sec; maximum: 460 Bq kg^{-1} sec), soit un facteur 5 (1·7 chez le Fucus).

Par ailleurs, des analyses effectuées sur des échantillons prélevés en 1977, 1982, 1983 ont donné les valeurs suivantes: 200, 270 et 214 Bq kg^{-1} de chair sèche, indiquant qu'il n'y a pas croissance de l'activité au cours du temps contrairement à ce qui est observé chez *Fucus* sp.

4.5. Transfert eau—Patelles

Comme pour le fucus on observe d'importantes fluctuations dans le rapport d'activité qui varie de 1000 à 7000, avec une moyenne de 3400 (Tab. VI); cette dispersion des résultats est liée aux fluctuations des concentrations de ^{99}Tc dans l'eau de mer et dans l'échantillon lui-même (cf. courbes, Fig. 6).

Tableau VI
Rapport des teneurs en ^{99}Tc: Patelles (chair)/eau de mer Goury: prélèvements effectués en 1983

Mois	*Eau de mer (mBq litre^{-1})*	*Chair (Bq kg^{-1} frais)*	*Rapport* $\frac{\textit{Patelles (chair)}}{\textit{Eau de mer}}$
Janvier	5	20 ± 1*	4 000
Fevrier	15	17 ± 1	1 133
Mars	16	$16 \pm 0{\cdot}5$	1 000
Avril	18	26 ± 2	1 440
Mai	—	53 ± 4	—
Juin	14	96 ± 3	6857
Juillet	19	37 ± 2	1 950
Août	9	67 ± 4	7 440
Septembre	12	51 ± 2	4 250
Octobre	12	54 ± 3	4500
Novembre	20	30 ± 2	1 500
Decembre	9	30 ± 2	3 330
Moyenne			3 400

* Incertitude sur le rendement chimique et le comptage ($p = 0{\cdot}95$ pour une mesure).

5. DISCUSSION

L'activité de ^{99}Tc, rejetée en mer par l'usine de La Hague, a été estimée pour l'année 1983 à 11·7 TBq.

Les teneurs de l'eau de mer prélevée sur la côte varient comme chez *Fucus* sp. selon un gradient décroissant de part et d'autre du point de rejet, de 20 mBq litre^{-1} à 2 mBq litre^{-1} environ. À 40 km au large, les teneurs sont de l'ordre du millibecquerel par litre (Ile de Jersey).

La valeur des rapports d'activité est très variable: de 20 000 à 90 000 chez *Fucus* sp. et de 1000 à 7000 chez *Patella* sp. Ces fluctuations sont dues aux variations des concentrations de ^{99}Tc dans l'eau de mer et dans l'échantillon biologique.

La comparaison de la moyenne des rapports observés 'in situ' avec celle observée en laboratoire, fait apparaître une différence importante [3, 10]—*Fucus* sp. 'in situ': 50 000, en laboratoire environ: 1500,—*Patella* sp. 'in situ': 3500, en laboratoire environ: 10. Cette différence est vraisemblablement due aux conditions de vie très différente dans les deux cas: activité physiologique, croissance, nourriture, etc.

En admettant une teneur moyenne de 400 Bq kg^{-1} frais de fucus et une masse alguale de 250×10^6 kg—entre Granville et Barfleur—la quantité de technétium 99 fixé est de 0·1 TBq; comparée aux 12 TBq rejetés dans l'année, on constate que la rétention à la côte, par la ceinture alguale, est très faible—de l'ordre du pourcent; les algues constituent le support qui fixe le plus ^{99}Tc et représentent la biomasse la plus importante; la majeure partie de ^{99}Tc reste donc dans la masse aqueuse.

Par ailleurs on observe à Goury une concentration moyenne dans l'eau de mer de 13·5 mBq litre^{-1} pour un rejet annuel moyen de 32 GBq j^{-1} soit pour 37 GBq j^{-1} (1 Ci j^{-1}) rejetés, une teneur de 15·6 mBq litre^{-1} (0·4 pCi litre^{-1}); cette valeur est conforme aux résultats obtenus par P. Gueguenjat sur d'autres radionucléides [11].

Le niveau de ^{99}Tc dû aux retombées atmosphériques se situe entre 0·5 et 1 Bq kg^{-1} sec de fucus. Ce niveau a également été observé sur la côte Est de la Suède et au Sud de l'Irlande [4]. La teneur de 7 Bq kg^{-1} sec, constatée à Roscoff, limite géographique de nos prélèvements, est environ dix fois supérieure. La présence persistante de ^{99}Tc à cette station montre la grande mobilité de l'ion pertechnétate en milieu marin. Compte tenu de la demi-vie de ce radionucléide ($2{\cdot}1 \times 10^5$ ans) et de l'effet cumulatif qui en résulte, il est souhaitable de poursuivre ces mesures afin d'en suivre l'évolution dans l'espace et le temps.

Le technétium 99 apparait comme un traceur intéressant qui devrait permettre de délimiter, en raison de sa mobilité, les masses d'eau soumises à la dispersion des rejets de l'usine de La Hague.

En ce qui concerne l'impact radiologique pour l'homme, résultant de l'ingestion de patelles, il peut être considéré comme négligeable; en effet, si l'on admet une consommation annuelle de 5·5 kg de chair de patelles et une concentration de 100 Bq kg^{-1} (valeur maximale à Goury), l'équivalent de dose effectif engagé est de $0{\cdot}2 \times 10^{-6}$ Sv an^{-1}, calculé suivant les données publiées par la CIPR 30 [12], soit 4×10^{-5} de la limite de dose annuelle.

REFERENCES

1. Dutton, J. W. R. et Ibbett, R. B. (1973). The determination of Technetium 99 in marine biological materials. Symposium on the determination of radionuclides in environmental and biological materials, London, CEGB, 2–3 April.
2. Jeanmaire, L., Masson, M., Patti, F., Germain. P. et Cappellini, L. (1981). Technetium 99 content in some marine biological materials. *Mar. Poll. Bull.*, **12**, 29–32.

3. APROSI, G. et MASSON. M. (1984). Bilan des études expérimentales de transfert de technétium à des sédiments et à des espèces marines benthiques et comparaison à des résultats 'in situ'. *Radioprotection*, **19**(2), 89–103.
4. HOLM, E., RIOSECO, J. et CHRISTENSEN, G. C. (1983). ^{99}Tc in Fucus from Norwegian waters. In: *International Symposium on the Behaviour of Long-lived Radionuclides in the Marine Environment*, La Spezia (Italy), 28–30 Sept. (in press).
5. GERMAIN, P., MASSON, M. et BARON, Y. (1979). Etude de la répartition de radionucléides émetteurs gamma chez des indicateurs biologiques littoraux des côtes de la Manche et de la Mer du Nord, de février 1976 à février 1978. Rapport CEA-R-5017.
6. PATTI, F., CAPPELLINI, L., JEANMAIRE, L. et MASSON, M. (1982). Détermination du technétium 99 dans les échantillons biologiques marins (IAEA-SM-252/49). In: *Methods of Low-level Counting and Spectrometry*, Berlin 6–10 April 1981, Vienne, IAEA.
7. PATTI, F. (1982). Détermination du technétium 99 dans les eaux résiduaires radioactives. Rapport DPr 82/4-SPS. CEN/Fontenay aux Roses, France.
8. PATTI, F., GARCET, M. et JEANMAIRE, L. (1983). Détermination du technétium 99 dans l'eau de mer. Note CEA-N-2347. CEN/Fontenay aux Roses, France.
9. MASSON, M., PATTI, F., CAPPELLINI, L., GERMAIN, P. et JEANMAIRE, L. (1983). Etude de la dispersion du technétium 99 sur les côtes françaises de la Manche à l'aide de deux indicateurs biologiques: *Fucus* sp. et *Patella* sp. *J. Radioanal. Chem.*, **77**, 247–55.
10. MASSON, M., APROSI, G., LA NIECE, A., GUEGUENIAT, P. et BELOT, Y. (1981). Approches expérimentales de l'étude des transferts du technétium à des sédiments et à des espèces marines benthiques (IAEA-SM-248/124). In: *Impacts of Radionuclide Release into the Marine Environment*, Vienne 6–10 oct. Vienne, IAEA.
11. GUEGUENIAT, P. Communication personnelle.
12. ICRP (1979). Limits for intakes of radionuclides by workers, ICRP Publication 30, supplement to Part 2, Oxford, Pergamon Press, p. 195.

5

Technetium-99 in Algae from Temperate and Arctic Waters of the North Atlantic

E. HOLM,[a] J. RIOSECO,[b] A. AARKROG,[c]
H. DAHLGAARD,[c] L. HALLSTADIUS,[b]
B. BJURMAN[b] and R. HEDVALL[b]

[a] *International Laboratory of Marine Radioactivity, IAEA, Musée Océanographique, Monaco.*

[b] *Department of Radiation Physics, Lund University, S-22 185 Lund, Sweden.*

[c] *Risø National Laboratory, DK-4000 Roskilde, Denmark.*

1. INTRODUCTION

Different species of algae were used as bioindicators for ^{99}Tc in the marine environment, close to the release sources [1–3]. It was then shown that different species of brown algae were the most suitable.

In 1983 results were presented for ^{99}Tc and the ^{99}Tc/^{137}Cs ratio in *Fucus vesiculosus* and *Ascophyllum nodosum* from Norwegian waters [4] contaminated by European reprocessing facilities. The study showed that this is an excellent way to follow the transport of not only radioactivity but also of other pollutants released into the North Sea.

The ^{99}Tc/^{137}Cs activity ratio in integrated global fallout is today $\sim 2 \times 10^{-4}$ [5]. In releases from, for example, the Sellafield reprocessing plants the corresponding ratio was between 5×10^{-2} and 2×10^{-3} during 1978–82 with values decreasing with time [6].

At long distances from a source the fallout from nuclear detonation tests would limit the possibilities of detecting the source of ^{137}Cs while the influence from fallout on ^{99}Tc is relatively much smaller.

2. SAMPLING AND ANALYSIS

Sampling of species of brown algae was performed along the Norwegian coast and Svalbard during 1980–82 [4]. Sampling expeditions were made in Iceland and the Faroes in 1981, and Scotland-England in 1982. Some limited sampling was performed in Ireland in 1983. From Greenland samples were obtained during 1969–82. There is of course a difficulty in comparing the results from the various places since yearly and seasonal variations play a role. To collect samples from such a large area at the same time is however rather difficult.

These samples, often in combination with water samples, were taken for studies of radiocaesium and transuranium elements. During most of these sample collections ^{99}Tc was not considered until later when a suitable analytical method was developed [7].

Briefly the method is based on extraction of technetium with TBP from a sulphuric acid medium and back extracted with sodium hydroxide solution. Technetium is finally electroplated onto stainless steel discs. The radiochemical yield is determined with ^{99}Tc^{m} ($T_{1/2} = 6{\cdot}0$ h, $E\gamma = 140$ keV). After the decay of ^{99}Tc^{m}, ^{99}Tc is measured with an anti-coincidence shielded multi-GM gas flow counter.

3. RESULTS AND DISCUSSION

In Fig. 1 the results available for *Fucus vesiculosus* are given as Bq per kg dry weight of ^{99}Tc. For Iceland results for *F. disticus* were also used. The ^{99}Tc/^{137}Cs activity ratio is also given after the ^{99}Tc value separated by a solidus. Some of the results for the Baltic Sea are also shown.

In 1978 Pentreath *et al.* [1] found ^{99}Tc activity concentrations of 6–21 kBq kg^{-1} wet weight (30–100 kBq kg^{-1} dry weight) in *Fucus vesiculosus* which is about 15–60 times higher than our values in 1982. This corresponds well with the decreased releases of ^{99}Tc from 180 Tbq in 1978 to 3·5 TBq in 1982.

Jeanmaire *et al.* [2] reported ^{99}Tc activity concentrations of 22–79 Bq kg^{-1} wet weight ($\sim$110–395 Bq kg^{-1} dry weight) in *Fucus serratus* at Cap de la Hague.

When regarding the results the transit times from Sellafield to the different areas must be considered in connection with release data. The transit times have been summarised by Dahlgaard *et al.* [8].

The transit time to the Scottish coast is about one year, to the North Sea $\sim$3 years, to the Baltic Sea $\sim$4 years, to the central Norwegian coast $\sim$3–4 years, to Svalbard $\sim$4–6 years and to the east coast of Greenland $\sim$6–8 years.

The results in Fig. 2 show that ^{99}Tc is released in a soluble form, enters

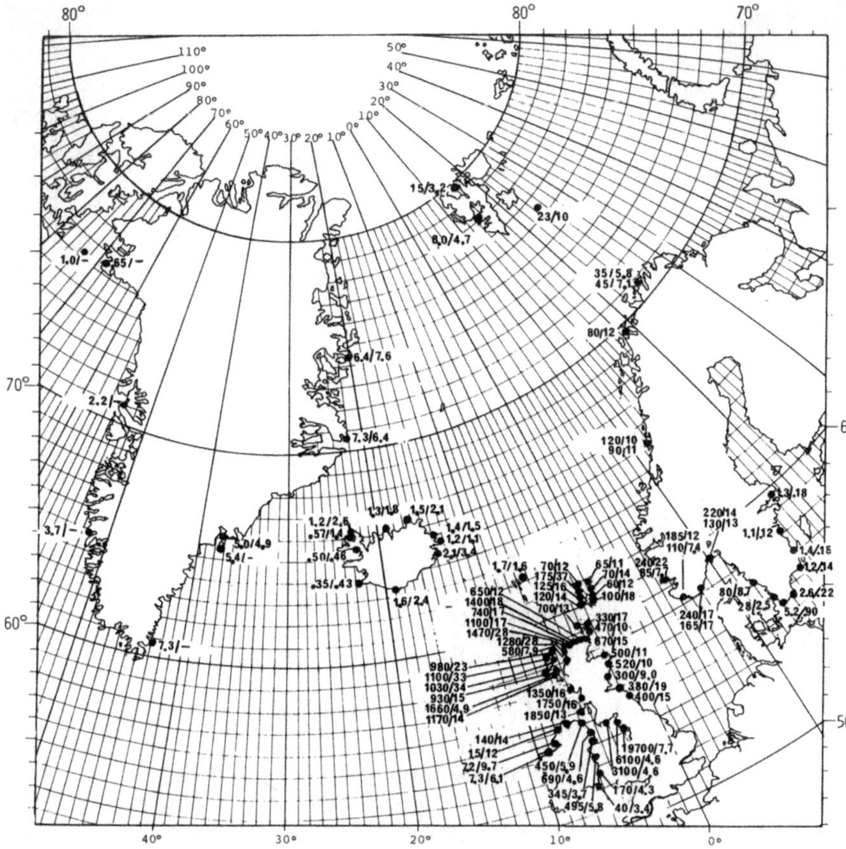

FIG. 1. Activity concentration of ^{99}Tc in *Fucus vesiculosus* (Bq kg^{-1} dry weight) from the North Atlantic and the Arctic Ocean. For the Faroes and Iceland results for *F. disticus* were also used. Results are from the Faroes and Iceland in 1981, from Norway in 1981 and 1982 (below), from Sweden in 1982, from England and Scotland in 1982, from Ireland in 1983 and from Greenland during 1979–82. The ^{99}Tc/^{137}Cs activity ratio is given after the solidus.

the North Sea and is effectively guided along the Norwegian coast by the Gulf current. It is also obvious that the Faroes and Iceland are little or not at all influenced by the releases from Sellafield. Here we find ^{99}Tc concentrations in the algae of 0·5–1·5 Bq kg^{-1} dry weight which we would like to designate as the North Atlantic fallout level. The ^{99}Tc/^{137}Cs activity ratios in samples from Iceland are 0·5–1·5. This could be compared to results from the east coast of Sweden in the Baltic where this ratio was

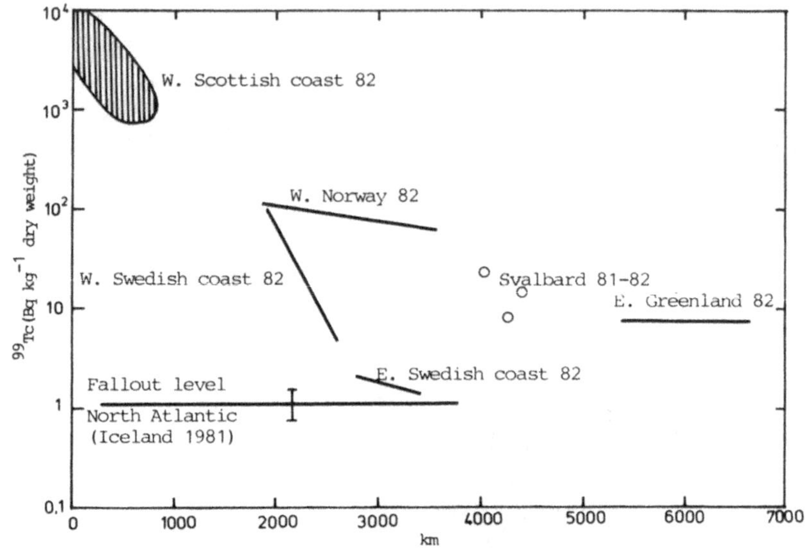

FIG. 2. ^{99}Tc activity levels in *Fucus vesiculosus* from different locations as a function of distance of route for transport from Sellafield.

around 0·15 at the same ^{99}Tc concentration level in the plants. This is because the observed concentration ratio *Fucus*/water is enhanced for ^{137}Cs but not for ^{99}Tc in low salinity water. The ^{137}Cs activity concentration in water from the Baltic Sea is about 14 Bq m^{-3} and for the area around Iceland 3–4 Bq m^{-3}. The Tc/Cs activity ratio in the water is expected to be higher than for integrated fallout due to the longer residence time in water of technetium than caesium [1].

The activity concentration decreased by only a factor of three for the years 1980 and 1982 along the Norwegian coast. In 1981 the decrease was stronger when the maximal levels occurred on the south coast of Norway. An observation similar to that made in 1980 was made for ^{137}Cs in seawater from the same area [9].

The activity ratio ^{99}Tc/^{137}Cs in the releases from Sellafield was varied as described in the introduction. Such ratios might be used to calculate transport times or differences in biological removal or sedimentation between technetium and caesium. Our results show that the major process causing decrease of concentration with distance from the North Sea along the Norwegian coast into the Polar Sea is dilution both for caesium and technetium. In the Baltic Sea [10] and in the Irish Sea [1] the other processes are important for caesium.

If we suppose that the activity ratio Tc/Cs in 1978–9 is that found at the south coast of Norway in 1981–2 and that the observed concentration ratios *Fucus*/water for caesium in the North Atlantic is ~260 [11] a concentration ratio *Fucus*/water for technetium of ~90 000 can be calculated. This would be an underestimation if we consider biological removal, sedimentation and physical decay of ^{137}Cs.

In Cattegat a concentration ratio of ~10^5 has been observed [12].

Very few measurements of ^{99}Tc have been performed on sea water. In an intercalibration sample from 1982 (IAEA, SW-N-2) collected off the NE Scottish coast, a ^{99}Tc activity concentration of 5 Bq m^{-3} was found. A concentration ratio of 85 000 *Fucus*/water can be calculated on the basis of these results. ^{99}Tc was also measured in a few samples of sea water collected off the NW coast of Norway during the Swedish YMER expedition in 1980. A concentration ratio of about 75 000 for *Fucus*/water was calculated on the basis of these results. In the Baltic Sea the concentration ratio was estimated to be 25 000 for lower salinity (~8 ‰) water [10]. Fallout levels of ^{99}Tc in the North Atlantic can be estimated on results from *Fucus* in Iceland at ~10 mBq m^{-3} and the activity ratio Tc/Cs at $2{\cdot}5 \times 10^{-3}$ in water which is 10 times higher than in integrated global fallout [5].

From some of the very same sampling sites brown algae other than *Fucus vesiculosus* were collected. It would of course be desirable to use many species as bioindicators and obtain comparable results. In Table I

Table I

^{99}Tc activity concentration and the ^{99}Tc/^{137}Cs activity ratio in some brown alga species relative to *Fucus vesiculosus*. The number of observations from the same sampling site is indicated

	Relative to Fucus vesiculosus			
	^{99}*Tc*		^{99}*Tc*/^{137}*Cs*	
Fucus serratus	0·56 ± 0·15	(1SD)	0·67 ± 0·24	(*n* = 9)
Fucus spiralis	0·54 ± 0·05		0·58 ± 0·10	(*n* = 3)
Fucus disticus	1·2 ± 0·6		1·3 ± 0·4	(*n* = 4)
Ascophyllum nodosum	2·1 ± 0·4		3·4 ± 1·3	(*n* = 18)
Laminaria digitata	0·26 ± 0·09		0·19 ± 0·03	(*n* = 2)
Halidrys siliquosa	0·74		0·98	(*n* = 1)
Pelvetia canaliculata	1·2 ± 0·4		2·7 ± 0·6	(*n* = 2)

the results are presented. From this table the uptake of ^{99}Tc by different species can be arranged as follows. For some species the number of our observations is however far too small. (*Ascophyllum nodosum* > *Pelvetia canaliculata* ~ *Fucus disticus* ~ *Fucus vesiculosus* > *Halidrys siliquosa* ~ *Fucus serratus* > *Fucus spiralis* > *Laminaria digitata*.)

This is different behaviour compared to ^{137}Cs where the difference in uptake is generally smaller.

A similar observation with uptake of ^{99}Tc in the order *Ascophyllum nodosum* > *Fucus serratus* > *Laminaria digitata* was also reported by Jeanmaire *et al.* [2].

ACKNOWLEDGEMENTS

This investigation was made possible by financial support from the Swedish Natural Science Research Council and from the CEC Radiation Protection Research Programme.

REFERENCES

1. Pentreath, R. J., Jefferies, D. F. and Lovett, M. B. (1980). In: Proceedings of Third NEA Seminar of Marine Radioecology, Tokyo, Japan, OECD, 1979, p. 203.
2. Jeanmaire, L., Masson, M., Patti, F., Germain, P. and Cappellini, L. (1981). *Marine Pollution Bulletin*, **12**, 29.
3. Masson, M., Aprosa, G., Lanièce, A., Guéguénia, P. and Belot, T. (1983). In: *Impacts of Radionuclide Releases into the Marine Environment*, Vienna, IAEA, p. 341.
4. Holm, E., Rioseco, J. and Christensen, G. C. (1983). In: Proceedings of International Symposium on the Behaviour of Long-Lived Radionuclides in the Marine Environment, La Spezia, Italy, CEC (in press).
5. Holm, E. and Rioseco, J. (1983). In: Proceedings of Seminar on the Transfer of Radioactive Materials in the Terrestrial Environment Subsequent to an Accidental Release to Atmosphere, Dublin, Ireland, CEC, p. 193.
6. ONFL (1978–1982). Annual report on radioactive discharge and monitoring of the environment. British Nuclear Fuels Limited.
7. Holm, E., Rioseco, J. and Garcia-Leon, M. (1984). *Nuclear Instruments and Methods in Physics Research*, **223**, 204.
8. Dahlgaard, H., Aarkrog, A., Hallstadius, L., Holm, E. and Rioseco, J. (1984). Paper presented at Symposium on Contaminated Fluxes through the Coastal Zone, Nantes, France.
9. Holm, E., Persson, B. R. R., Hallstadius, L., Aarkrog, A. and Dahlgaard, H. (1983). *Oceanologica Acta*, **6**, 457–62.

10. HOLM, E., RIOSECO, J. and MATTSSON, S. Technetium-99 in the Baltic Sea (this volume, pp. 61–8).
11. AARKROG, A., DAHLGAARD, H., HALLSTADIUS, L., HOLM, E. and LIPPERT, J. (1982). Risø Report R-470.
12. AARKROG, A., DAHLGAARD, H., HALLSTADIUS, L., HOLM, E., MATTSSON, S. and RIOSECO, J. Time trend of ^{99}Tc in seaweed from Greenland Waters (this volume, pp. 69–78).

6

Technetium-99 in the Baltic Sea

E. Holm,[a] J. Rioseco[b] and S. Mattsson[c]

[a]*International Laboratory of Marine Radioactivity, IAEA, Musée Océanographique, Monaco*

[b]*Department of Radiation Physics, Lund University, S-22 185 Lund, Sweden*

[c]*Department of Radiation Physics, University of Gothenburg, S-413 45 Gothenburg, Sweden*

1. INTRODUCTION

Technetium is released to the biosphere from nuclear facilities, from nuclear detonation tests and by the use of $^{99}Tc^{m}$ in nuclear medicine.

The only important isotope for environmental concern is ^{99}Tc which is considered as one of the most important radionuclides in the long-term storage of nuclear waste.

^{99}Tc ($T_{1/2} = 2{\cdot}1 \times 10^5$ a) is a pure β-emitter ($E\beta^{max} = 292$ keV) and is the daughter product of ^{99}Mo ($T_{1/2} = 67$ h) which is produced both as fission and activation products.

The fission yield of ^{99}Mo is about equal to that of ^{137}Cs. An activity ratio of about $1{\cdot}45 \times 10^{-4}$ (^{99}Tc/^{137}Cs) can be calculated.

^{99}Tc in the Baltic Sea originates from various sources such as fallout, inflow of water from the North Sea contaminated from European reprocessing plants and releases from local nuclear power plants.

It is well known that species of brown algae are excellent bioindicators for various radionuclides including technetium.

In samples of *Fucus vesiculosus* collected close to a nuclear power plant on the southwest coast of Sweden, ^{99}Tc activity concentrations of about 10 Bq kg^{-1} dry weight were observed.

This was much higher than expected from global fallout and led us to believe that there were significant releases of ^{99}Tc from the power plant. However, the activity concentrations increased with distance towards the north of the power plant which encouraged us to perform further investigations on source distribution and pathways for ^{99}Tc in the Baltic Sea. For this purpose, the brown alga *Fucus vesiculosus* played an important role as bioindicator.

2. SAMPLING AND ANALYSIS

Occasional collection of *Fucus vesiculosus* and *Fucus serratus* has been carried out along the west coast of Sweden at various places since 1979.

At one well-defined site at Sardal (56·76°N, 12·63°E) collection started as early as 1967 and is still going on. From this site more frequent sampling has been performed since 1978. These samples have been used for studies of long-term and seasonal variations for different radionuclides.

During October 1982 samples of *Fucus vesiculosus* were collected at various sites along the east and south coasts of Sweden.

During May 1983 sampling of *Fucus* was extended to cover the west coast of Sweden and at the same time 10 water samples (50–100 litres) were collected in the southern Baltic Sea during the 'Gauss' expedition organised by DHI, Hamburg.

The analysis of ^{99}Tc was carried out according to a method described earlier [1]. Briefly, technetium is extracted after ashing or precipitation (sea water) with TBP from sulphuric acid medium. Technetium is back extracted with sodium hydroxide solution and electroplated onto stainless steel discs. The radiochemical yield is determined by use of $^{99}Tc^{m}$ ($T_{1/2} = 6{\cdot}0$ h) and ^{99}Tc counted by an anti-coincidence shielded GM gas flow counter.

3. RESULTS AND DISCUSSION

3.1 Source term variation

The results for ^{99}Tc in *Fucus* collected in October 1982 along the east coast of Sweden are shown in Fig. 1. The results for the $^{99}Tc/^{137}Cs$ activity ratio are also given. Except for the most southern part the activity concentrations of ^{99}Tc are very constant: $1{\cdot}3 \pm 0{\cdot}1$ (1 SE) Bq kg^{-1} dry weight, as is the activity ratio Tc/Cs $0{\cdot}15 \pm 0{\cdot}04$.

On the south coast we see clearly higher values with ^{99}Tc concentrations $\sim 4{\cdot}5$ Bq kg^{-1} and Tc/Cs activity ratios of about 0·7. This trend of increasing values is very pronounced on the west coast of Sweden. There

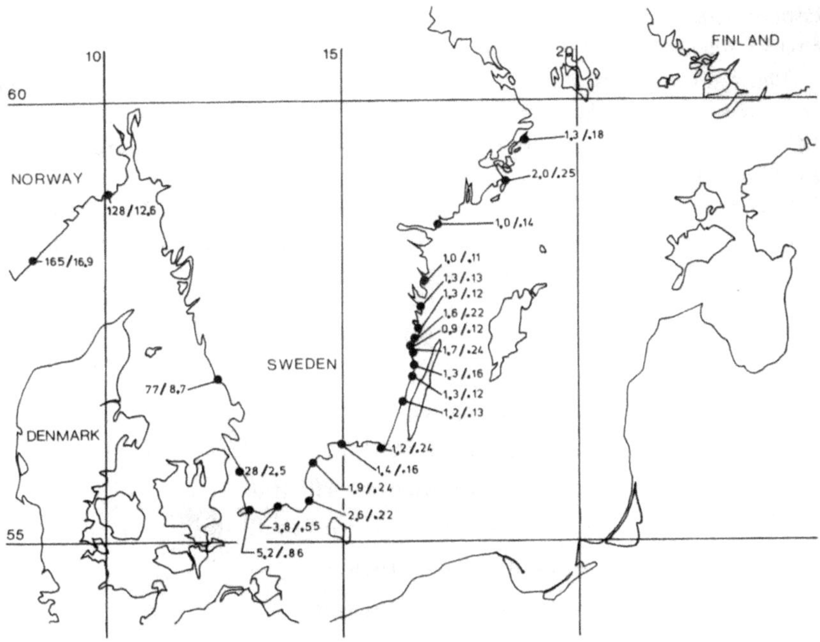

FIG. 1. Activity concentration of ^{99}Tc in *Fucus vesiculosus* (Bq kg^{-1} dry weight) during October 1982. The activity ratio ^{99}Tc/^{137}Cs is given after the solidus.

are few results from the west coast and southern Norway [2] for October 1982; although more results are available they are not easy to use here due to the large seasonal variation (which will be discussed later).

The corresponding results for May 1983 are shown in Fig. 2. Also included are results from the islands of Gotland, Oland and Aland.

The results for 1983 show the same overall tendency as those for 1982, although the ^{99}Tc concentration was $3{\cdot}0 \pm 0{\cdot}3$ Bq kg^{-1} and the ^{99}Tc/^{137}Cs activity ratio $0{\cdot}27 \pm 0{\cdot}02$ on the east coast of Sweden. The concentrations were also higher on the west coast in 1983, showing that we had a high inflow from the North Sea during spring 1983. It is obvious that the impact from different sources varies along the coast.

On the east coast fallout from nuclear detonation tests is certainly the most important source of ^{99}Tc. The integrated fallout was determined to be 800 ± 300 mBq m^{-2} in central Sweden (62·3°N, 12·4°E) by analysing carpets of lichen and soil [3]. The present activity ratio ^{99}Tc/^{137}Cs was determined to be 2×10^{-4}. We cannot detect any increase in the ^{99}Tc concentrations in *Fucus* in the vicinity of the nuclear power plants on the east or west coast of Sweden.

The releases from nuclear power plants were investigated during 1983. From one power plant on the southwest coast (Barseback, 2 BWR 570 MWe) the total releases to the water recipient during 1983 were measured as 0·86 MBq of ^{99}Tc.

The decrease of ^{99}Tc activity with distance, going from the north to the south of the west coast, which depends more on dilution than biological removal and sedimentation, can be described as

$$A(x) = A(0)\exp - (k_1 + k_2 + k_3) \times \exp\left(-\left(\lambda \frac{x}{v}\right)\right)$$

where

x = distance in km
k_1 = dilution coeff. km^{-1}
k_2 = biological removal coeff. km^{-1}
k_3 = sedimentation removal coeff. km^{-1}
λ = physical decay constant
v = velocity of current km a^{-1}.

We found a total removal coefficient for the transport of ^{99}Tc from the North Sea into the Baltic Sea of 0·010 in October 1982 and a value of 0·009 for May–June 1983.

In May 1983 during the 'Gauss' expedition 10 water samples were taken in the Baltic Sea (between 13–26°E and 55–64°N) and analysed for ^{99}Tc. For surface water we found a mean value of 68 ± 11 mBq m^{-3} and for bottom water 95 ± 11 mBq m^{-3}. The observed activity concentration ratio *Fucus*/water can be calculated to be in the order of 40 000 in the Baltic Sea. This is lower than observed for other areas such as Cattegat [4], i.e. 5×10^4–10^5. Our water samples were taken in the open sea and at different sites from where the *Fucus* samples were taken, which might lead to higher concentrations of ^{99}Tc in water and lower estimates of transfer factors to *Fucus*.

The activity ratio ^{99}Tc/^{137}Cs in sea water (surface) was 4–5×10^{-3} which is 20 times higher than in integrated fallout. More water samples have to be analysed in order to verify these results.

These higher activity ratios are partly dependent on the longer residence time of technetium in water compared to that of caesium, especially in the Baltic Sea where runoff from land is significant. The area where the water samples were taken is also to some extent contaminated with ^{99}Tc

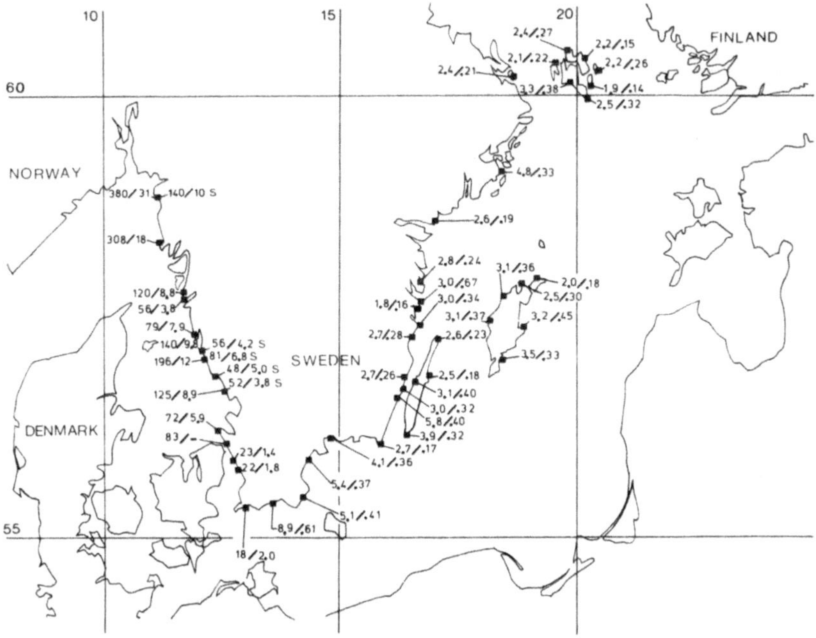

FIG. 2. Activity concentration of ^{99}Tc (Bq kg^{-1} dry weight) in *Fucus vesiculosus* during May–June 1983. Samples of *Fucus serratus* are marked with s. The activity ratio $^{99}Tc/^{137}Cs$ is given after the solidus.

and ^{137}Cs from European reprocessing plants where the $^{99}Tc/^{137}Cs$ activity ratios in the releases are as high as 0·02–0·04.

The $^{99}Tc/^{90}Sr$ activity ratio in the water is calculated to be $\sim 3{\cdot}5 \times 10^{-3}$ for the same samples. Data for ^{90}Sr were obtained from Kautsky *et al.* [5].

A theoretical $^{99}Tc/^{90}Sr$ activity ratio in the oceans would be $3{\cdot}1 \times 10^{-4}$ in 1982 with global fallout as the only source, which is 10 times lower than that which we found.

3.2. Seasonal and long-term variation

The long-term variation of ^{99}Tc in *Fucus* was investigated at a well-defined site, Sardal (56·76°N, 12·63°E), where sampling has been performed since 1967. The results are given in the lower part of Fig. 3.

The activity concentration has increased from ~ 3 to ~ 150 Bq kg^{-1} dry weight since 1972. The transit time from the Sellafield reprocessing plant to this site is about 4 years [6, 7]. A sharp increase in the ^{99}Tc

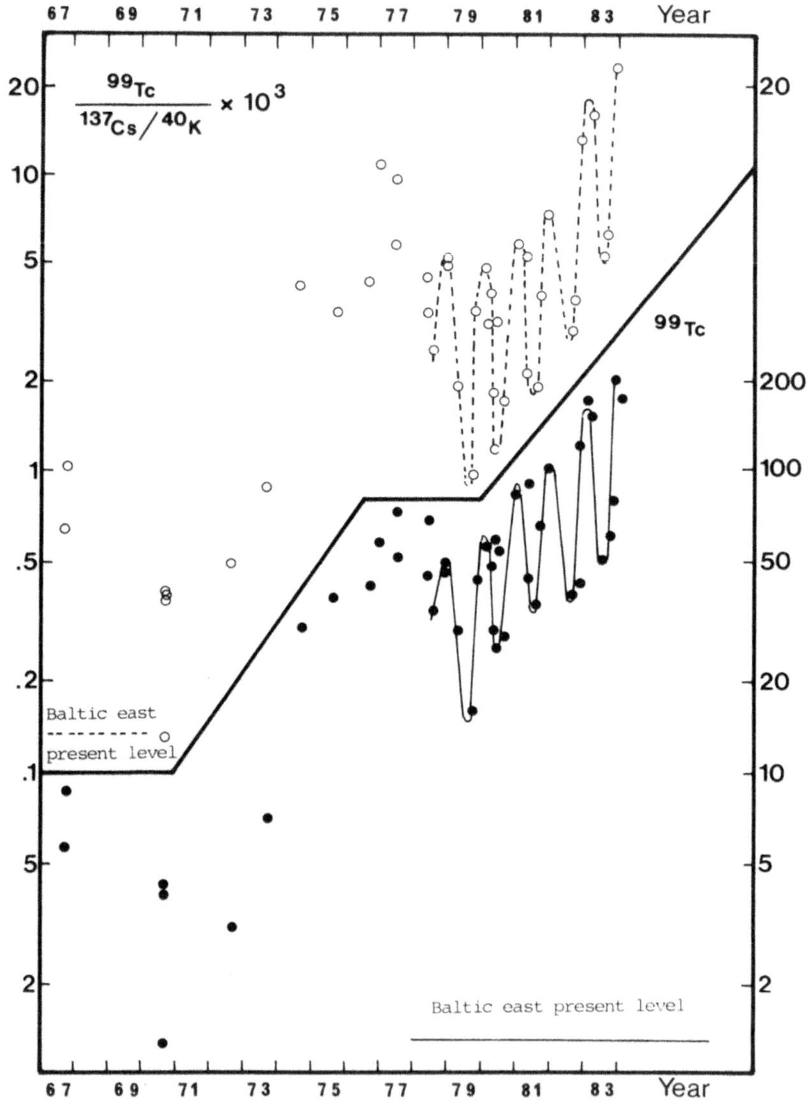

FIG. 3. Activity concentration of ^{99}Tc in *Fucus* at the west coast of Sweden (56·76°N, 12·63°E) during 1967–84. In the upper part of the figure the ^{99}Tc/(^{137}Cs/^{40}K) ratio is given.

concentration in *Fucus* occurred during 1972–3. Data for releases of ^{99}Tc from the reprocessing plant have only been reported since 1978 while figures are available since 1952 for ^{137}Cs [8].

Our results indicate significant increases in the release of ^{99}Tc from European reprocessing plants around 1970 and 1975.

In a closer look at our results for ^{99}Tc in *Fucus* we observe a seasonal variation which is rather significant. Maximal ^{99}Tc concentrations appear during winter and minimal values during summer. An explanation for this is the higher inflow of contaminated water from the North Sea during winter and the higher outflow from the Baltic Sea during spring–summer.

However, the opposite seasonal variation is observed for ^{137}Cs and the ^{137}Cs/^{40}K activity ratio for the same samples [9].

Our results show that the uptake of technetium is not metabolically controlled or salinity-dependent, in contrast to caesium. For ^{137}Cs the amplitude between summer and winter values is concealed by higher water concentrations during winter.

Since the seasonal behaviour of ^{99}Tc is opposite to that of ^{137}Cs the ^{99}Tc/(^{137}Cs/^{40}K) ratio is given in the upper part of Fig. 3. A better seasonal relationship was obtained by using this ratio instead of the ^{99}Tc/^{137}Cs activity ratio. The average ^{40}K activity concentration was 817 ± 33Bq kg^{-1} in the *Fucus* samples during 1978–83.

The winter values for the ^{99}Tc/(^{137}Cs/^{40}K) ratios are a factor of 4 higher than the summer values. The corresponding difference for ^{99}Tc alone is about 2·8. For ^{137}Cs/^{40}K the ratio between the summer and the winter values is about 1·5.

Indicated in Fig. 3 are the present levels for the east coast of Sweden, where ^{99}Tc and ^{137}Cs mainly originate from global fallout. The indicated levels are lower than the earlier results from 1967–70 for the west coast. The fallout levels were higher at that time but as the present Tc/(Cs/K) level on the east coast is also lower it shows that ^{99}Tc from European reprocessing facilities was present on the west coast in 1967.

The reported releases of ^{137}Cs during 1963–7 (4 years earlier) were about 10–15 times lower than during 1973–9. Our results show that this was also the case for ^{99}Tc.

ACKNOWLEDGEMENTS

This investigation has been made possible by financial support from the Swedish Natural Science Research Council and the Swedish Radiation Protection Institute.

REFERENCES

1. Holm, E., Rioseco, J. and Garcia-Leon, M. (1984). *Nuclear Instruments and Methods in Physics Research*, **223**, 204–7.
2. Holm, E., Rioseco, J. and Christensen, G. C. (1984). In Proceedings of Symposium on the Behaviour of Long-Lived Radionuclides in the Marine Environment, La Spezia, Italy, CEC (in press).
3. Holm, E. and Rioseco, J. (1983). In Proceedings of Seminar on the Transfer of Radioactive Materials in the Terrestrial Environment subsequent to an Accidental Release to the Atmosphere. Dublin, Ireland, CEC, pp. 193–203.
4. Aarkrog, A. (1984). Personal discussion.
5. Kautsky, H., Wedekind, C. and Eicke, H.-F. (1984). Results presented at Final Research Co-ordination Meeting on the Study of Radioactive Materials in the Baltic Sea, Helsinki, Finland, IAEA.
6. Kautsky, H. (1980). In Proceedings of the Third NEA Seminar of Marine Radioecology, Tokyo, Japan, pp. 283–9.
7. Aarkrog, A., Bøtter-Jensen, L., Dahlgaard, H., Hansen, H., Lippert, J., Nielsen, S. P. and Nilsson, K. (1980). Risø Report 421.
8. NRPB R171. 1984.
9. Mattsson, S. (1984). In Proceedings of 6th International Congress of IRPA, Berlin, FRG, pp. 901–4.

7

Time Trend of ^{99}Tc in Seaweed from Greenland Waters

A. Aarkrog,[a] H. Dahlgaard,[a] L. Hallstadius,[b] E. Holm,[c] S. Mattsson[d] and J. Rioseco[b]

[a] *Risø National Laboratory, DK-4000 Roskilde, Denmark*
[b] *Department of Radiation Physics, University of Lund, S-22 185 Lund, Sweden*
[c] *International Laboratory of Marine Radioactivity, IAEA, Musée Océanographique, Monaco*
[d] *Department of Radiation Physics, University of Gothenburg, S-413 45 Gothenburg, Sweden*

1. INTRODUCTION

In the past few years the development of a suitable analytical method [1] for determining ^{99}Tc in low-level concentrations has intensified the study of this radionuclide in the environment [2–4].

^{99}Tc is a fission product and the theoretical fission yield is approximately 0·06 atom per fission for nuclear weapons as well as for thermal fission [5]. The fission yield activity ratio $^{99}Tc/^{137}Cs$ is $1{\cdot}44 \times 10^{-4}$. The half-life of ^{99}Tc is $2{\cdot}1 \times 10^5$ y, and it is a pure β-emitter ($E\beta_{max} = 292$ keV).

The global inventory of ^{99}Tc from nuclear weapons testing in the atmosphere is calculated from the ^{137}Cs production by 1980 [6] to be 0·14 PBq. The annual production in civil nuclear reactors (273 $GW_e\,a^{-1}$) [7] is 3 PBq ^{99}Tc (1985). The purpose of this study has been to follow the levels of ^{99}Tc in the marine environment and from this to estimate the discharge pattern of ^{99}Tc from the Sellafield reprocessing plant in the UK prior to 1978, which is the first year in which discharge data are available [8].

2. METHODS

The materials used for this study have been preferentially *Fucus vesiculosus* samples collected in Greenland (Fig. 1), Swedish and Danish coastal

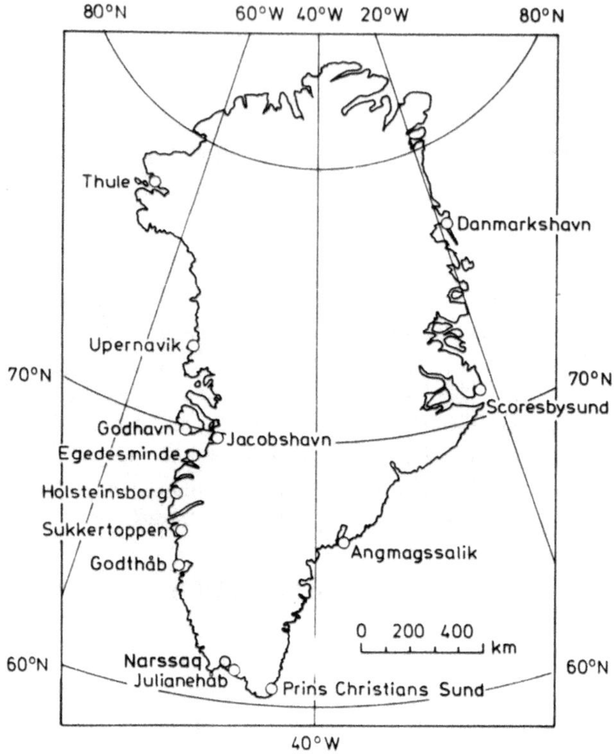

FIG. 1. Sampling locations in Greenland.

waters (Fig. 2) since 1967 [9]. However, some of the Greenland fucoids have been *Fucus disticus*. In a comparative study carried out on samples collected in Iceland in 1981 we have not, however, seen any significant difference between the two fucoids with regard to ^{99}Tc concentrations (Risø-R470) [9].

The samples have been analysed radiochemically for ^{99}Tc on either ash or dried material. We have found that the ^{99}Tc content of ashed samples is $0{\cdot}92 \pm 0{\cdot}05$ (1SD, $N = 4$) times that of dried samples. Hence, we have corrected the analyses on ash by dividing by 0·92. Futhermore, in cases where no dry matter measurements were available we have assumed an ash/dry matter ratio of 0·34, which is the mean obtained from measurements.

In the radiochemical method [1] ^{99}Tc^{m} ($t_{1/2} = 6{\cdot}0$ h) is used as radiochemical yield monitor. Technetium is extracted from sulphuric acid solution

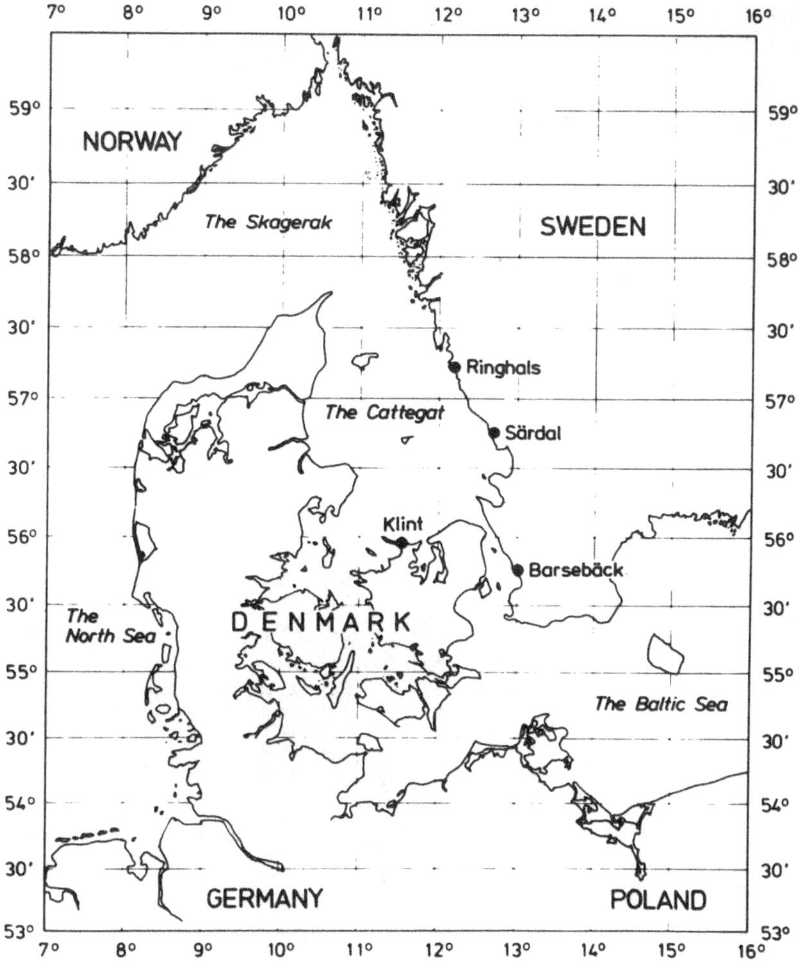

FIG. 2. Sampling locations in the Danish Straits.

with tributylphosphate and back-extracted with sodium hydroxide solution. Prior to the measurement, technetium is electroplated from 2 M sodium hydroxide solution onto stainless steel discs.

The yield is determined by gamma spectrometry using the 140 keV transition from $^{99}Tc^{m}$. After decay of $^{99}Tc^{m}$ (5 days), ^{99}Tc is measured by an anti-coincidence shielded multi-GM gasflow counter. The radiochemical yield is typically about 70%. Double determinations show a relative SD of 3–4%.

Table I

Technetium-99 in *Fucus vesiculosus/disticus* collected in Greenland and Swedish coastal waters 1963–83 (summer period) (Unit: Bq ^{99}Tc kg^{-1} dry weight; ± 1SD due to counting statistics)

Year	Danmarkshavn	East Greenland coastal waters			West Greenland coastal waters			The Cattegat	
		Scoresby Sund	*Angmaggsalik*	*Prins Chr. Sund*	*Narssaq-Godthåb*	*Godhavn-Upernavik*	*Thule*	*Ringhals*	*Särdal*
1963			4·73 ± 0·23						
64							2·72 ± 0·04		
65					2·82 ± 0·14				
66			2·79 ± 0·05		3·16 ± 0·07				
67				3·91 ± 0·09; 2·93 ± 0·06	3·82 ± 0·08				5·64
68	2·55 ± 0·06			2·40 ± 0·06					
69	1·75 ± 0·15	2·06 ± 0·16; 1·93 ± 0·13		1·75 ± 0·18					
70	3·13 ± 0·47			2·06 ± 0·05					4·05; 3·91; 3·92; 1·29
71	2·68 ± 0·21								
72	1·40 ± 0·13; 2·15 ± 0·17		2·91 ± 0·14						3·02
73	3·65 ± 0·09; 2·44 ± 0·13			2·83 ± 0·12					6·94
74				2·98 ± 0·16					29·8
75				3·24 ± 0·31					36·7
76								77·7 ± 0·8	40·4; 54·7
77			8·06 ± 0·31	6·81 ± 0·20				80·8 ± 1·0	72·0
78		4·71 ± 0·35; 4·07 ± 0·20						40·9 ± 0·4	33·8
79		8·96 ± 0·29		10·51 ± 0·40	1·86 ± 0·10		0·96 ± 0·38; 0·65 ± 0·18	59·4 ± 0·7	29·7; 16
80				7·31 ± 0·19				34·3 ± 0·3	59·0; 29·0; 52·5; 27·2
81						2·25 ± 0·13		77·8 ± 1·4	90·3; 35·3
82	6·42 ± 0·26	7·29 ± 0·28; 12·61 ± 0·18	7·43 ± 0·37 9·37 ± 0·56; 8·66 ± 0·52		3·78 ± 0·09; 3·71 ± 0·11				39·2; 42·8
83						3·65 ± 0·14			152; 49·6; 59·9

3. RESULTS AND DISCUSSION

The East Greenlandic samples (cf. Table I and Fig. 3) from the beginning of the 1960s show ^{99}Tc concentrations of about 3–5 Bq kg^{-1} dry weight. Later on until the middle of the 1970s the concentrations are 2–3 Bq ^{99}Tc kg^{-1}. In 1977 we notice a steep increase to 7–8 Bq ^{99}Tc kg^{-1} and this higher level is kept for the rest of the period of observation.

The samples from West Greenland (cf. Table I and Fig. 3) also show enhanced concentrations in the latest 3 years, but the increase is less pronounced than that in East Greenland.

The higher concentrations in the beginning of the 1960s were due to the enhanced fallout levels at that time. To compensate for this effect on the ^{99}Tc data, we have in Fig. 4 shown the ratio of ^{99}Tc in *Fucus* to ^{90}Sr in coastal sea water [9] collected in the same region as the fucoids. The ^{90}Sr concentrations have been corrected for radioactive decay of ^{90}Sr. We have applied UNSCEAR's [6] ^{90}Sr data on annual deposition and

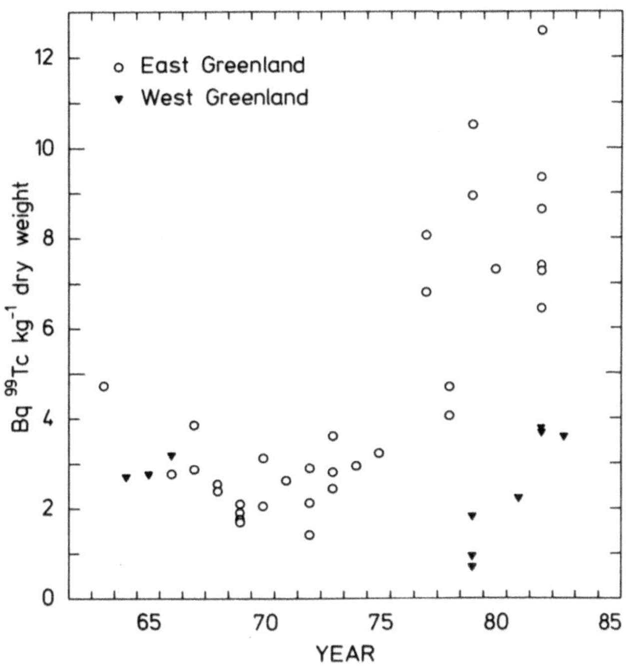

FIG. 3. Technetium-99 in *Fucus vesiculosus/disticus* collected in Greenland 1963–83.

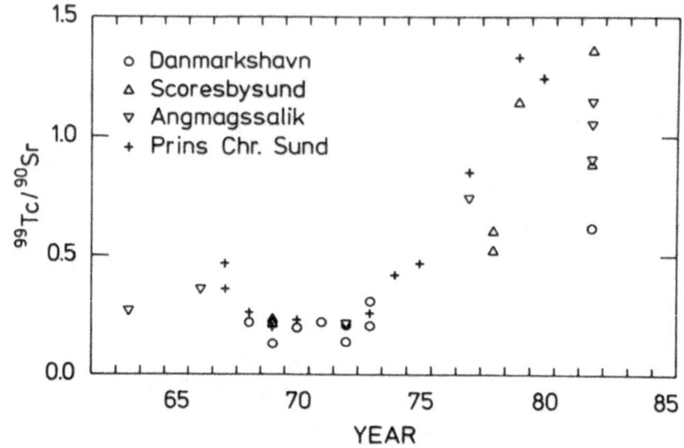

FIG. 4. [^{99}Tc in *Fucus* (Bq kg^{-1} dry weight)/^{90}Sr in seawater (Bq m^{-3})]. Collected at four locations in East Greenland 1963–82. The ^{90}Sr concentrations have been corrected for decay, i.e. we have assumed no decay of the activities considered.

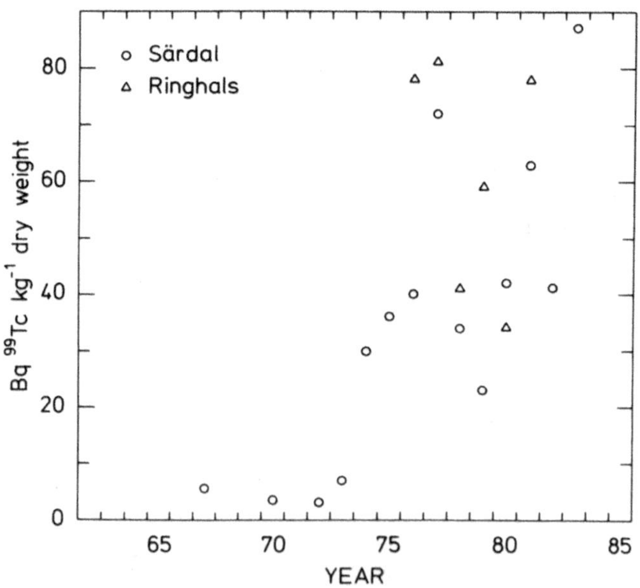

FIG. 5. Technetium in *Fucus vesiculosus* collected at two locations at the Swedish West coast (cf. Fig. 2) 1967–83.

cumulative deposit in the Northern Hemisphere throughout the years, so that the corrected ^{90}Sr concentrations will be proportional to a fallout radionuclide with essentially no radioactive decay, as, for example, ^{99}Tc. In other words, if we assume that in 1962 we had the theoretical activity ratio [5] of ^{99}Tc to ^{90}Sr of $2{\cdot}08 \times 10^{-4}$ in the ocean, we would have a ratio of $3{\cdot}1 \times 10^{-4}$, i.e. approximately 50% higher, in 1982, due to the decay of ^{90}Sr. Figure 3 shows that we still see a marked increase in the relative ^{99}Tc content of fucoids in 1977. In a similar way we may correct the West Greenland data. In this case we observe the $^{99}Tc/^{90}Sr$ ratios in 1981–3 to be twice those in 1964–7.

If we look at the data from Swedish and Danish waters (Fig. 5) in Table I, we see that the levels are low (1–6 Bq $^{99}Tc\,kg^{-1}$) from 1967 to 1972, but in 1974 the concentration increases to 30 Bq $^{99}Tc\,kg^{-1}$. This level was further enhanced in the years that followed.

In our studies [10] of radiocaesium from Sellafield in the Danish Straits we have seen that it took four years before the increased discharges from that installation appeared in the bottom water and a little longer before the surface water also showed enhanced concentrations. The increase observed at Särdal (Fig. 2) in 1974 may thus be due to an increased discharge of ^{99}Tc from Sellafield in 1969–70. From our studies [11] of radiocaesium in Arctic waters we know that it takes approximately 7 years for the effluent from Sellafield to reach the East Greenland Polar Current. Hence we would expect that enhanced levels in fucoids from East Greenland would appear in 1976–7. This is what we actually observe.

If we assume that the mean concentrations we have observed in *Fucus vesiculosus* collected at Särdal in 1982 and 1983 are due to the discharges from Sellafield in 1978 and 1979 (cf. Table II), we can estimate a transfer factor of 1100 Bq $^{99}Tc\,kg^{-1}$ in *Fucus* (dry weight) per PBq $^{99}Tc\,a^{-1}$. Futhermore, if we assume that the concentration factor for *Fucus* is equal to the ratio of ^{99}Tc in *Fucus* to that of sea water found in the Danish

Table II
Annual discharge rates of ^{99}Tc from Sellafield [8]

Year	*TBq $^{99}Tc\ a^{-1}$*
1978	179
1979	44
1980	57
1981	5·8
1982	3·6

Straits (Klint, cf. Fig. 2) in 1984, i.e. 10^5, we get a transfer factor to sea water of 11 Bq m^{-3} per PBq a^{-1}. This figure may be compared with those determined for ^{90}Sr, ^{134}Cs, and ^{137}Cs in the Cattegat, which were 16, 13 and 7 respectively [10]. We would have expected the transfer factor for ^{99}Tc to have been similar to that of ^{90}Sr because both these nuclides are sedimented only to a very small extent. The uncertainty of the estimate does not, however, preclude such an agreement. If we use the above transfer factor of 1100 Bq kg^{-1} per PBq a^{-1}, from the Ringhals and Särdal data in Table I we may estimate the annual mean discharge rate of ^{99}Tc from Sellafield to have been of the order of 0·04 PBq during the period 1970–7. In the next years when we get more *Fucus* data from the Danish Straits, for the period corresponding to the discharge years 1980–2 we may get a more reliable determination of the transfer factor, which admittedly at present is very poorly known.

The transfer factor for ^{99}Tc to *Fucus* in East Greenland is estimated as

$$\frac{8-2}{0{\cdot}04} = 150\ \text{Bq kg}^{-1}\ \text{per PBq a}^{-1}$$

and in West Greenland at

$$\frac{3{\cdot}7-2}{0{\cdot}04} = 40\ \text{Bq kg}^{-1}\ \text{per PBq a}^{-1}$$

(an estimated fallout background of 2 Bq ^{99}Tc kg^{-1} *Fucus* was applied in these calculations).

If we have the same concentration factors between *Fucus* and sea water in Greenland as in the Danish Straits, i.e. 10^5, the transfer factors to Greenland coastal waters become 1·5 for East and 0·4 Bq ^{99}Tc m^{-3} per PBq ^{99}Tc a^{-1} for West Greenland. These values are within the range of those calculated for ^{90}Sr in the western and eastern branches of the East Greenland Current, i.e. 0·1 and 3, respectively.

4. CONCLUSION

Technetium-99 may be used as a long distance (10^3–10^4 km) tracer for Sellafield effluents in Arctic waters when *Fucus vesiculosus/disticus* is used as a bioindicator. We have confirmed a transport time from Sellafield to the East Greenland Current of approximately 7 years. Furthermore, we have shown that enhanced ^{99}Tc discharges began about 1970 from

Sellafield. The annual mean discharge rates from 1970 to 1977 have been approximately 0·04 PBq ^{99}Tc and the transfer factor to *Fucus* in East Greenland waters is 150 Bq ^{99}Tc kg^{-1} dry weight per PBq ^{99}Tc a^{-1}.

ACKNOWLEDGEMENT

Martin Munch Hansen, the Greenland Fisheries and Environmental Research Institute kindly provided us with *Fucus* samples from West Greenland in 1982–3.

REFERENCES

1. HOLM, E., RIOSECO, J. and GARCIA-LEÓN, M. (1984). Determination of ^{99}Tc in environmental samples. *Nuclear Instruments and Methods in Physics Research* (in press).
2. HOLM, E. and RIOSECO, J. (1983). Tc-99 in carpets of lichen and soil. Seminar on the Transfer of Radioactive Materials in the Terrestrial Environment Subsequent to an Accidental Release to the Atmosphere. Dublin, 11–15 April, 1983.
3. HOLM, E., RIOSECO, J. and CHRISTENSEN, G. C. (1983). Technetium-99 in *Fucus* from Norwegian waters. International Symposium on the Behaviour of Longlived Radionuclides in the Marine Environment. Forte St. Teresa, 28–30 Sept., 1983.
4. DAHLGAARD, H., AARKROG, A., HALLSTADIUS, L., HOLM, E. and RIOSECO, J. (1984). Radiocesium transport from the Irish Sea via the North Sea and the Norwegian Coastal Current to East Greenland. Symposium on Contaminant Fluxes through the Coastal Zone. Nantes, France, 14–16 May, 1984.
5. HARLEY, J. H. (Ed.) (1972). HASL procedures manual, HASL-300, 602 pp.
6. UNSCEAR (1982). *Ionizing Radiation: Sources and Biological Effects*, New York, United Nations, 773 pp.
7. IAEA (1984). Reference Data Series No. 2. Nuclear Power Reactors in the World.
8. BNFL (1978–82). Annual report on radioactive discharges and monitoring of the environment. Risley, Warrington, Cheshire, UK, British Nuclear Fuels Limited.
9. AARKROG, A. *et al.* Environmental Radioactivity in Denmark/the Faroes/Greenland in 1962–82. Risø-R-Nos. 63, 64, 65, 85, 86, 87, 107, 108, 109, 130, 131, 132, 154, 155, 156, 180, 181, 182, 201, 202, 203, 220, 221, 222, 245, 246, 247, 265, 266, 267, 291, 292, 293, 305, 306, 307, 323, 324, 325, 345, 346, 347, 361, 362, 363, 386, 387, 388, 403, 404, 405, 421, 422, 423, 447, 448, 449, 469, 470, 471, 487, 488 and 489. Annual reports from Risø National Laboratory.

10. AARKROG, A., DAHLGAARD, H. and BOELSKIFTE, S. (1984). Transfer of radiocesium and ^{90}Sr from Sellafield to the Danish Straits. Cooperative Programme on the Study of Radioactive Materials in the Baltic Sea. IAEA, Vienna.
11. AARKROG, A., DAHLGAARD, H., HALLSTADIUS, L., HANSEN, H. and HOLM, E. (1983). Radiocesium from Sellafield effluent in Greenland waters. *Nature*, **304**, 49–51.

8

Behaviour of Technetium in Freshwater Environments

B. G. BLAYLOCK, M. L. FRANK, F. O. HOFFMAN
and D. L. DEANGELIS

Environmental Sciences Division, Oak Ridge National Laboratory, Oak Ridge, Tennessee 37831, USA

1. INTRODUCTION

Currently the major sources of ^{99}Tc released to the environment in the United States are gaseous diffusion plants that enrich reprocessed nuclear fuel. At present these facilities are located on freshwater streams into which liquid effluents are released. Of the total amount of ^{99}Tc that escapes these facilities more than 90% may enter the liquid pathway [1]. Information on the behaviour of technetium in aquatic ecosystems is limited; most data on the uptake of technetium by aquatic organisms are restricted to marine environments.

Most of the information available on technetium in freshwater ecosystems is based on data from an experimental pond study with ^{95m}Tc [2, 3]. Concentration factors of 11, 75 and 121 were reported for carp (*Cyprinus carpio*), mosquitofish (*Gambusia affinis*) and snails (*Helisoma* sp.), respectively [2]. Recently, concentration factors were reported for other freshwater organisms based on laboratory studies with ^{95m}Tc. The freshwater mussel (*Margaritifera margaritifera*) had a whole-body concentration factor of 0·9; the newt (*Taricha granulosa*) had a whole-body concentration factor of 11 [4]. In similar studies with the crayfish *Pacifactacus leniusculus* and the snail *Juga silicula*, whole-body concentration factors were 1·6 and 41, respectively [5].

In previous pond studies [2] periphyton and macrophytes accumulated technetium rapidly and contributed to the concentration of ^{95m}Tc in higher trophic levels via the food chain pathway. Concentrations in zooplankton and aquatic insects were not measured because these populations were kept at low densities by fish predation.

One objective of the current study was to determine the concentration of ^{95m}Tc in zooplankton, aquatic insects and other lower-trophic-level components of an experimental pond. These data were incorporated with data from previous experimental pond studies to obtain a better understanding of the bioaccumulation of technetium in freshwater organisms. A second objective was to estimate variability by comparing the bioaccumulation of technetium in freshwater organisms with that of marine organisms.

2. MATERIALS AND METHODS

An acute release of ^{95m}Tc in the form of a pertechnate was made into a small, experimental, freshwater pond. The experimental pond, described previously [3], was approximately 4 m on each side and had a maximum depth of approximately 1 m. The pond was lined with a plastic (hypolon) to prevent seepage. The ecosystem in the pond was established by covering the bottom with 15 cm of sediment taken from a nearby pond and then filling it with spring water. The system was then allowed to develop naturally for approximately one year before it was spiked with ^{95m}Tc. Fish were excluded from the pond to allow for the development of denser zooplankton populations. Only one species of macrophyte (*Elodea canadensis*) developed in this pond, whereas at least three species developed in earlier experimental pond studies.

Sampling techniques for biotic and abiotic components of the pond were similar to those described previously [3]. Collections were made at 1 and 4 h after the release of ^{95m}Tc to the pond and at days 1, 2, 3, 6, 13 and 20. The following components of the pond were sampled: water, sediments, macrophytes, invertebrates, periphyton, algae and zooplankton. Zooplankton was collected using a plankton net (mesh size 153 μm). Samples were separated according to size by passing the zooplankton through a series of sieves (size 500, 250, 150 and 45 μm). The dominant species in the different size groups were then identified. Samples were oven dried and weighed, and the activity of ^{95m}Tc was measured in a gamma spectrometer. Aquatic insects, both mature and immature, were collected using a long-handled net. The sampling for each collection period was limited to prevent undue disturbance of the pond; therefore, only two families of insects were present for each sampling period.

3. RESULTS

One hour after the release of ^{95m}Tc to the pond, the water contained $0{\cdot}75 \pm 0{\cdot}10$ Bq/ml. The concentration decreased approximately 20% to $0{\cdot}58 \pm 0{\cdot}06$ Bq/ml after 4 h and then gradually over the next 20 d to $0{\cdot}21 \pm 0{\cdot}14$ Bq/ml (Fig. 1). Less than 2% remained on the particulate matter when the water was filtered through a 0·4 μm Nuclepore filter; thus, most of the ^{95m}Tc remained dissolved in the water.

The mean ($n = 4$) concentrations of ^{95m}Tc in sediment core samples are also shown in Fig. 1. The sediment did not accumulate ^{95m}Tc rapidly but continued to accumulate it throughout the experiment, with the highest concentrations occurring on day 20. The percentage of organic matter in the sediment samples averaged 7·7% and ranged from 4·7 to 14·5%; however, in this short-term experiment, the concentration of ^{95m}Tc in the sediment was not directly related to the percentage of organic matter in the sample.

Zooplankton accumulated ^{95m}Tc rapidly. One hour after ^{95m}Tc was

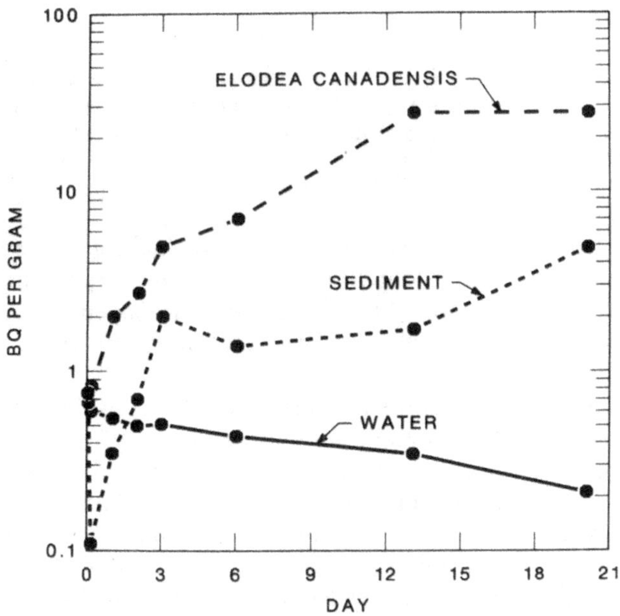

FIG. 1. The concentrations of ^{95m}Tc in water (Bq/ml), sediments (Bq/g dry wt) and *Elodea canadensis* (Bq/g wet wt) are shown over time after an acute release of $^{95m}TcO_4^-$ into an experimental freshwater pond.

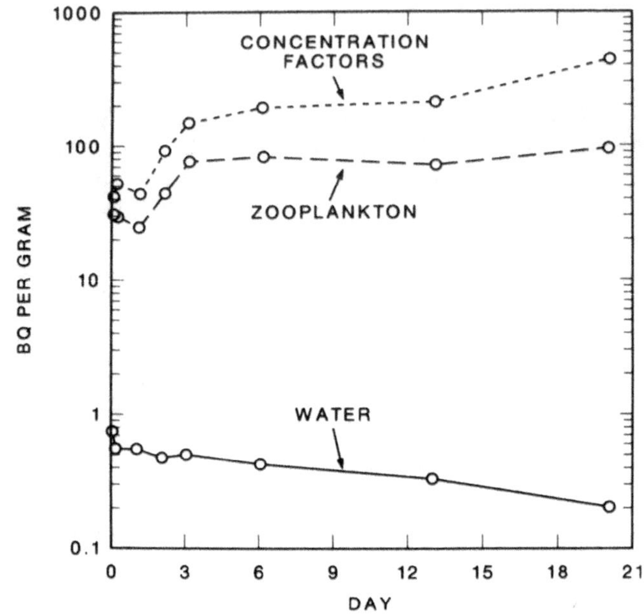

FIG. 2. The concentrations of ^{95m}Tc in zooplankton (Bq/g dry wt) and water (Bq/ml) and concentration factors (Bq/g dry wt zooplankton/Bq/ml water) are shown for different sampling periods after an acute release of $^{95m}TcO_4^-$ into an experimental freshwater pond. The wet weight to dry weight ratio for zooplankton in the experimental pond was 1:12·5.

released to the pond, zooplankton contained 31 Bq/g dry weight. The concentration of ^{95m}Tc in zooplankton and the concentration factor based on dry weight (Bq/g zooplankton/Bq/g water) are given in Fig. 2. A dry weight to wet weight ratio of 1:12·5 was determined for the zooplankton in the experimental pond.

Diaptomus reighardi was the dominant zooplankton species in the 500 to 250 μm size classes and made up approximately 95% of the biomass within these size classes. Ostracods (40%), *Diaptomus* (30%) and *Difflugia* (30%) were the dominant taxa in the 150 μm size class. A considerable amount of variability, but no consistent relationship, was observed in the concentrations of ^{95m}Tc among the different size classes of zooplankton. Data on the different size classes were combined and the mean concentrations are shown in Fig. 2.

Periphyton was not sampled until day 3 of the experiment. At this time the periphyton scraped from the sides of the pond contained 326·9 ± 20·8

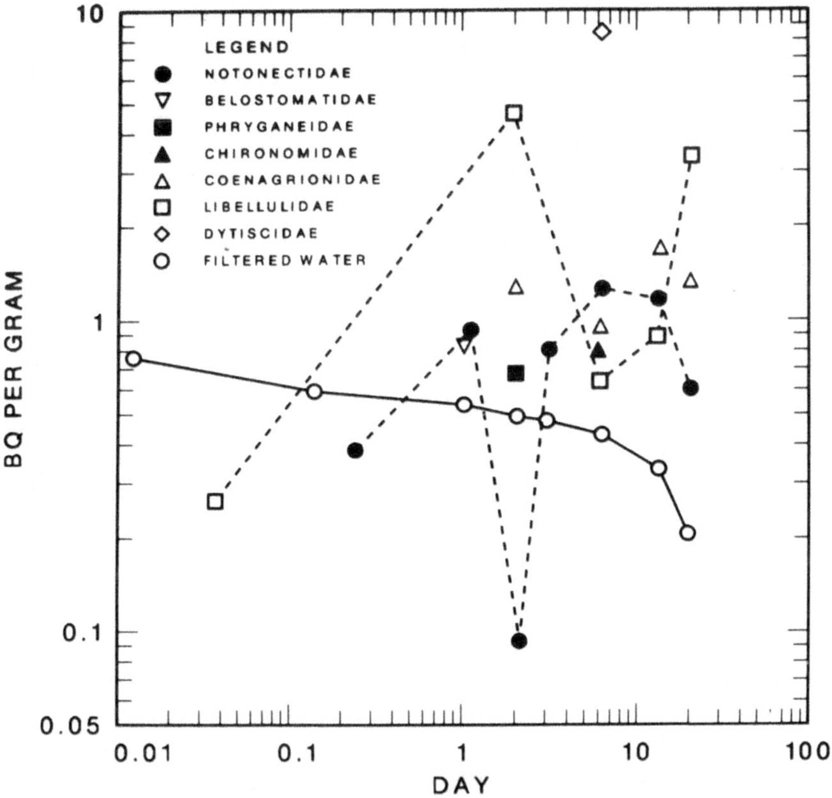

FIG. 3. Concentrations of ^{95m}Tc are shown in seven families of mature and immature insects after an acute release of $^{95m}TcO_4^-$ into an experimental freshwater pond. The adult backswimmer (*Notonectidae*) and the predaceous dragonfly nymph (*Libellulidae*) were collected the most consistently. The range of concentrations of ^{95m}Tc observed in these two families of insects included most of the concentrations measured for all the other families. In most cases, the samples were composite samples of insects.

Bq/g dry weight. A dry weight to wet weight ratio of 1:13·8 was determined for the periphyton in the experimental pond. By day 13 the concentration of ^{95m}Tc in periphyton had decreased to 138·3 ± 80·3 Bq/g.

Aquatic insects, both mature and immature stages, concentrated ^{95m}Tc by factors of 3–8. Five orders of insects were collected from the experimental pond: Coleoptera, Diptera, Hemiptera, Odonata and Trichoptera. The concentrations of ^{95m}Tc were measured in nine families of insects belonging to these five orders (Fig. 3). Only adult backswimmers

(Notonectidae) and dragonfly nymphs (Libellulidae) were collected consistently because of the restrictions on sampling to prevent undue disturbance of the pond system. The highest mean concentration of ^{95m}Tc, $4{\cdot}7 \pm 1{\cdot}4$ Bq/g wet wt, was observed in Libellulidae (dragonfly) nymphs at day 2. Concentrations of ^{95m}Tc for the other families of insects are also shown in Fig. 3. Most of these samples were composite samples made up of several insects. Although the number of insects was limited, concentrations in all the insects were within the range obtained for adult Notonectidae and the benthic, predaceous nymph belonging to the family Libellulidae. Concentrations in the nymph appeared to be slightly higher than those in the adult.

The snail (*Helisoma* sp.) and the larva (tadpole) of the bullfrog (*Rana catesbeiana*) represent organisms at the secondary trophic level in the experimental pond. The snails and larvae accumulated ^{95m}Tc gradually and reached their highest concentrations of $33{\cdot}5 \pm 22{\cdot}5$ and $18{\cdot}8 \pm 3{\cdot}2$ Bq/g, respectively, at day 20.

4. DISCUSSION

In the experimental pond, organisms are not only exposed to the ^{95m}Tc in the water but also obtain ^{95m}Tc via their natural food chain pathway. As a result, concentration (bioaccumulation) factors obtained from pond studies would be more realistic than concentration factors obtained from direct uptake studies conducted in the laboratory.

Since the experiment was conducted for a period of only 20 d following an acute release of ^{95m}Tc to the pond, organisms that have a relatively long biological half-life for technetium would not be expected to reach equilibrium conditions. A continuous decrease in the concentration of ^{95m}Tc in the water would also prevent equilibrium conditions from being attained; however, in this experiment, ^{95m}Tc in the form of a pertechnetate remained dissolved in the water and decreased by less than a factor of three over a 20-d period. Thus, near-equilibrium conditions should be attained for organisms such as phytoplankton, zooplankton and algae, which have relatively short biological half-lives and short life spans.

In the present study, the concentration factor was determined by dividing the highest average concentration of ^{95m}Tc in the organisms in a particular sampling period by the time-averaged water concentration. These concentration factors are probably near-equilibrium concentration factors for most organisms in the pond. For example, in a previous study

[2] an equilibrium concentration factor of 121 was calculated for the snail (*Helisoma* sp.). This number is in relatively good agreement with the concentration factor of 75 that was measured in the present experiment.

The initial decrease of ^{95m}Tc in pond water was accompanied by a concomitant increase in organisms composing the lower trophic levels. Zooplankton and periphyton accumulated ^{95m}Tc very rapidly. After 1 h of exposure zooplankton accumulated ^{95m}Tc by a factor of 3·4. Although the concentration of ^{95m}Tc in water gradually decreased for the next 20 d, the concentration in zooplankton continued to increase with the highest concentration factor, 36, occurring on day 20. Apparently the initial uptake of ^{95m}Tc by zooplankton was directly from water, but as the experiment proceeded, ^{95m}Tc was probably also taken up through the food chain.

The concentration factor for periphyton decreased with time, reflecting the decrease in water concentration. In an earlier study [3], periphyton taken from slides had accumulated ^{95m}Tc by a factor of three after 4 h of exposure. Thus, we concluded that zooplankton and periphyton accumulate technetium very rapidly directly from water and that technetium has a short residence time in periphyton.

Information on the early uptake of technetium by freshwater algae is unavailable, but in an earlier pond study [3] the green algae *Spirogyra* had a maximum concentration factor of 10 after 22 d of exposure. The initial uptake of ^{95m}Tc by *Spirogyra* would probably be similar to that by periphyton.

Accumulation of ^{95m}Tc by *Elodea canadensis* was similar to the results obtained previously [3]. For the first 4 h of exposure *Elodea* contained approximately the same concentration of ^{95m}Tc as did the water, but after 1 d *Elodea* had accumulated ^{95m}Tc by a factor of three. Accumulation of ^{95m}Tc continued throughout the experiment with the maximum concentration factor, 62, occurring on day 20. This concentration factor is near the upper end of the range of concentration ratios reported for terrestrial plants [1]. While ^{95m}Tc was increasing in the sediments, *Elodea* was probably obtaining ^{95m}Tc from the sediment as well as from the water.

Organisms grazing on periphyton accumulated ^{95m}Tc for the duration of the experiment. The snail (*Helisoma* sp.) and the larva of the bullfrog (*Rana catesbeiana*) have similar uptake patterns (Fig. 4). The highest concentrations of ^{95m}Tc were observed in these organisms. After 20 d of exposure, snails and frog larvae had accumulated ^{95m}Tc by a factor of 75 and 42, respectively. The concentration in snails is relatively close to the equilibrium concentration factor calculated by Blaylock *et al.* [2]. It is

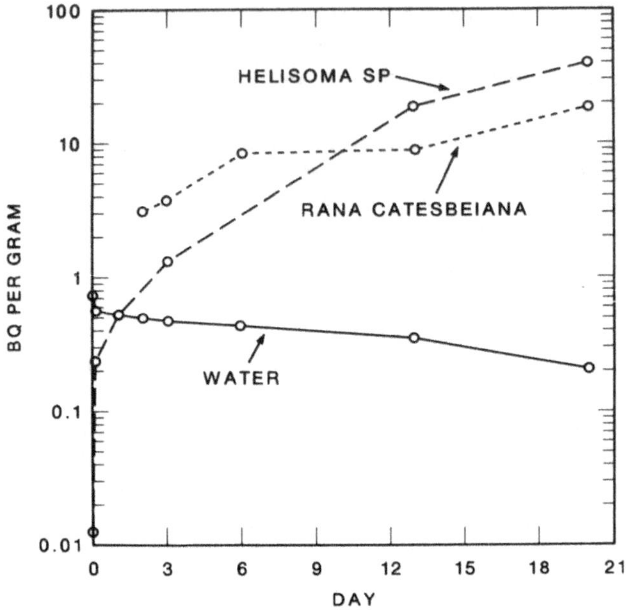

FIG. 4. The concentrations of ^{95m}Tc in the snail (*Helisoma* sp.), the larva of the frog (*Rana catesbeiana*) in Bq/g wet wt and water (Bq/g) are shown after an acute release of $^{95m}TcO_4^-$ into an experimental freshwater pond.

apparent that the food chain of these organisms contributes significantly to their body burden of ^{95m}Tc.

In Table I a comparison of the concentration factors in marine and freshwater organisms is made. With the exception of two laboratory studies, data on freshwater organisms were from experimental pond studies. In most cases, the data for marine organisms were obtained from laboratory studies that excluded the process of food chain uptake. Brown algae and lobster have the highest concentration factors in the marine ecosystem, their values exceeding any of the concentrations found in freshwater organisms. In the freshwater system, snails and macrophytes have the highest concentration factors. Mosquitofish have a much higher concentration factor than any of the other freshwater fish or any marine fish. In comparison, the carp has a concentration factor similar to that of marine fish. However, despite differences between the concentration factors in freshwater and marine organisms, there is good agreement in the values obtained for similar groups of organisms.

Table I
Comparison of technetium concentration factors (CF) for freshwater and marine organisms

Freshwater	*CF*[a]	*Marine*	*CF*
Periphyton		Phytoplankton	
Slide samples	40	*Heterocapsa pygmaca* [6]	17
		Mixed flagellates [7]	1
		Phytoflagellates [8]	1
Algae		Algae	
Green algae	1–10	Brown algae	250–2 500
		Red and green algae [9–14]	1–20
Macrophytes			
Elodea canadensis	61		
Zooplankton		Zooplankton	
Diaptomus reighardi	36	*Artemia salina* [15]	3
Difflugia		Euphausiid [16]	1
Ostracods			
Other crustaceans		Other crustaceans	
		Shrimp crab, isopod amphipods [7, 8]	2–12
Crayfish			
Pacifastacus leniusculus [5]	1·6	Lobster [8, 10, 17]	1 000–1 400
Cambarus	30	*Hormarus gammarus*	
Molluscs		Molluscs	
Margaritifera margaritifera [4]	0·9	*Mytilus edulis*	2
Corbicula manilensis	1	*Mytilus galloprovinciallis*	
		Mytilus californianus	
		Crassostrea gigas [8, 10, 18]	
Snails		Red abalone	
Helisoma sp.	121	*Haliotis rufescens* [14, 19]	100–200
Juga silicula [5]	41	Gastropod	
		Aporrnais pespelicana [20]	3
Aquatic insects	3–8		
Notonectidae			
Belostomatidae			
Chironomidae			
Dytisidae			
Tricoptera			
Odonata			
Amphibians			
Taricha granulosa [4]	11		
Fishes		Fishes	
Cyprinus carpio	11	*Pleuronectes platessa* [11]	9
Gambusia affinis	75	*Raja clavata* [17]	8
		Blennius pholis [10]	2

[a] Freshwater data except those that have specific references were obtained from experimental ponds.

5. CONCLUSIONS

Technetium in the form of a pertechnetate released into an experimental freshwater pond had a very low affinity for particulate matter and sediments. Most of the technetium remained dissolved in the water column. Technetium was accumulated rapidly from the water by periphyton, zooplankton and algae. Other organisms, such as snails and mosquitofish, accumulated technetium slowly but attained some of the highest whole-body concentrations. Aquatic insects concentrated technetium by factors ranging from 3 to 8 times greater than the concentration in water. Macrophytes very rapidly reached tissue concentrations of technetium equivalent to the water concentrations and continued to accumulate technetium until the concentration factor was equal to the upper range of those reported for terrestrial plants.

A comparison of concentration factors for marine and freshwater organisms shows that, with the exception of brown algae and lobsters, the bioaccumulation of technetium by similar organisms is in relatively good agreement despite marked differences in environmental and experimental conditions.

ACKNOWLEDGEMENT

Research sponsored by the Office of Health and Environmental Research, US Department of Energy, under Contract No. DE-ACO5-840R21400 with Martin Marietta Energy Systems, Inc.

REFERENCES

1. Till, J. E., Shor, R. E. and Hoffman, F. O. Environmental effects of the uranium fuel cycle—A review of data for technetium. Oak Ridge National Laboratory, Oak Ridge, TN. NUREG/CR-3738 (in press).
2. Blaylock, B. G., Frank, M. L. and DeAngelis, D. L. (1982). Bioaccumulation of ^{95m}Tc in fish and snails. *Health Physics*, **42**, 257–66.
3. Blaylock, B. G. and Frank, M. L. (1982). Bioaccumulation and distribution of ^{95m}Tc in an experimental freshwater pond. In: *Environmental Migration of Long-lived Radionuclides* (Proc. Symp. Knoxville, 1982), IAEA, pp. 451–64.
4. Willis, D. L. and Hevland, M. E. (1984). Radiotechnetium metabolism in two freshwater species, the mussel *Margaritifera margaritifera* and the newt *Taricha granulósa* (abstract). *Health Physics*, **47**, 209.
5. Willis, D. L. and McKenzie-Carter, M. (1984). Radiotechnetium metabolism in two freshwater species, the crayfish *Pacifastacus leniusculus* and the snail *Juga silicula* (abstract). *Health Physics*, **47**, 209.

6. FISHER, N. (1982). Bioaccumulation of technetium by marine phytoplankton. *Env. Sci. Tech.*, **16**, 579–81.
7. FOWLER, S. W., BENAYOUN, G., PARSI, P., ESSA, M. W. A. and SHULTE, E. H. (1981). Experimental studies on the bioavailability of technetium in selected marine organisms. In: *Impacts of Radionuclide Releases into the Marine Environment* (Proc. Symp. Vienna, 1980), IAEA, pp. 319–39.
8. BEASLEY, T. M., GONOR, J. J. and LORZ, H. V. (1982). Technetium: Uptake, organ distribution and loss in the mussel *Mytilus californianus* (Conrad) and the oyster *Crassostrea gigas* (Thunberg). *Mar. Env. Res.*, **7**, 103–16.
9. PENTREATH, R. J., JEFFERIES, D. F., LOVETT, M. B. and NELSON, D. M. (1980). The behavior of transuranic and other long-lived radionuclides in the Irish Sea and its relevance to the deep sea disposal of radioactive wastes. In: *Marine Radioecology* (Proc. 3rd NEA Seminar, Tokyo, 1979), OECD, pp. 203–21.
10. MASSON, M., APROSI, G., LANIECE, A., GUEGUENIAT, P. and BELOT, Y. (1981). Approches experimentales de l'etude des transfers du technetium a des sediments et a des especes marine benthiques. In: *Impacts of Radionuclide Releases into the Marine Environment* (Proc. Symp. Vienna, 1980), IAEA, pp. 341–59.
11. PENTREATH, R. J. (1981). Biological availability to marine organisms of transuranium and other long-lived nuclides. In: *Impacts of Radionuclide Releases into the Marine Environment* (Proc. Symp. Vienna, 1980), IAEA, pp. 241–72.
12. JEANMAIRE, L., MASSON, M., PATTI, F., GERMAIN, P. and CAPPELLINI, L. (1981). Technetium-99 content in some marine organisms collected near La Hague, France. *Mar. Poll. Bull.*, **12**, 29–32.
13. BEASLEY, T. M., LORZ, H. V. and GONOR, J. J. (1982). Biokinetic behavior of Tc in the red abalone *Haliotis rufescens*: A reassessment. *Health Physics*, **43**, 501–7.
14. TOPCUOGLU, S. and FOWLER, S. W. (1984). Factors affecting the biokinetics of technetium (^{95m}Tc) in marine macroalgae. *Mar. Env. Res.*, **12**, 25–43.
15. SCHULTE, E. H. (1981). In: *Biogeochemical Studies of Technetium in Marine and Estuarine Ecosystems.* Progress Report to US Dept. of Energy, DOE/EV-10251-3, Washington, DC, 75 pp.
16. FOWLER, S., HEUSSNER, S., LAROSA, J. and TOPCUOGLU, S. (1983). The biokinetic behavior of technetium in marine organisms. In: Biennial Report, 1981–1982, International Laboratory of Marine Radioactivity, IAEA TECDOC-297.
17. PENTREATH, R. J. (1981). In: *Biogeochemical Studies of Technetium in Marine and Estuarine Ecosystems.* Progress Report to US Dept. of Energy, DOE/EV-10251-3, Washington, DC, 75 pp.
18. FOWLER, S. W., ASTON, S. R., BENAYOUN, G. and PARSI, P. (1983). Bioavailability of technetium from artificially labelled northeast Atlantic deep-sea sediments. *Mar. Env. Res.*, **8**, 87–100.
19. SPIES, R. B. (1975). Uptake of technetium from seawater by red abalone, *Haliotis rufescens. Health Physics*, **29**, 695–9.
20. FOWLER, S. W. (1981). In: *Biogeochemical Studies of Technetium in Marine and Estuarine Ecosystems.* Progress Report to US Dept. of Energy, DOE/EV-10251-3, Washington, DC, 75 pp.

9

Role of Organic Matter as a Geochemical Sink for Technetium in Soils and Sediments

M. STALMANS, A. MAES and A. CREMERS
Laboratorium voor Colloïdale Scheikunde, Fakulteit der Landbouwwetenschappen, KU Leuven, Kardinaal Mercierlaan 92, B-3030 Leuven, Belgium

1. INTRODUCTION

The fate of a radionuclide when introduced (accidentally or otherwise) into a soil or sediment (marine or geological) is ruled by a combination of chemical, physico-chemical and biological factors (pH, redox conditions, sediment composition, complexing agents, microflora). The understanding of the behaviour of a radionuclide in such systems is needed on two main accounts: bioavailability and geochemical mobility.

An expedient and purely empirical way of dealing with the problem is based on K_D measurements which enable an estimate to be made of radionuclide distribution between particulate and solute phases. However, in recent years it has been realised that the speciation of the element is of crucial importance, not only in the liquid phase but in the solid phase as well. In general, it is a not too difficult problem to establish speciation in the liquid phase, as evidenced by a wealth of literature data. The methods used consist of (a combination of) ultrafiltration, equilibrium dialysis, chromatography and, if available, complex stability constants and hydrolysis data. Nevertheless, in the quantitative interpretation of the data, some caution is warranted in view of possible adsorption phenomena on membranes or chromatographic materials and shifting equilibria in chromatographic separations.

The speciation of radionuclides in the solid phase, soil or sediment, is a much more difficult problem which has not been solved satisfactorily so far. This is one of the main topics which is being addressed in this paper. Its purpose is somewhat limited in scope in attempting a quantitative assessment of the radionuclide fraction in just one of the geochemical phases: organic matter. In view of possible relevance of organic–metal associations to bioavailability in aquatic organisms and plant roots, this may be a worthwhile objective.

2. PROBLEM STATEMENT

Whatever the speciation of a radionuclide in a liquid phase extract, it may be present in the solid phase in a variety of forms, such as:

— simple ions in the ion exchange complex (aluminosilicates);
— insoluble precipitate (hydroxide, sulphide) or coprecipitate;
— complexed with high MW (insoluble) organic matter;
— surface complexes with hydrous oxides;
— associations with micro-organisms.

Ideally, the quantitative assessment of the (possibly transient) geochemical phase associations of radionuclides should be attempted under conditions approximating as closely as possible to those prevailing in situ. However, the methods currently in use are quite far removed from such conditions. The most popular method is based on sequential extraction procedures of the sediment, aiming at selectively and sequentially dissolving the various geochemical phases. The method of Tessier *et al.* [1], or some modification of it, has found widespread use in this respect. There has recently been some tendency to overinterpret the results of such procedures in that the metal extracted by the sequential and selective dissolution of some geochemical phase can, in fact, be assigned to that phase in the original sample. This is far from certain in view of a statement by Tessier *et al.* [1] that 'it is obvious that the distribution of a given metal between various fractions does not necessarily reflect the relative scavenging action of discrete sediment phases but rather should be considered as *operationally* defined by the method of extraction'.

The main advantage of these techniques is that 'they simulate to a certain extent various environmental conditions to which a sediment may be subjected' [1] and as such may reflect on the degree of bioavailability [2]. Perhaps one example may suffice to illustrate the point. Prior to extracting the organic fraction with H_2O_2, the sample is subjected to an acidified hydroxylamine treatment (pH 2–3) intended to extract the 'reducible phases' (Mn oxides, Fe oxyhydrates) [3]. It is possible that, under such extreme conditions and in view of the pronounced pH dependency of metal–organic complex formation, a significant fraction of the metals, associated with the organic phase at the original 'in situ' conditions, is being co-extracted in such a treatment. Alternatively, a fraction of metal, associated in situ with the oxide fraction, may be readsorbed on to the OM fraction, following the hydroxylamine treatment. In being forced to disturb the sediment sample in the process of carrying out some form of analysis, it is doubtful that (except in rare cases) conditions can ever be exactly identical to those occurring in situ. It will be shown below that minor changes in pH, about one unit, may already cause a substantial shift of metals from one geochemical sink to another.

In any case, we propose to examine the possibilities of using mild extraction procedures for the purpose of making quantitative assessments of metal–organic associations in sediments.

3. TECHNETIUM BEHAVIOUR IN SOILS AND SEDIMENTS

It has been established in recent years that the rate of soil to plant transfer of technetium, when applied as TcO_4^- to a soil, decreases over time [4–6], the effect being particularly pronounced under flooded conditions [7]. These effects can be rationalised in terms of speciation changes from the highly soluble and readily available TcO_4^- to less available and highly insoluble forms such as TcO_2, sulphides and high MW organic complexes. Evidently, the condition responsible for these effects is anaerobicity leading to the reduction of Tc to lower oxidation states (IV). It may seem surprising that Tc can be reduced in soils which are aerobic. However, it is well known that soils cannot be considered as uniformly oxic and that anaerobic centres can develop in apparently well-drained soils [8–10]. This is caused by oxygen depletion through microbial growth in small water-filled pores, resulting in a sluggish, diffusion-controlled replenishment of oxygen. Similar phenomena may take place in marine sediments in which, under seemingly oxic conditions as evidenced by E_h readings, microbial H_2S production may occur in reducing microniches [11].

From the seemingly persistent nature of the phenomenon of Tc immobilisation in soils, it would furthermore appear that, in otherwise oxic conditions, the reduced forms of Tc are fairly stable towards re-oxidation in soils.

It has also been reported that the immobilisation of technetium is more pronounced in soils containing high levels of organic matter [5, 12, 13]. Moreover, it appears that the microflora plays a key role in establishing the proper redox conditions, conducive to the reduction of technetium [14], such conditions leading in turn to relatively high K_D ($\sim 10^3$) values in marine sediments [15]. The role of the microflora in these processes is of course indirect and it is not clear to what extent micro-organisms may quantitatively affect the fate of Tc in environmental conditions in more direct ways. It has been demonstrated that technetium (TcO_4^-) is accumulated by the microflora [16, 17], but the relative importance of these effects is not yet clear. It is to be expected, however, that where Tc is present as an insoluble precipitate or an HWM humic acid complex the rate of uptake will be severely impeded. Moreover, upon decay of the microflora the biomass-associated (and reduced) fraction of Tc will eventually be returned to the soil OM fraction.

In some cases, when TcO_4^- is introduced into geological formations which have reducing properties *per se*, it may be reduced directly, as demonstrated by Bondietti and Francis [18]. Another example is the TcO_4^- reduction in Boom clay, where it can be shown that the Tc in the extract is quantitatively present as an OM complex [19].

A particularly interesting finding is that (in contrast to other complex-forming radionuclides or transition metals) reduced Tc species (precipitates or humic acid complexes) are not resolubilised by complexing agents [20] such as EDTA and DTPA, known to form stable Tc complexes [21–24]. This would indicate that either Tc complexes are relatively 'inert' in displaying quite sluggish ligand exchange rates (a matter to be discussed in some detail in the experimental section) or that EDTA and DTPA complexes cannot match the stability of the Tc humate complexes.

At present, there appear to be no suitable methods that enable quantitative distinction to be made between the various possible insoluble forms (precipitates, insoluble HA complexes). Usually, rather harsh procedures are used, such as peroxide treatment, which may re-oxidise all reduced forms. By the same token, it has not been possible to show if, in environmental conditions, some of these immobilised forms are transient and that some of these forms are eventually converted into other more stable forms.

It appears in any case well established that humic acids (HA) may play a significant role in that they may lock reduced Tc into relatively stable complexes. In fact, it was recently shown that when hydroxyl groups are adjacent to carboxylic groups, conditions are quite favourable for the formation of stable Tc complexes [25]. Evidently, these are conditions which are abundantly met in the organic matter fractions of soils and sediments. In the following pages, we describe a novel approach which is liable to enable quantitative assessments to be made of the role of the OM fraction in the environmental chemistry of technetium and other radionuclides.

4. RADIONUCLIDE–ORGANIC MATTER ASSOCIATION: A QUANTITATIVE METHOD

Let the amount of a radionuclide present in a soil or sediment be represented by Q (conveniently expressed per unit mass of solid) which is partitioned among different geochemical sinks (reducible oxides, precipitate, humics, microflora . . .). At the outset, we shall assume that only a negligibly small fraction of water-soluble radionuclide (such as TcO_4^-) is present. If not, such a fraction can be eliminated and measured beforehand by a prior extraction step at low pH (5–6), avoiding possible loss of the OM-associated radionuclide. Let the amount of radionuclide associated with the soil or sediment organic fraction (humics) be represented by Q_A, the objective being the evaluation of Q_A/Q, the fraction of radionuclide bound to the OM sink.

The principle of the method is to isolate a small but measurable fraction of the OM sink, using conditions as close as possible to the in situ situation, ensuring no changes in speciation and leaving the other geochemical sinks unaltered in the process. This may often be difficult to achieve unless the extraction pH (necessary for isolating a sizeable fraction of OM) happens to coincide with the sediment pH, as in the case of Boom clay [19]. In most cases, such as marine sediments and soils, pH conditions of extraction are bound to differ from in situ values (which may lead to a partial radionuclide shift from one sink to another). In view of the fact that Tc humic acid complexes exhibit quite slow ligand exchange rates (as shown below), negligibly small changes in speciation are ensured, at least on the time scale of the extraction process. Therefore, pH conditions of extraction may differ from those in situ, making Tc an ideal element for testing the method. It is furthermore imperative that the extractant used is in no way competitive with the OM extracted (and with any other geochemical sink), the purpose in mind being to avoid speciation changes. In practice, a dilute solution of sodium acetate at a pH of 8–9 can be used. Eventually, $NaHCO_3$ may be used provided it is verified that OM levels in the extract are sufficiently high to prevent the formation of carbonate complexes. Extractions are preferably carried out under inert atmosphere so as to minimise alkaline hydrolysis [26] which might generate additional functional groups.

In the extract, obtained by high-speed centrifugation, the amount of radioactivity is assayed and expressed per equivalent of organic matter, the latter being measured by a method presented at this seminar [27]. Let q represent the (OM-associated) radionuclide in the extract and n the amount of OM, both being expressed per unit volume. We assume that the amount of radionuclide scavenged per equivalent OM in the extract is identical to the value in the equilibrium sediment, i.e.

$$\frac{q}{n}=\frac{Q_A-qv}{N-nv} \tag{1}$$

where v is the volume of extractant used and N the amount of organic matter (in equivalent terms) in the sample. Therefore, Q_A-qv and $N-nv$ represent the amount of OM-associated radionuclide, or organic matter, per unit amount of sediment, after extraction equilibrium (qv and nv referring to total amounts in the extract). Multiplying both sides of eqn (1) by $(Q-qv)$, i.e. the total (organic-bound + other sinks) amount of radionuclide remaining in the sediment, the expression can be rearranged into the form

$$\frac{Q_A-qv}{Q-qv}=\frac{N-nv}{n}\Bigg/\frac{Q-qv}{q}=\text{EFAR} \tag{2}$$

The left-hand side is the fraction of OM-associated radionuclide in the sediment after extraction. The right-hand side is the equilibrium ratio of K_D values of OM and radionuclide, a quantity which we define as EFAR or 'Equilibrium Fractional Activity Ratio', i.e.

$$\text{EFAR}=\frac{q}{n}\Bigg/\frac{Q-qv}{N-nv} \tag{3}$$

The fraction of OM-associated nuclide in the original sample (prior to extraction), Q_A/Q, is obtained by rearranging eqn (2) into the form

$$\frac{Q_A}{Q}=\text{EFAR}+\frac{qv}{Q}(1-\text{EFAR}) \tag{4}$$

Obviously, if $qv \ll Q$, i.e. the fraction of radionuclide in the extract is very small, eqn (4) takes the simple form

$$Q_A/Q=\text{EFAR} \tag{5}$$

Therefore, the fraction of radionuclide associated with the OM sink is directly obtainable from K_D values of the radionuclide and the organic matter. There are at present no suitable methods for the quantitative measurement of the ion capacity of the OM fraction in organic–inorganic mixtures such as soils and sediments [26]. Such methods are now being investigated in our laboratory. However, reasonable estimates can be made on the basis of overall OM content, using capacity values of 3–4 meq/g [26].

The assumption embodied in eqn (1) requires some comment. It is *a priori* not known whether metals might not exhibit some preference for the highly soluble and low molecular weight organics which are being selectively extracted in the lower pH range. Such an effect, if any, could perhaps be accounted for on the basis of a higher functional group density (i.e. precisely the reason of higher solubility). In view of the fact that the level of radionuclides is in effect expressed with respect to equivalent OM concentration, the assumption does not seem to be unreasonable. In any case, it remains subject to experimental verification, e.g. by performing extractions in a broad pH range. Moreover, if such an assumption is only partially fulfilled, the method could still be of use in studying possible transient effects of metal–organic matter associations. The method is illustrated in the experimental section for the case of Tc binding in a podzol soil.

A particularly interesting feature of this method is that it enables one to study the partitioning of radionuclides between organic matter and well-defined adsorbants (clays, oxides), even in cases which are complicated by adsorption phenomena of the organic phase. In such systems, differentiation between the two sinks is difficult to achieve since changes in metal concentrations in the liquid phase may be due either to a shift of the metal species towards the competitive solid or to adsorption of the (metal-bearing) OM on to the solid [28]. The method presented here enables such questions to be settled: in the first case, the amount of radionuclide per unit OM will be decreased whereas, in the second case, it shall remain invariant. This conclusion is validated by some recent findings [29], indicating that adsorption of fulvic acid does not result in a loss of its complexing properties and that the characteristics of the complexation reaction appear to be similar in the adsorbed state and the solution.

This particular aspect of the methodology is illustrated by an example on the partition of Eu^{152} between α-Fe_2O_3 (haematite) and a commercial humic acid extract (Fluka). Haematite dispersions are mixed with Eu^{152}-labelled humic acid solutions at an initial pH of about 8. Haematite content is fixed at 10 g/litre and OM concentrations are 2 and 0·2 g/litre. The ionic strength is adjusted to 0·1 M with $NaNO_3$. The systems are

Table I
Europium partitioning between haematite and humic acid

Initial HA concentration: 2 g/litre (8·07 meq/litre); haematite concentration: 10 g/litre; initial Eu^{152} activity: $20·5 \times 10^3$ cpm/meq HA

pH	Eu^{152} *activity (cpm/ml)*	*HA conc. (meq/litre)*	Eu^{152} *activity (cpm/meq HA)*	*EFAR*	log K_D
8·00	131·0	7·41	$17·7 \times 10^3$	0·86	1·42
8·44	127·2	7·42	$17·1 \times 10^3$	0·83	1·52
8·94	123·2	7·73	$15·9 \times 10^3$	0·78	1·62
9·46	119·0	7·85	$15·2 \times 10^3$	0·74	1·72

Initial HA concentration: 0·2 g/litre (0·844 meq/litre); haematite concentration: 10 g/litre; initial Eu^{152} activity: $1·5 \times 10^5$ cpm/meq HA

pH	Eu^{152} *activity (cpm/ml)*	*HA conc. (meq/litre)*	Eu^{152} *activity (cpm/meq HA)*	*EFAR*	log K_D
7·47	44·5	0·70	$6·36 \times 10^4$	0·42	2·26
7·97	42·1	0·72	$5·85 \times 10^4$	0·39	2·35
8·54	39·9	0·73	$5·47 \times 10^4$	0·36	2·43
9·13	35·7	0·72	$4·96 \times 10^4$	0·33	2·55

shaken for 30 min and a sample of the homogenised slurry is shaken overnight and centrifuged at high speed. The supernatant liquid is assayed for Eu^{152} activity and the OM content is measured, using the Ag-thiourea method [27]. The remaining slurry is adjusted to a slightly higher pH by addition of 1 M NaOH, another sample shaken overnight and analysed, the procedure being repeated four times, covering a pH range of 7·5–9·5. The results are summarised in Table I. It appears that the fraction of Eu in the OM phase amounts to about 0·75–0·85 at high OM content (2 g/litre) and 30–40% at low OM content (0·2 g/litre). Decreasing OM levels to 0·02 g/litre (data not shown) leads to a 100% association of Eu with the oxide phase. It may seem surprising that for a fifty-fold excess of oxides over OM about one-third of the Eu remains associated with the OM fraction. Of course, these results should not be extrapolated to natural conditions because (a) of possible irreversible sorption of Eu on the organic phase and (b) ferric oxyhydroxides in natural sediments are often quite amorphous and in environmental conditions are often derived from oxidation and hydrolysis of Fe(II) [30]. Surface areas and porosities, and therefore sorption properties, may be quite different. In any case, the method may offer some perspectives in studying the partition of metals

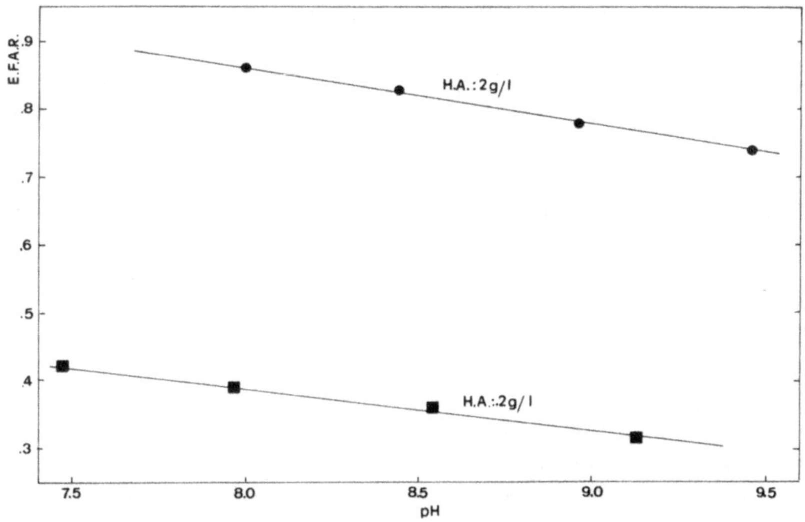

FIG. 1. Effect of pH on the fraction of Eu in humic acid at concentrations of 2 g/litre (●) and 0·2 g/litre (■) in the presence of haematite (10 g/litre).

between oxides and organic matter under well-defined conditions approximating those in natural environments. Examination of the data in Table I also shows a distinct effect of pH on the partitioning of Eu between oxide and organic phases. It is seen that at the high OM content, the fraction of OM-associated Eu decreases from 86 to 74% when increasing the pH from 8 to 9·5. At low OM content, the change occurs from 42 to 33%. The results are summarised in Fig. 1. The shift of Eu towards the oxide phase is probably related to the higher pH dependency of the oxide charge (as compared to OM): the zpc (zero point of charge) value of haematite is about 8·5, i.e. the oxide phase acquires an increasingly negative charge at higher pH, as reflected in the decreasing adsorption trend of OM at the high OM contents.

One final comment is in order. It is obvious that K_D values alone (also given in Table I), in the absence of quantitative data on solid–liquid partitioning of OM, are of limited use in assessing the distribution of the radionuclide between the oxide and the organic sink. In fact, the increase in K_D with higher pH is the combined result of two processes having opposite effects: firstly, the desorption of OM at higher pH, which would lead to a lower K_D; secondly, the shift of Eu—the most important effect—towards the oxide phase which becomes more competitive at increasing pH.

5. EXPERIMENTAL SECTION

The purpose of this section is to present and discuss some recent data, all of which relate to the possible role of OM as a geochemical sink for Tc.

5.1. Fate of Tc in the presence of humic acids (HA) in reducing conditions generated by micro-organisms

A 200-ml growth medium (glucose: 1 g/litre; KH_2PO_4: 10 g/litre; yeast extract: 10 g/litre; Trypton: 10 g/litre) was first saturated with N_2 and inoculated with *E. coli* in an air-tight glass vessel. The system was equipped with a set of electrodes (glass, Pt, Ag–AgCl), enabling continuous pH and E_h monitoring. In addition, the system is continuously stirred by a slow nitrogen flow, evacuated by way of a capillary, connected to a 'water lock'. The complete set-up is shown schematically in Fig. 2. Within about 4–5 h, E_h values drop to about −300 mV (H^+/H_2 scale). After 15 h, the system is heavily stirred by a short but high nitrogen flow (leading to a sharp increase in E_h to about −100 mV, probably the result of the

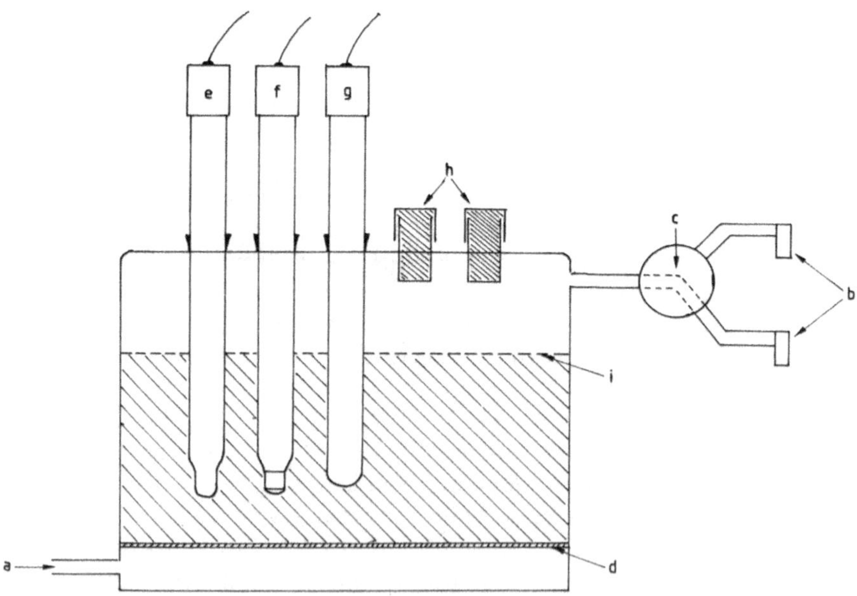

FIG. 2. Reactor vessel for reducing TcO_4^- by microbial action: (a) nitrogen inlet, (b) water lock, (c) 3-way valve, (d) sintered glass, (e) combined glass electrode, (f) Pt electrode, (g) reference electrode, (h) septa, (i) fluid level.

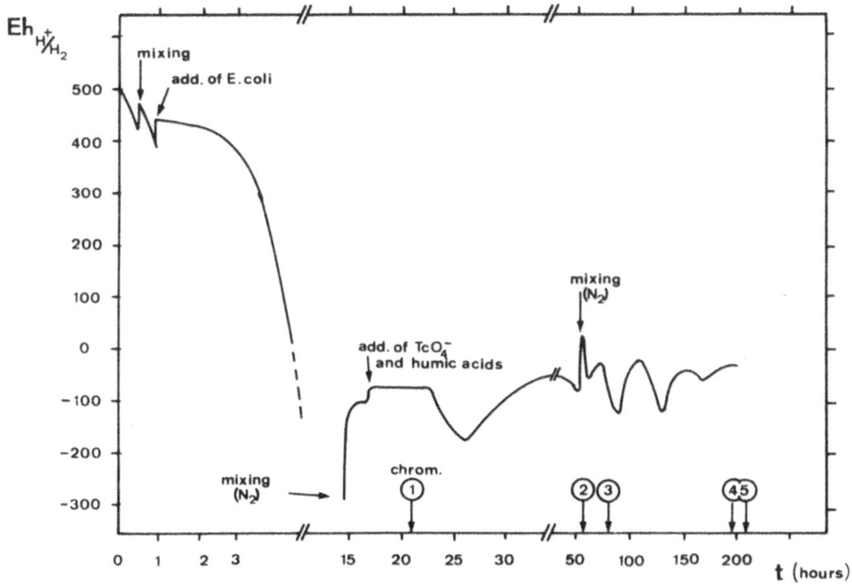

FIG. 3. Redox potential and sequence of operations in the microbial reduction and complexation of technetium.

evacuation of some volatile fermentation products). By way of a septum, 30 ml of a 10^{-6} M solution of TcO_4^- (^{99}Tc, labelled ^{95m}Tc) in a high molecular weight fraction (> 10 000) of HA (≃300 ppm) is injected into the medium. The HA was extracted from Boom clay at pH = 9 and the high molecular weight fraction obtained by ultrafiltration. Throughout the course of the experiment, pH stabilised around 5·9–6·2 and E_h values fluctuated between 0 and −100 mV, conditions which are within the stability domain of the Tc(IV) oxidation state [31]. The liquid is sampled aseptically at regular time intervals and analysed by gel exclusion chromatography and γ-scintillation counting. The overall procedure is summarised in Fig. 3.

A series of chromatograms of filtered (0·2 μm) samples taken between 20 and 200 h are shown in Figs 4(a–d). Examination of Fig. 4(a) shows clearly that the HA is well separated from the growth medium organics: the HA peak occurs at $V_e = 10$ ml (void volume) whereas the other organics cover a broad range of elution volumes between 10 and 40 ml, a distinct peak occurring at about $V_e = 20$. Consequently, a fairly reliable assignment can be made of Tc distribution between the two 'sinks'. A

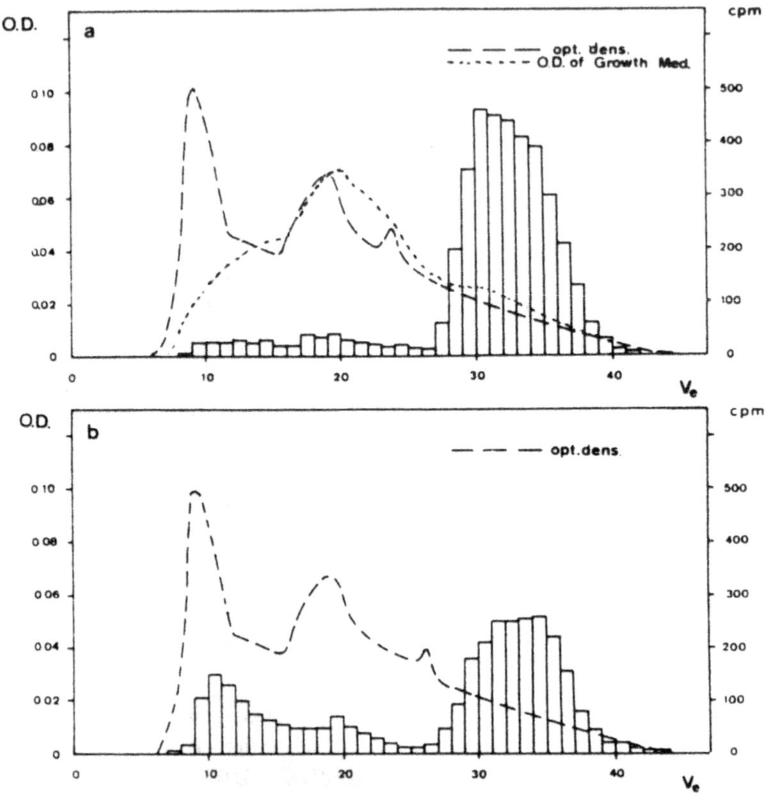

FIG. 4. Gel chromatograms (Sephadex G-25) and Tc distribution in growth medium filtrate, taken 5 h after humic acid injection (a), after 45 h (b), after 64 h (c), after 184 h (d) (G-15).

comparison of the chromatogram shows that after 21 h, i.e. 5 h after TcO_4^- injection, only a small fraction is reduced and eluted as a complex. This fraction gradually increases and after a week Tc is quantitatively reduced, as evidenced by the chromatogram in Fig. 4(d) (the elution volume of TcO_4^- in Sephadex G-15 is 68 ml). Moreover, it can be estimated from Fig. 4(c) that about 50–60% of the Tc in the filtrate is associated with the HA, the remaining part being bound to the growth medium organics. This may seem surprising in view of the fact that, on a weight percentage basis, the content represents only about 0·2–0·3% of the other organics (yeast extract, Trypton), a finding which points to the extremely high stability of Tc–HA complexes.

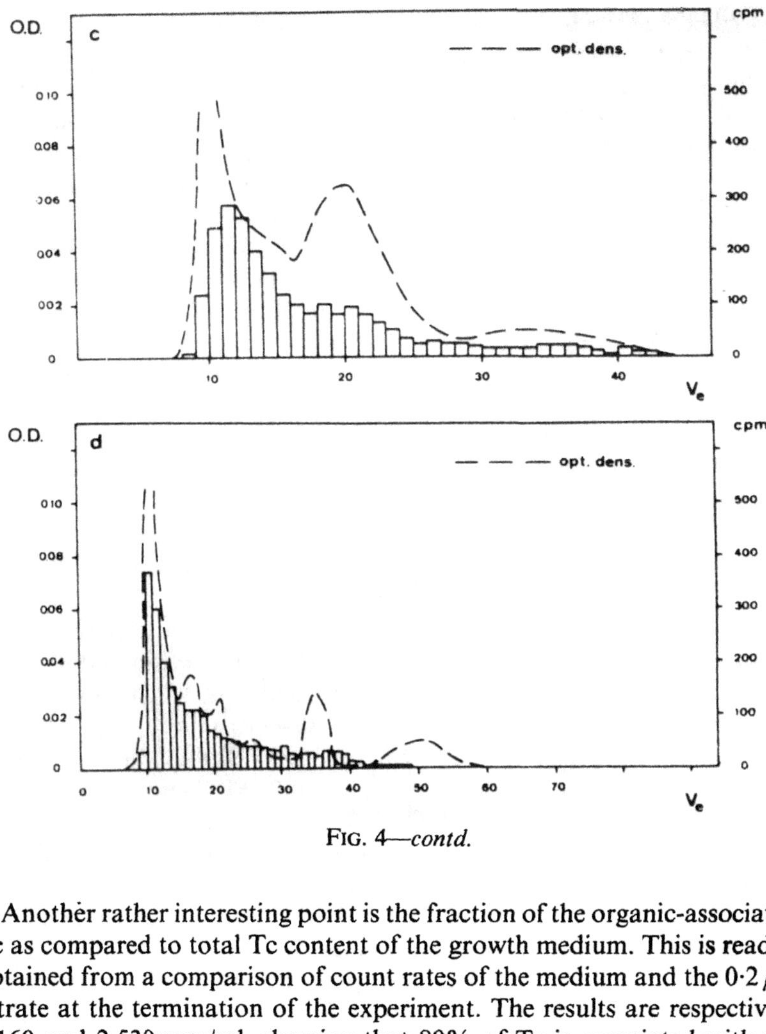

FIG. 4—*contd.*

Another rather interesting point is the fraction of the organic-associated Tc as compared to total Tc content of the growth medium. This is readily obtained from a comparison of count rates of the medium and the 0·2 μm filtrate at the termination of the experiment. The results are respectively 3·160 and 2·530 cpm/ml, showing that 80% of Tc is associated with the liquid phase organic fraction (i.e. HA, growth medium compounds and possibly microbial metabolites), the remaining 20% being either associated with the microflora or present as insoluble compounds (oxide, sulphide). It is obvious that these results can hardly be extrapolated to environmental conditions but they do clearly show that, when appropriate redox conditions prevail, HA complexation may represent an important geochemical sink for Tc.

5.2. Ligand exchange phenomena in Tc–HA complexes

It was stated above that technetium–organic complexes appear to be characterised by slow ligand exchange rates. This is qualitatively confirmed by a ligand exchange experiment involving the Tc transfer from a high molecular weight (HMW) to a low molecular weight (LMW) HA fraction. The HMW fraction was prepared by overnight extraction of a 100 g podzol (Kalmthout-Belgium) sample in 200 ml of NaOH (0·5 M). After high-speed centrifugation, Ha is precipitated at pH = 2, using HNO_3, and dialysed (Visking, Medicell Int. Ltd) free of salt. In the process, the LMW compounds (arbitrarily defined as the fraction which permeates the membrane) are eliminated. The pH of the HA is raised to 7 and the system is again dialysed against distilled water (discarding again the LMW compounds in the dialysate). Finally, the HMW fraction is treated twice with $CaCl_2$ 1 M (converting the exchange complex into the Ca form) and dialysed until salt-free. The last step is taken to avoid coagulation of the OM in the process of Tc reduction, using Sn^{2+}.

The LMW fraction is prepared by extraction of a batch of the same soil with NaOH (0·1 M). The pH of the supernatant is lowered to 7 and dialysed against distilled water. The dialysate, containing a LMW fraction (fulvic + humic), is collected, concentrated by freeze-drying and submitted five times to a dialysis, each time discarding the residual fraction remaining in the dialysis bag. The result is a LMW fraction which should freely migrate through the membrane. $^{99}TcO_4^-$ is then introduced ($\sim 5 \times 10^{-8}$ M) in the HMW fraction and labelled with $^{95m}TcO_4^-$. Sn^{2+} is added (5×10^{-5} M) and the system is shaken for 24 h. These conditions are adequate for a quantitative reduction and complexation of Tc, as shown in another paper at this seminar [32]. A chromatogram showed all Tc to be associated with the organic fraction, eluted at the void volume in a Sephadex G-25 column; 15 ml of the HMW Tc–HA complex is then introduced into a dialysis bag and immersed in an equal volume of the LMW fraction, allowing its free transport to the inner compartment of the dialysis bag and possible transfer of Tc from the HMW to the LMW fraction of HA. The process is monitored by sampling the LMW fraction at regular time intervals and analysis by gel chromatography and γ-scintillation. The pH of the system is about 7. The appearance of Tc in the LMW fraction is illustrated in Figs 5(a–e). The chromatographic data for the LMW fraction are given in Fig. 5(a), showing two distinct peaks at elution volumes of 15 and 25 ml. It is seen that after 13 days (Fig. 5(b)) some Tc has shifted towards the LMW fraction (showing the same OD pattern as the original sample) and increases with time (Figs 5(c, d)). It appears, however, that after a month some slight changes occur in the OD pattern of the LMW fraction, corresponding to a small shift towards lower MW values. In addition, it appears that in all cases some slight re-oxidation has occurred, as evidenced by the Tc activity increase at 37–38 ml, the elution volume of TcO_4^-.

At the termination of the experiment, the HMW fraction in the dialysis bag was coagulated: the chromatogram of the supernatant in the bag, shown in Fig. 5(e), is quite similar to the one in Fig. 5(d) and relates to a LMW fraction identical to the one outside the membrane.

The overall result can be summarised by stating that, after 46 days, 52% of the activity remains in the HMW fraction in the sediment within the bag, the remainder having shifted towards the LMW fraction. It cannot be stated that LMW complexes are more stable since no efforts were made to relate Tc distribution to functional group concentrations in the two fractions. It is shown, however, that ligand changes do occur but that the process is rather slow. The important implication of this finding is that within the time scale (hours) used for determining Tc associations with OM by extraction techniques (see below) no changes in speciation are expected to occur. Moreover, such a result may explain also why extraction of soils with EDTA or DTPA (provided of course such complexes can match the stability of HA complexes), using relatively short equilibration times, are unsuccessful.

5.3. Extraction of Tc–HA complexes from soil

In attempting to assess the role of the OM sink to radionuclide immobilisation at in situ conditions, it is imperative that, in the course of an extraction of OM, no change in speciation (or a radionuclide shift from one geochemical sink to another) occurs. Such conditions appear not to be satisfied in the case of Europium, as demonstrated above by the distinct shift from the OM to the oxide phase at increasing pH. In the case of Tc, however, speciation changes are unlikely during extraction, on account of the slow rate of ligand exchange.

We here present an attempt to quantify the role of HA on Tc immobilisation in a podzol, in which Tc has quantitatively been reduced by microbial action. The soil used is the B_1 horizon from a podzol (Kalmthout), a strongly leached and very acid soil ($pH(H_2O) = 3{\cdot}72$; $pH(KCl) = 3{\cdot}19$; clay content = 0·014%; %C = 2·53). Its low pH indicates that, using the Henderson Hasselbach equation and a pK of 4·6–4·7 [26], at least some 90% of carboxylic groups occur as COOH.

Prior to use, the soil batch is neutralised to pH = 6 by $CaCO_3$ addition.

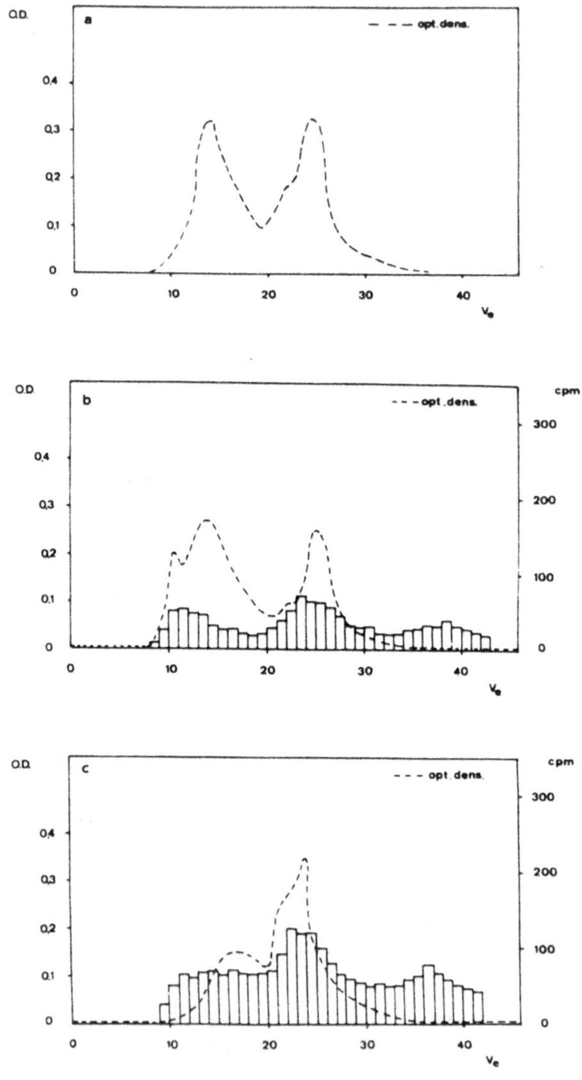

FIG. 5. Tc–ligand exchange rates (see text) from HMW to LMW humic acids, as measured by gel chromatography (G-25); (a), (b), (c) and (d) show the OD of and Tc activity in the LMW fraction after 0, 13, 31 and 44 days, (e) shows the LMW fraction in the dialysis bag after 46 days.

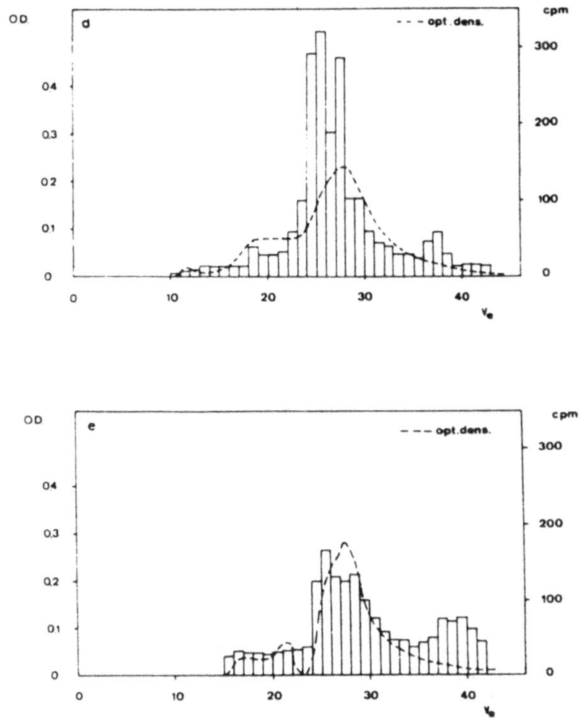

FIG. 5—*contd.*

For purposes of stimulating microbial development, some glucose (3 ml of a 1% solution/30 g soil) and a nitrogen source (3 ml 0·02 M NH_4NO_3/30 g soil) are added to the soil. After a few days $^{95m}TcO_4^-$ is added to the system, which is shut off from the air (N-atmosphere) and kept under a saturated water atmosphere. An aqueous extract (pH = 6·5) showed that, after a week, Tc is completely reduced, as evidenced by the minute levels of Tc found in the extract. Proper conditions being established for quantitative immobilisation of Tc (in HA, oxides, microflora ...), the effect of OM was investigated by submitting a series of ± 2 g batches to such treatment, followed by extraction in a pH range 7–9 under nitrogen atmosphere. Such conditions are relatively mild so as to exclude alkaline hydrolysis [26] which would generate additional carboxyl groups during extraction.

Extractions are carried out by dispersing the soil samples in 20 ml of distilled water–NaOH mixtures of varying ratios so as to cover a pH

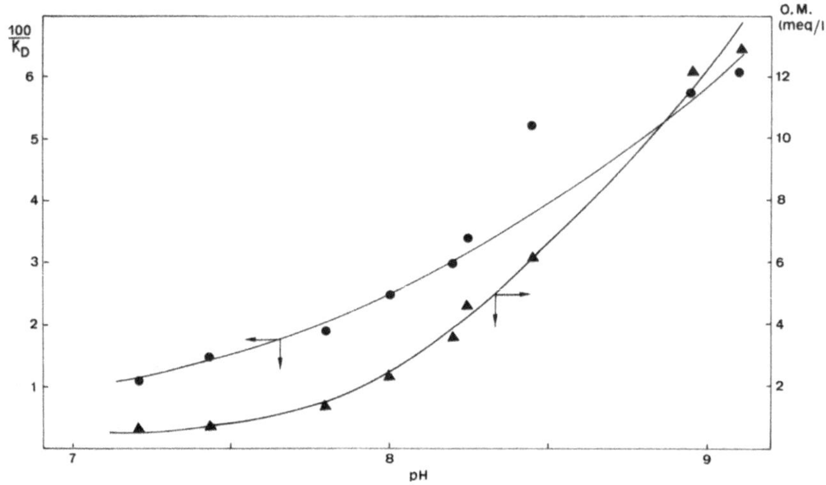

FIG. 6. Effect of extractant pH on reciprocal K_D–Tc and organic extraction yield (in meq/litre) in a podzol soil.

range of 7–9. In so doing, increasing levels of organic matter are extracted. The liquid phase is separated by high-speed centrifugation and supernatants are assayed for Tc activity, pH and HA concentration, using the Ag–thiourea method [27]. Four samples were controlled by gel chromatography and confirmed the absence of TcO_4^- . The results are summarised in Fig. 6, showing the effect of pH on HA concentration and reciprocal K_D values for Tc.

With K_D values ranging from 90 (at pH 7·2) to 16 (at pH 9·1), reciprocal K_D values are a direct measure of the fraction of Tc in the liquid phase. Examination of Fig. 6 shows that, as expected, increasing pH leads to higher yields of OM in the extracts: at pH = 7·2, OM concentrations are 7×10^{-4} eq/litre and, at pH = 9·1, $1{\cdot}2 \times 10^{-4}$, i.e. an increase by a factor of 17. The increase in OM concentration is accompanied by a decrease in K_D(Tc), clearly indicating Tc–OM association; the effect is, however, not proportional. Reciprocal K_D values (i.e. $100/K_D$) increase from 1·12 to about 6 within the same pH range, i.e. a factor of 5–6. This is, of course, a direct indication that the amount of Tc, associated per unit amount of OM, is significantly higher at low pH (7·2), roughly by a factor of 5. This is nicely illustrated in Fig. 7, showing the Tc levels in the extracts expressed per microequivalent OM. This effect may be due to an intrinsically higher stability of LMW Tc complexes (preferentially extracted at lower pH) or an overestimate of the Tc/OM ratio (perhaps

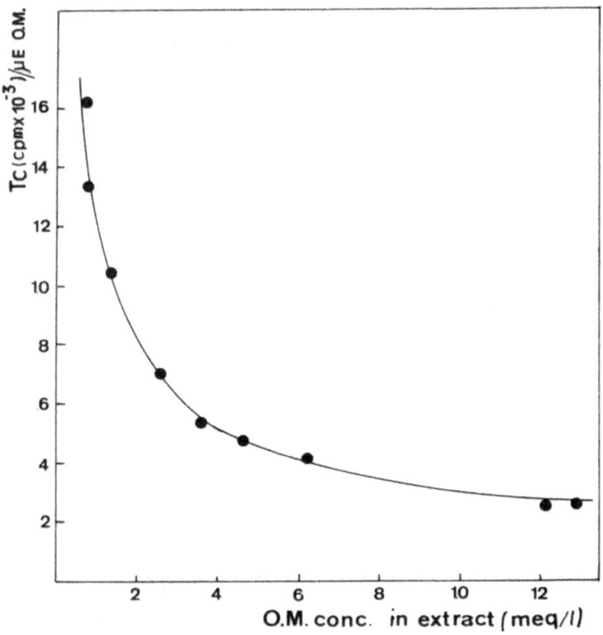

FIG. 7. Effect of organic matter concentration in the extract on Tc activity, expressed per unit amount of OM, extracted from a podzol soil.

due to the fact that a number of functional groups, such as amine or phenolic functions or perhaps microbial metabolites, are not being taken by the Ag–thiourea). Whatever the reason for this effect, it appears that at higher pH values, corresponding to increasingly lower contributions of LMW compounds to the overall functional group capacity, Tc/OM ratios level off to some constant value, where the enrichment effect of the LMW fraction is 'swamped' by the high excess of HMW compounds. The next step is the quantitative estimate of the contribution of the soil OM sink to Tc immobilisation. This requires the use of the EFAR concept, defined above, and the amount of functional (COOH) groups in the soil as a whole. Measurement of the exchange capacity of the bulk soil, using the Ag–thiourea method, yields a value of 7·6 meq/100 g. It would appear that such a value is strongly underestimating the number of carboxylic groups on several counts. Firstly, on the basis of organic carbon content (2·53%), one would estimate the OM content at about 4·5%, which would correspond to a value of 18 meq/100 g, if we take a COOH capacity of 4 meq/g OM. Secondly, the amount of OM extracted at pH = 9 is in the range of 8–9 meq/100 g, as determined by the Ag–thiourea method on the dilute extracts. These amounts exceed or are similar to the CEC value (7·6 meq/g), a fact which is related of course to the release of blocked sites, as discussed in another paper at this seminar [27]. The third reason is based on titration data of the acid soil [33]. The amount of NaOH or $Ba(OH)_2$ required to raise the system to a pH of 7–7·5 (i.e. the pH value generally considered to correspond to complete neutralisation of the COOH groups) amounts to 16 or 20 meq/100 g. The fourth reason is based on the Ca acetate method [33], considered to yield an estimate of the number of COOH groups. The result is 2·7 meq/100 g. We therefore consider a value of 18 meq/100 g to be a more realistic estimate of the number of carboxyl groups in the soil.

In any case, we shall calculate EFAR values, using two alternative values for functional group densities: 7·8 and 18 meq/100 g. Of course, when using the lower value, we are restricted to the systems corresponding to OM extraction yields less than a value of 7·8 meq/100 g. Depending on the choice made for the number of functional groups in the soil, two sets of EFAR values are obtained for the range of OM extraction yields investigated. From these EFAR values, the fraction of Tc in the OM sink, Q_A/Q, can be obtained, using eqn (4). The results are summarised in Fig. 8, showing Q_A/Q as a function of OM levels in the extract.

It is seen, as could be expected on the basis of the Tc enrichment in the LMW fraction extracted at low pH (see Fig. 7), that Q_A/Q values appear to be higher than unity. In other words, it would seem that the fractional Tc loading of the LMW fraction is not representative of the scavenging power of the soil OM fraction as a whole and the assumption contained in eqn (1) is not valid. However, at high OM concentrations in the extract, where the HMW fraction by far exceeds the LMW contribution to the overall binding capacity, the enrichment effect of the LMW is 'drowned' and Q_A/Q values level off to some constant value. It would therefore appear that the assumption contained in eqn (1), i.e. identical relative 'loadings' of Tc in the OM fraction in the extract and the solid, is more likely to be fulfilled. Depending on the choice made for the number of functional groups in the soil (8 or 18 meq/100 g), Q_A/Q values level off at values of 0·4 or 0·6, which corresponds to about 50% of Tc being locked in the OM sink. The remaining part may be associated with any of the other possible sinks.

Perhaps some comment is in order with regard to the nature of the functional groups involved in Tc–humic acid binding. It has been argued [27] that, using the Ag^+–TU as index cation, no other groups besides

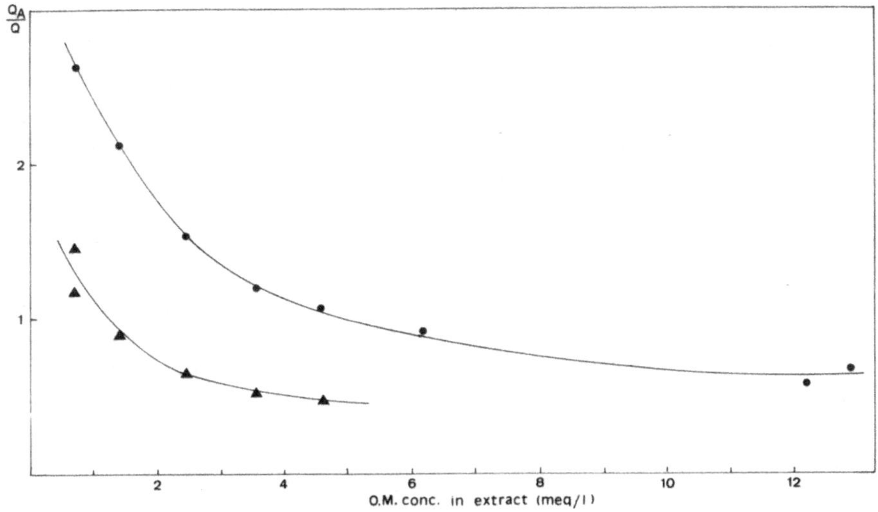

FIG. 8. Fraction of Tc in the organic matter phase of a podzol, as calculated from eqn (4), using either 78 (▲) or 180 (●) μeq/g for the number of functional groups per gram of soil.

COOH are measured in the quantitative evaluation of HA concentration. It is, however, quite likely that Tc complexing also involves other groups (OH, NH_2). In as much as the method is able to produce quantities which are directly proportional to the Tc/OM ratio (OM being quantified in terms of just one of its functional groups), this is not a serious problem, provided the abundance of COOH groups, relative to the values for the other groups, is not significantly different in the various MW fractions.

6. CONCLUDING COMMENTS

On the basis of various evidence, we have demonstrated that the organic matter fraction in soils constitutes an important geochemical sink for technetium. Tc–humic acid complexes were shown to be quite stable and characterised by rather slow rates of ligand exchange. Furthermore, a new method was presented, based on straightforward dilution principles, enabling quantitative assessments to be made of the fraction of Tc, and possibly other radionuclides, associated with the OM fraction of a soil or sediment.

Although the results are of a preliminary nature, they do nevertheless indicate the potential use of the method for studying sediment–radionuclide interactions. Moreover, such an approach might enable the study of the dynamics of radionuclide–soil or radionuclide–sediment associations, preferably relying on field studies. In particular, the question is to be resolved whether some given distribution of Tc between HA and the other sinks evolves in time.

The question whether the enrichment of Tc in the low molecular weight fraction is a real effect also deserves further study and could be resolved by quantitative Tc distribution measurements between well-defined LMW and HMW fractions. The relevance is rather obvious in that the answer has ramifications on the two issues emphasised in the introductory part: bioavailability and geochemical mobility, both in soils and in geological HLW repositories.

ACKNOWLEDGEMENT

This work has been carried out under contract with the Commission of the European Communities, Radiation Protection Program, Contract No. BIO-B-531-83-B. One of us (M.S.) acknowledges the receipt of a doctoral fellowship from the IWONL.

REFERENCES

1. TESSIER, A., CAMPBELL, P. G. C. and BISSON, M. (1974). *Anal. Chem.*, **51**, 844.
2. VANGENECHTEN, J. H. D., ASTON, S. R. and FOWLER, S. W. (1983). *Mar. Ecol. Progr. Ser.*, **13**, 219.
3. CALMANO, W. (1983). *Sci. Total Envir.*, **28**, 77.
4. HOFFMAN, F. O., GARTEN, C. T., LUCAS, D. M. and HUCKABEE, J. W. (1982). *Envir. Sci. Technol.*, **16**, 214.
5. MOUSNY, J. M. and MYTTENAERE, C. (1982). In: *Environmental Migration of Long-lived Radionuclides*, Vienna, IAEA, p. 353.
6. SAAS, A., GRAUBY, A., COLLE, C., DENARDI, J. L., NOMURA, T. and JOYER, P. (1982). In: *Environmental Migration of Long-lived Radionuclides*, Vienna, IAEA, p. 263.
7. VANDECASTEELE, C. M., MOUSNY, J. M. and LIETART, E. A. (1983). IAEA–CEC Seminar, Environmental Transfer to Man of Radionuclides Released from Nuclear Installations, Brussels (in press).
8. CURRIE, J. A. (1962). *J. Sci. Fd Agric.*, **13**, 380.
9. GREENWOOD, D. J. and GOODMAN, D. (1964). *J. Sci. Fd Agric.*, **15**, 579.
10. ROWELL, D. L. (1981). In: *The Chemistry of Soil Processes* (Ed. D. J. Greenland and M. H. B. Hayes), Chapter 7, New York, Wiley-Interscience.

11. JORGENSON, B. B. (1977). *Mar. Biol.*, **41**, 7.
12. LANDA, E. R., THORVIG, L. J. and GAST, R. G. (1977). *J. Envir. Qual.*, **6**, 181.
13. WILDUNG, R. E., MCFADDEN, K. M. and GARLAND, T. R. (1979). *J. Envir. Qual.*, **8**, 156.
14. MASSON, M., APROSI, G., LANIECE, A., GUEGENIAT, P. and BELOT, Y. (1981). In: *Impacts of Radionuclides into the Environment*, Vienna, IAEA, p. 341.
15. HARVEY, B. R. and KERSHAW, P. J. In: *The Behaviour of Long-lived Radionuclides in the Marine Environment* (Ed. C. Myttenaere), La Spezia (in press).
16. GEARING, P. J., BAALEN, C. VAN and PARKER, P. L. (1975). *Plant Physiol.*, **55**, 240.
17. VANDECASTEELE, C. M., DELMOTTE, A., ROUCOUX, P. and HOVE, C. VAN (1982). In: *Environmental Migration of Long-lived Radionuclides*, Vienna, IAEA, p. 275.
18. BONDIETTI, E. and FRANCIS, C. (1979). *Science*, **203**, 1337.
19. CREMERS, A. and HENRION, P. N. (1984). In: *CEC Int. Seminar on the Behaviour of Radionuclides in Estuaries*, Renesse (in press).
20. MARTIN, L. Y. and FRANZ, J. A. (1980). *Trans. ANS*, **35**, 53.
21. SCHWOCHAU, K. (1983). *Radiochim. Acta*, **32**, 139.
22. GORSKI, B. and KOCH, H. (1970). *J. Inorg. Nucl. Chem.*, **32**, 3831.
23. NOLL, B., SEIFERT, S. and MUENZE, R. (1983). *Int. J. appl. Radiat. Isot.*, **34**, 581.
24. RUSSELL, C. D., CRITTENDON, R. C. and GASH, A. G. (1980). *J. Nucl. Med.*, **21**, 354.
25. GROSSMAN, B. and MUENZE, R. (1982). *Int. J. appl. Radiat. Isot.*, **33**, 189.
26. STEVENSON, F. J. (1982). *Humus Chemistry*, New York, Wiley-Interscience.
27. STALMANS, M., KEIJZER, S. DE, MAES, A. and CREMERS, A. Measurement of the complexation capacity of organic matter in dilute extracts of soils and sediments (this volume, pp. 155–67).
28. DAVIS, J. A. and LECKIE, J. D. (1978). *Envir. Sci. Technol.*, **12**, 1309.
29. DALANG, F., BUFLE, J. and HAERDI, W. (1984). *Envir. Sci. Technol.*, **18**, 135.
30. CROSBY, S. A., GLASSON, D. R., CUTTLER, A. H., BUTLER, I., TURNER, D. R., WHITFIELD, M. and MILLWARD, G. E. (1983). *Envir. Sci. Technol.*, **17**, 709.
31. JENSSEN, B. S. (1982). *Migration Phenomena of Radionuclides into the Geosphere*, New York, Harwood Academic.
32. LOON, L. VAN, STALMANS, M., MAES, A., CREMERS, A. and COGNEAU, M. Soil–humic acid complexes of technetium: Synthesis and characterisation (this volume, pp. 143–53).
33. LOON, L. VAN (1982). Fixatie van zink in humusrijke bodems, Thesis, K.U. Leuven.

10

Technetium Sorption in Surface Soils

R. E. Wildung, T. R. Garland, K. M. McFadden and C. E. Cowan

Pacific Northwest Laboratory,
Richland, Washington 99352, USA

1. INTRODUCTION

Technetium from a number of sources has the potential for entering soils in initially volatile or solubilised forms [1, 2]. Several assessments [3] have suggested that subsequent soil solubility and uptake by plants will be important factors in governing Tc radiation dose to man. However, prediction of Tc behaviour in soils and availability to plants is complicated by its complex chemistry relative to soil processes.

Technetium is known to exist in all valence states from Tc^{+7} to Tc^{-1} and in chemical species representing a wide range in volatilities and solubilities [4–10]. The solubility of Tc in soil will be a function of Tc chemical speciation and distribution between the solution and solid phase. Soil factors which have been implicated in Tc sorption have been the degree of aeration, the nature and quantity of sorptive surfaces and microbial activity.

Under aerobic conditions the most soluble chemical species in water is the pertechnetate ion (TcO_4^-), which is highly stable over broad pH concentration ranges. When trace quantities of the pertechnetate ion (in 0·01 M $CaCl_2$) were equilibrated with 22 soils for 24 h under aerobic conditions, K_d values (pCi sorbed per g soil/pCi solute per ml equilibrating solution) ranged from 0·007 to 2·8 [11], indicating high solubility relative to hydrolysable elements such as Pu ($K_d > 1000$).

In another soil equilibration study over longer time periods (2–5 weeks), 8 of 11 surface soils sorbed 98% of pertechnetate from solution [12]. The longer equilibration times also provided an opportunity for microbial processes to play a role. However, this experiment was conducted in a closed, unbuffered dialysis system that likely led to reducing conditions. Concentrated solutions of pertechnetate have exhibited chemical reduction in the presence of dust and organic vapours [5, 7]. The reducing environment in the dialysis systems may have resulted in the formation of highly insoluble $TcO_2{\cdot}2H_2O$, sulphides or coprecipitates of Tc(IV) with other insolubles such as Fe sulphides. When the experiment was repeated on several of the soils under aerobic conditions, sorption occurred, but not to the same extent and at a slower rate. Thus, as predicted from its basic chemistry, pertechnetate remained relatively soluble under aerobic conditions, at least over short equilibration times; but redox conditions may be important in influencing Tc solubility in soil. Surface agricultural soils are not generally anaerobic for extended periods and the importance of this effect, as well as the limited sorption observed under aerobic conditions, will depend on soil properties, topography, climatic conditions, and agricultural use as well as the kinetics of mobilisation/immobilisation.

The sorption studies by Wildung *et al.* [11] encompassed soils representing a range in physicochemical properties. Sorption that occurred was directly correlated with organic C content and inversely correlated with pH. Thus, retention appeared to be closely related to the positive charge on the organic matter fraction. However, the oxidic Fe and Al fraction, which exists as amorphous coatings on crystalline minerals, may have also played a key role in some soils. Positive charge on both organic and mineral components may be expected to increase with decreasing soil pH. A lack of sorption by low organic matter and clay soils, reduction in sorption following H_2O_2 digestion of soil, and recovery of sorbed Tc by NaOH extraction also implicated the organic fraction in sorption [12]. Using soil thin-layer chromatography (TLC) to characterise Tc mobility in a range of Minnesota soils, Balogh and Grigal [13] also implicated the soil organic matter in Tc retention and suggested that weak binding mechanisms were involved. Although the TLC method did not accommodate highly organic soils, all the studies to date suggest the soil organic fraction will play a role in immobilisation of Tc, possibly through sorption by ion exchange mechanisms, reduction to less soluble forms, and complexation. Complexation of Tc might be expected to influence Tc mobility in soils in two ways [1, 2]. Low molecular weight organic ligands may serve to increase the solubility of reduced forms of Tc, whereas complexation reactions with the higher molecular weight, predominantly insoluble, soil humic fraction may lead to immobilisation.

Investigation of sterilisation [12] and temperature effects [14] on sorption of Tc implicated the soil microflora in Tc retention by soils equilibrated for periods longer than 2 days. However, it is also well established that sterilisation procedures alter the chemistry of soil surfaces and the soil solution. Interpretation is further complicated in studies in which soils are incubated in moist conditions in pots or columns because anaerobic conditions may exist within soil beds, even under well-drained conditions, perhaps leading to insolubilisation through chemical reduction.

The current studies were undertaken to (1) aid in resolving the anomalies arising from sorption and incubation studies conducted over different periods of time under aerobic and anaerobic conditions; and (2) to further examine the effects of the amorphous mineral fraction in conjunction with organic matter on Tc solubility in soil. The solubility of Tc (initially added as TcO_4^-) was determined under strictly aerobic conditions over a 1050 h period in 30 surface soils representing a wide range in genesis and physicochemical properties, including muck and peat soils of high organic C content. Solubility was related to soil properties (particle size distribution, elemental composition, mineralogy, amorphous Fe, Al, Si and Mn) and solution parameters (E_h, pH, organic C, cations and anions).

2. MATERIALS AND METHODS

2.1. Soil sampling and preparation

To provide a range in soil genesis and physicochemical properties, surface (0–15 cm) soils were sampled throughout the United States at locations in the states of Washington, Oregon, Colorado, Illinois and Minnesota under arid, semi-arid and humid climates and under sagebrush, grassland, pastureland, intensive agricultural and forest vegetation. After sampling, soils were air-dried ($\sim 8\%$ moisture), screened (< 50 mm), thoroughly mixed, quartered and stored at 25°C. Samples utilised in subsequent sorption studies were obtained by aggregating subsamples from each quarter.

2.2. Soil properties

Soil analyses were directed toward those properties with potential importance in Tc sorption (Tables I and II).

2.3. Soil equilibration

Sorption of Tc was determined by equilibration of $^{95m}TcO_4^-$ in distilled water with soil (10:1 water:soil) at tracer concentrations (0·03 μCi/g soil)

Table I

Summary of the properties of soils employed in studies of technetium behaviour: bulk physicochemical properties

Soil property	*Range*
Particle size distribution (%)[1,2]	
Sand	1·8–84·1
Silt	8·5–75·8
Clay	3·2–50·6
Primary clay minerals (% of clay fraction)[1,3]	<2–>20%
Composition (%)	
Organic carbon[4]	0·38–22·3
Total nitrogen[5]	0·050–1·42
Total sulphur[6]	0·013–0·57
pH[1]	4·7–8·3
Cation exchange capacity (meq/100 g)[7]	5·5–123

[1] Jackson [15].
[2] Summary of two clay fractions, three silt fractions and five sand fractions.
[3] Summary for individual measurement of montmorillonite, illite, kaolinite, quartz, feldspar, vermiculite, chlorite, interstratified constituents.
[4] Wildung *et al.* [16].
[5] Welcher [17].
[6] Bardsley and Lancaster [18].
[7] Routson *et al.* [19].

for a period of 1050 h at 26°C with gentle agitation (60 rpm) in an incubator-shaker.

2.4. Solution chemistry

Solution parameters pertinent to understanding Tc sorption with time were measured periodically during soil equilibration after removal of an aliquot of the equilibrating solution and filtration ($<0{\cdot}22$ μm).

Technetium was detected by gamma spectroscopy (Packard Model 5230). Major inorganic cations (Ca, K, Mg and Na) and trace elements (Al, As, B, Ba, Cd, Co, Cr, Cu, Fe, Li, Mn, Mo, Ni, Pb, S, Sr, Ti, Zn) were analysed directly or after concentration (lyophilisation) using a Jarrell Ash Model 975 inductively-coupled plasma emission spectrometer (Fisher Scientific Company).

Major inorganic anions (Cl^-, F^-, NO_3^-, NO_2^-, PO_4^{3-} and SO_4^{2-}) were analysed using a Dionex Model 16 ion chromatograph equipped with an AGI guard column, ASI anion separator column, eluent suppressor, and conductivity detector. Samples were eluted with 3 mM $NaHCO_3 + 2{\cdot}4$ mM Na_2CO_3 (2·3 ml/min); normal injection volume was 100 μl. Leachates

Table II

Summary of the properties of soils employed in technetium behaviour: oxidic silicon, aluminium, iron and manganese

Extractant and element	*Range* (*μg/g*)
Potassium hydroxide[1]	
Silicon	750–16 300
Aluminium	80–2 550
Citrate-bicarbonate-dithionite[2]	
Iron	390–8 000
Aluminium	50–11 700
Sodium pyrophosphate[3]	
Iron	10–12 300
Aluminium	20–3 940
Ammonium oxalate[4]	
Iron	220–10 900
Aluminium	162–14 100
Acidic hydroxylamine[5]	
Manganese	30–5 210

[1] Jackson [15].
[2] Mehra and Jackson [20).
[3] McKeague *et al.* [21].
[4] Schwertmann [22].
[5] Chao [23].

were periodically analysed to ensure the absence of thiocyanate (SCN^-) and thiosulphate ($S_2O_3^{2-}$) using Dionex brine (L20) columns. Fluoride was also periodically analysed using an ion-selective electrode (Orion Model 96-09).

Organic and inorganic carbon concentrations were determined using a Dohrmann DC-80 carbon analyser that consists of a UV-enhanced persulphate oxidation system and a CO_2 detector (Horiba Model PIR-2000).

Routine water quality parameters including conductivity, alkalinity, hardness, pH, E_h, and total dissolved solids were determined by standard methods [24].

3. RESULTS AND DISCUSSION

The primary objectives of the current studies were to determine Tc sorption under strictly aerobic conditions and to establish, with time, the

Table III
Summary of properties distinguishing soil groups and technetium sorption[1]

Primary vegetation	*Soil orders represented*	*Organic C content (%)*	*pH*	*Soluble nitrate-nitrogen*	*Ammonium oxalate extractable amorphous oxides*		*Technetium sorption*	
					Iron (μg/g)	*Aluminium (μg/g)*	*48 h (%)*	*1050 h (%)*
Grassland-shrub (15 soils)	Entisols Aridisols Mollisols	< 2·1	> 5·4	> 2·7	< 1 300	< 1 190	< 1–5·7	< 1–12·7
Forest-marshland (15 soils)	Entisols Inceptisols Spodosols Alfisols Ultisols Histosols	> 1·8	< 6·1	< 0·4[2]	> 1 610	> 1 230[3]	5·1–31·0	8·3–46·6

[1] See Tables I and II for ranges in soil properties.
[2] 14 soils.
[3] 13 soils.

relationships between sorption and soil properties of known importance in ion retention.

3.1. Soil properties

Soil organic, pH-dependent sites have previously been implicated in Tc sorption and particular attention was given in this study to the selection of soils that would extend the range in organic matter content (Table I) and allow examination of pH-dependent inorganic sites. To achieve a range in physicochemical properties, it was necessary to examine soils representing a range in parent material and genesis. Many of the properties of soils are therefore strongly interrelated and a selection of one set of variables often results in a range in associated properties reflecting soil-forming processes. Thus, the soils employed in these investigations also exhibited a wide range in particle size distribution, clay mineralogy and ion exchange capacities (Table I).

The 30 soils ultimately employed in these investigations encompassed 9 of the 10 soil orders of the United States comprehensive system (Table III). In particle size distribution (Table I), the soils ranged from sands to clays (3·2–50·6% clay). Montmorillonite was the major (> 20%) clay mineral in most soils, but illites also comprised a major component of several mollisols; kaolinite dominated the ultisols; vermiculite, along with montmorillonite, was a major component of the entisols and chlorite and interstratified minerals predominated in the inceptisols (Table I). The wide range of organic C contents, pH values, particle size distributions and mineralogies was reflected in major differences in reactive surface chemistry as exemplified by the cation exchange capacities (Table I), which were generally a function of increasing clay and organic matter content and pH.

In view of the known role of oxidic materials in ion sorption, Si, Al, Fe and Mn were extracted using a range of solvents designed to characterise the continuum of amorphous materials present in soil (Table II).

The KOH extraction has been employed to remove amorphous aluminosilicates. The citrate-bicarbonate-dithionite extraction has been commonly used to estimate crystalline oxides and amorphous material; sodium pyrophosphate provides a measure of organic forms; ammonium oxalate extracts primarily amorphous materials, likely products of recent weathering; acidic hydroxylamine has been used to estimate manganese oxides. All of the extractants remove significant organic C when applied to surface soils. As in the case of the soil properties, the extractability of these materials differed by orders of magnitude depending on the soil, and when used together in the present study, provided insights into the nature of soil development and types of reactive surfaces present.

3.2. Technetium sorption

Most surface soils of the world are well-drained for a significant portion of the year, but prior to this study, Tc sorption under strictly aerobic conditions had not been measured for extended periods. Furthermore, an evaluation of earlier studies suggested that in cases in which Tc solubility significantly decreased with time, chemical reduction of the pertechnetate ion may have been an important factor in retention, although this is difficult to separate from microbial processes. To accomplish unqualified aerobic conditions (solution $E_h > +375$) over extended time periods, shaking, batch-type equilibrations were employed. These systems had the additional advantage that solution properties could be fully characterised and related to the kinetics of Tc sorption over extended periods. The integrated effect of initial soil properties on Tc sorption were examined by empirical and correlative methods.

3.2.1. Effect of soil properties

Technetium, initially added as the pertechnetate ion, generally remained in solution for extended periods in a diversity of soils. The sorption that occurred was related to soil genesis and physicochemical properties.

The soils could be classified in two groups based on Tc sorption characteristics (Table III). Soils developed primarily under grassland-shrub vegetation and arid to semi-arid conditions sorbed less than 5·7% Tc from solution after 48 h equilibration. In contrast, soils developed in forests and marshlands under humid and subhumid conditions exhibited up to 31% Tc sorption over the same period. The two groups of soils were distinguished by markedly different (t-test, $p < 0{\cdot}05$) physicochemical properties, with the forest-marshland soils exhibiting higher organic C and amorphous Fe, Al contents and lower pH values than grassland-shrub soils (Tables I–III).

When soil properties were compared to Tc sorption in all soils at the 48 h equilibration time, sorption was highly correlated ($p < 0{\cdot}01$) to organic C, total N, extractable Fe and Al, several clay and silt fractions and pH (Table IV). The latter was inversely correlated to sorption substantiating the importance of pH-dependent sorption sites, which would tend to increase in positive charge with decreased pH on both the

Table IV
Significant simple correlations ($p < 0{\cdot}01$) relating properties of all the soils investigated to technetium sorption

Soil Property	*Correlation coefficient*[1] *for equilibration times of*	
	48 h	*1050 h*
Organic carbon	0·67	0·85
Total nitrogen	0·71	0·87
Extractable iron		
Ammonium oxalate	0·74	0·62
Citrate-bicarbonate-dithionite	0·73	
Extractable aluminium		
Ammonium oxalate	0·72	0·51
Citrate-bicarbonate-dithionite	0·78	0·55
Particle size distribution		
Fine clay	0.58	0·71
Coarse clay	0·71	0·61
Medium silt	0·58	0·49
pH	−0·56	−0·62

[1] $n = 25$; $p < 0{\cdot}01 = 0{\cdot}49$; $R = 0{\cdot}36$ for citrate-bicarbonate-dithionite extractable iron at 1050 h.

organic and oxidic mineral fractions. There was no significant correlation between Tc sorption and other soil parameters measured. Highest correlation coefficients occurred with organic C, total N, extractable Fe and Al and the coarse clay fraction. The higher (or equivalent) correlations of Tc sorption with the coarse clay and silt size fractions relative to the fine clay fraction may reflect higher concentrations of amorphous materials in these fractions. After 1050 h of equilibration, the same properties (except citrate-bicarbonate-dithionite extractable Fe) were highly correlated ($p < 0{\cdot}01$) with Tc sorption, but the organic C and N components increased in importance whereas the amorphous mineral fraction decreased in importance.

Separate examination of the grassland-shrub and forest-marshland soil groups provided further insight into the effects of soil properties on sorption (Table V). Although organic C and fine clay approached significance, only the citrate-bicarbonate-dithionite extractable Al was significantly correlated ($p < 0{\cdot}05$) to sorption for the grassland-shrub soils. Technetium sorption in the forest-marshland soils had a major effect on the relationships established for all soils (Table IV), and was predominantly

Table V
Simple correlations relating properties of grassland-shrub and forest-marshland soils to technetium sorption

Soil property	*Correlation coefficient*[1] *for equilibration times of*	
	48 h	*1050 h*
Grassland-shrub[2]		
Organic carbon	0·47	0·46
Total nitrogen	0·50	0·51
Extractable aluminium		
Citrate-bicarbonate-dithionite	0·66*	0·53
Particle size distribution		
Fine clay		0·50
Medium silt		0·41
Forest-marshland[3]		
Organic carbon		0·78**
Total nitrogen	0·55**	0·83**
Extractable iron		
Ammonium oxalate	0·65**	
Citrate-bicarbonate-dithionite	0·76**	
Extractable aluminium		
Ammonium oxalate	0·69**	
Citrate-bicarbonate-dithionite	73**	
Particle size distribution		
Fine clay		0·55*
Coarse clay	0·61*	0·58*
Fine silt		0·54*
Medium silt	0·57*	

[1] Single and double asterisks denote significance at $p < 0{\cdot}05$ and $p < 0{\cdot}01$ levels, respectively.
[2] $n = 10$; $p < 0{\cdot}05 = 0{\cdot}58$; $p < 0{\cdot}01 = 0{\cdot}71$; all properties not listed, $R < 0{\cdot}4$.
[3] $n = 13$; $p < 0{\cdot}05 = 0{\cdot}51$; $p < 0{\cdot}01 = 0{\cdot}64$; all properties not listed, $R < 0{\cdot}5$.

correlated ($p < 0{\cdot}01$) with the amorphous mineral fraction at 48 h equilibration and with the organic fraction after 1050 h. The clay and silt fraction was also implicated but was of lesser importance. The pH was significantly correlated with Tc sorption only when all soils were considered (Table IV).

3.2.2. *Effect of time*

The ranges in Tc sorption in both the grassland-shrub and forest-marshland soils increased over the equilibration time from 48 h to 1050 h

(Table III). However, comparison (*t*-test) of the means ± SD for Tc sorption at all time points for the periods < 48 h and 48–1050 h indicated that sorption increased significantly with time only in eight soils. Equilibrium in these soils was attained after approximately 300 h. As a group these soils did not exhibit unique properties and included three soils (aridisol, mollisol) from the grassland-shrub group and five soils from the forest-marshland group (histosol, inceptisols, entisol, alfisol). This overall increase with time, as well as the maintenance of Tc attenuation with time in other soils (sorption did not decrease in any soil with time) would appear to be attributable to a change in importance of initial soil properties (Tables IV and V) from predominant control by amorphous mineral components in the early phases to organic matter controls in the latter phases. This may have resulted from alteration of mineral surfaces with time, or perhaps increased importance of microbial activity. The 48 h period was chosen for comparison because prior studies with selected soils indicated that mineral dissolution and microbial effects were minimal over this time period.

The equilibration systems were not maintained in a sterile condition, and after 48 h progressive decreases in soluble organic C occurred. This was accompanied by a decrease in NO_2^- (if it was present) and progressive increases in NO_3^- after approximately 300 h in most soils, perhaps reflecting the increased activities of heterotrophic and nitrifying micro-organisms. Increases in Tc sorption with time may therefore have resulted from microbial uptake. Related studies in which nutrients were not limiting (unpublished) indicate that micro-organisms isolated from soil are capable of removing up to 60% of the Tc from solution and that exocellular metabolites may reduce Tc sorption by up to 10%.

It is also possible that increases in Tc sorption reflected changes, possibly microbially mediated, in reactive surface sites with time. The order of magnitude increases in NO_3^- that occurred during the equilibration period originated from soil organic N, which was strongly implicated (Table IV) in Tc sorption. The possible roles in Tc sorption of organic N and other functional groups or of amorphous Fe in the reduction of Tc to less soluble forms also remain to be elucidated. The aerobic conditions maintained in these systems do not totally preclude chemical reduction at micro-sites on particle surfaces, suggested as a possibility by Balogh and Grigal [13]. A detailed examination is underway of the solubility controls on Tc behaviour using the complete solution compositions determined in this investigation with time and geochemical modelling techniques.

Table VI
Stepwise multiple regressions relating selected soil properties to technetium sorption

Soil property	*Square of multiple*[1,2,3] *correlation coefficient*	
	48 h	*1050 h*
All soils		
Organic carbon	44·3**	71·6**
Extractable iron		
Citrate-bicarbonate-dithionite	80·0**	75·0**
pH	82·1**	80·5**
Fine clay	82·6**	81·2**
Grassland-shrub soils		
Organic carbon	21·3	21·5
Extractable iron		
Citrate-bicarbonate-dithionite	41·1	22·8
pH	49·1	23·0
Fine clay	69·6	52·2
Forest-marshland soils		
Organic carbon	22·9	61·5**
Extractable iron		
Citrate-bicarbonate-dithionite	78·7**	64·7**
pH	80·1**	83·3**
Fine clay	82·9**	89·3**

[1] Double asterisks denote significance (F test) at the $p < 0{\cdot}01$ level.
[2] Properties were entered into the regression in the order given.
[3] The multiple regression equation describing Tc sorption for all soils at the 48 h time period is as follows:

$$\% \text{ adsorption at } 48\,\text{h} = 6{\cdot}70 + 0{\cdot}478\,(\text{organic C}) + 0{\cdot}002\,52\,(\text{dithionite Fe}) - 1{\cdot}06\,(\text{pH}) + 0{\cdot}112\,(\text{fine clay}).$$

3.2.3. Multiple regression analyses

In order to evaluate the combined effects of key soil properties, Tc sorption was subject to stepwise regression against organic C, citrate-bicarbonate-dithionite extractable Fe, pH and fine clay. These properties were selected because they generally provided the best simple correlations (Tables IV and V) and represent different intercorrelated groups of soil properties. Organic C was selected over total N because it was highly correlated ($R = 0{\cdot}99$) to organic N and represented a less labile property. Similarly, citrate-bicarbonate-dithionite extractable Fe has been well-documented and evaluated in the literature and was correlated to ammonium oxalate extractable Fe ($R = 0{\cdot}72$) and Al extractable by both solvents ($R > 0{\cdot}84$). The stepwise correlation was forced in the order given.

The results of the multiple regression analyses (Table VI) tended to verify the simple correlations, indicating the major influence on Tc sorption of oxidic Fe at earlier times and organic C at later times during the equilibration period. These properties accounted for 80% and 75% of the variability in sorption at 48 h and 1050 h, respectively, for all soils investigated. The addition of pH and fine clay increased these values only slightly. Similar relationships existed for the two groups of soils, except that less variability could be accounted for in the grassland-shrub soils and the fine clay fraction was quite influential for these soils at both time periods. In the forest-marshland soils, pH was a significant factor in explaining sorption after 1050 h. The combined properties accounted for approximately 90% of the variability between soils of the forest-marshland group after 1050 h equilibration time.

4. SUMMARY AND CONCLUSIONS

The results of these studies verify the relatively low sorption of Tc (initially as TcO_4^-) under aerobic conditions, even under extended time periods and in soils of elevated organic matter and clay contents. Sorption that occurred was related to soil genesis and soil properties. Sorption was more pronounced in forest-marshland than grassland-shrub soils and was related primarily to pH-dependent sites on both the oxidic Fe and Al and organic matter fractions. The former was most important at time periods < 48 h; the latter was most important at the time periods > 48 h.

Additional sorption (up to 100%) previously observed in dialysis and column systems, in which soil was incubated in a moistened state, likely resulted from reduction of TcO_4^- to the much less soluble Tc^{+4} and/or microbial uptake. The current investigations suggest that chemical reduction phenomena would be most important as microbial uptake of TcO_4^- in conjunction with TcO_4^- sorption would not account for all of the Tc removed from solution in these studies.

Future investigations should address the relationships of Tc sorption to soil properties under conditions of reduced oxygen characteristic of poorly drained surface soils and subsoils.

ACKNOWLEGEMENTS

This work was supported by the US Department of Energy under contract DE-AC06-76RLO 1830. The authors sincerely appreciate the assistance of Sharon Lepel in the preparation of this manuscript.

REFERENCES

1. KEENEY, D. R. and WILDUNG, R. E. (1977). Chemical properties of soils. In: *Soils for Management of Organic Wastes and Waste Waters* (Ed. L. F. Elliott and F. J. Stevenson), Chapter 4. Madison, Wisconsin, American Society of Agronomy.
2. CATALDO, D. A. and WILDUNG, R. E. (1983). The role of soil and plant metabolic processes in controlling trace element behavior and bioavailability to animals. *The Science of the Total Environment*, **28**, 159–68. Amsterdam, The Netherlands, Elsevier Science Publishers.
3. BURKHOLDER, H. C., CLONINGER, M. O., BAKER, D. A. and JANSEN, G. (1975). Incentives for partitioning high-level waste. USERDA Rep., BNWL-1927. Springfield, Virginia, National Technical Information Service.
4. KENNA, B. T. (1962). The search for technetium in nature. *Journal of Chemical Education*, **39**, 436–42.
5. KOTEGOV, K. V., PAVLOV, O. N. and SHVEDOR, V. P. (1968). Technetium. In: *Advances in Inorganic Chemistry and Radiochemistry, Vol. II*, New York, Academic Press, pp. 1–90.
6. BOYD, G. E. (1959). Technetium and promethium. *Journal of Chemical Education*, **36**, 3–14.
7. RULFS, C. L., PACER, R. A. and HIRSCH, R. F. (1967). Technetium chemistry, oxidation states and species. *Journal of Inorganic Chemistry*, **29**, 681.
8. ANDERS, E. (1960). Radiochemistry of technetium. USAEC Rep., NAS-NS-3021, Springfield, Virginia, National Technical Information Service.
9. COBBLE, J. W. (1964). Technetium. In: *Treatise on Analytical Chemistry. II: Analytical Chemistry of the Elements* (Ed. I. M. Kolthoff and J. Elving), New York, Interscience, pp. 407–434.
10. KORKISCH, J. (1969). *Modern Methods for Separation of Rarer Metal Ions*, New York, Pergamon Press, pp. 504–523.
11. WILDUNG, R. E., ROUTSON, R. C. SERNE, R. J. and GARLAND, T. R. (1974). Pertechnetate, iodide and methyl iodide retention by surface soils. In: Pacific Northwest Laboratory Annual Report for 1974. BNWL-1950, Pt. 2. Pacific Northwest Laboratory, Richland, Washington, pp. 37–40. Also In: Agronomy Abstracts, ASA, SSSA and CSSA Annual Meeting, Chicago, Illinois, November 10–15, 1974.
12. LANDA, E. R., THORVIG, L. J. and GAST, R. G. (1977). Effect of selective dissolution, electrolytes, aeration and sterilization on technetium-99 sorption by soils. *Journal of Environmental Quality*, **6**, 181–7.
13. BALOGH, J. C. and GRIGAL, D. F. (1980). Soil chromatographic movement of technetium-99 through selected Minnesota soils. *Soil Science*, **130**, 278–82.

14. Mousny, J. M. and Myttenaere, C. (1981). Absorption of technetium by plants in relation to soil type, contamination level and time. *Plant and Soil*, **61**, 403–12.
15. Jackson, M. L. (1974). Soil Chemical Analysis—Advanced Course, 2nd edn, published by author, Madison, Wisconsin, Department of Soil Science, University of Wisconsin.
16. Wildung, R. E., Chesters, G. and Behmer, D. E. (1970). Alkaline nitrobenzene oxidation of plant lignins and soil humic colloids. *Plant and Soil*, **32**, 221–37.
17. Welcher, F. J. (Ed.) (1966). *Standard Methods of Chemical Analysis, Vol. III, Instrumental Methods*, Princeton, New Jersey, D. Van Nostrand Company, Inc.
18. Bardsley, C. E. and Lancaster, J. D. (1965). Sulfur. In: *Methods of Soil Analysis* (Ed. C. A. Black), Madison, Wisconsin, American Society of Agronomy, pp. 1102–16.
19. Routson, R. C., Wildung, R. E. and Serne, R. J. (1972). A column cation-exchange-capacity procedure for low exchange capacity sands. *Soil Science*, **115**, 107–12.
20. Mehra, O. P. and Jackson, M. L. (1960). Iron oxide removal from soils and clays by a dithionite-citrate system buffered with sodium carbonate. *Clays and Clay Minerals*, **7**, 317–27.
21. McKeague, J. A., Brydon, J. E. and Miles, W. M. (1971). Differentiation of forms of extractable iron and aluminum in soils. *Soil Science of America Proceedings*, **35**, 33–8.
22. Schwertmann, U. (1973). Use of oxalate for Fe extraction from soils. *Canadian Journal of Soil Science*, **53**, 244–6.
23. Chao, T. T. (1972). Selective dissolution of manganese oxides from soils and sediments with acidified hydroxylamine hydrochloride. *Soil Science Society of America Proceedings*, **36**, 704–68.
24. American Public Health Association (APHA) (1980). *Standard Methods for the Examination of Water and Waste Water*, 15th edn, Washington, DC, American Public Health Association.

11

Technetium Behaviour in Soils of the Canadian Precambrian Shield

MARSHA I. SHEPPARD and S. C. SHEPPARD
Environmental Research Branch,
Whiteshell Nuclear Research Establishment,
Atomic Energy Canada Limited,
Pinawa, Manitoba, Canada R0E 1L0

1. INTRODUCTION

Technetium-99, produced by the fission of ^{235}U and ^{239}Pu, has a radiological half-life of $2{\cdot}15 \times 10^5$ years and has been dispersed in the environment via weapons testing and releases from establishments using it for medical or industrial purposes. Technetium could also reach the environment via releases from nuclear power plants, nuclear fuel processing facilities or nuclear fuel waste disposal vaults. Environmental impact assessments associated with the licensing of such nuclear facilities usually include analysis of radiation exposure pathways that involve the transport of radionuclides through agricultural soils and their uptake by crop plants. The environmental assessment of the Canadian concept for the permanent disposal of nuclear fuel waste in plutonic rock differs from the assessments required for surface nuclear facilities in that the source of radionuclides in the fuel waste disposal vault is located 500–1000 m deep in a rock mass. Thus, the models for the assessment of this concept must predict radionuclide migration up from the vault, through the geosphere and into the biosphere, and the eventual radiological dose to man [1]. Only radionuclides with very long half-lives could reach the biosphere and, of those, only the most mobile nuclides could escape from the vault and the surrounding rock. Technetium has both of these properties and so is important in the research supporting the Canadian Nuclear Fuel Waste Management Program.

The Canadian reference concept involves placement of the nuclear fuel waste in a relatively homogeneous plutonic rock body located somewhere on the Precambrian Shield. It seems most likely that the first appearance of a radionuclide, possibly ^{99}Tc, in the biosphere would be in a low-lying swamp or bog. For this reason, organic soils, which are commonly drained for agricultural production, such as market gardening and sod-farming, or processed commercially as soil amendments for gardens and nurseries, must be included in the terrestrial exposure pathways.

The ability of plants to extract Tc from mineral and organic soils, including quantitative removal from some soils, has been demonstrated previously [2–11]. The migration of Tc in the soil then becomes the critical step in the entry of Tc into the terrestrial foodchain. Numerous experiments have been carried out to study the soil sorption characteristics of Tc on disturbed mineral and organic soil horizons and geological materials, under laboratory conditions [5, 7, 12–15]. Few experiments have been carried out with undisturbed complete soil profiles [3], mineral or organic, or under the anoxic conditions expected in shallow water-table systems.

In our Tc experiments, we use sandy soils, the predominant soil type covering 55% of the Precambrian Shield, as well as highly organic mineral soils and purely organic soils. We study them under the influence of a shallow water table and for both oxic and anoxic conditions. Three of these experiments are discussed in this paper.

2. EXPERIMENTAL

2.1. First experiment

In this experiment, field lysimeters (15 cm in diameter × 40 cm in length) were filled with an Orthic Black Chernozem, a clay loam soil with an organic matter content of 0·073 mg/kg and a pH of 8·1 (see Table I). The soil was amended with wood sawdust (2·5% w/w). Technetium, as NH_4TcO_4, was uniformly dispersed in a 4-cm layer of soil at a concentration of 1 μg ^{99}Tc/g dry soil, labelled with 84·0 Bq ^{95m}Tc/μg ^{99}Tc. This was a total of 440 μg ^{99}Tc/lysimeter. Only one contaminated layer was used in each lysimeter, located 16–20, 24–28 or 32–36 cm below the soil surface. The water table was held fixed 26 cm from the soil surface in half the lysimeters, but fluctuated weekly between 18 and 34 cm in the other lysimeters. Barley was grown for 53 days in all lysimeters, until it was fully in head. The plants were collected and the soil columns sampled. The Tc content of the plant shoots was determined. The soil columns were sectioned in 2-cm slices, which were analysed for total Tc, 0·02

Table I
Selected soil properties

Property	*Orthic Black Chernozem*	*Gleyed Dystric Brunisol*			
		Horizon			
		Organic	*Ae*	*B*	*C*
Bulk density, ρ_b (g/cm^3)	1·01	0·29 ± 0·04	1·34 ± 0·02	1·47 ± 0·03	1·44 ± 0·00
pH	8·1	5·1 ± 0·1	5·1 ± 0·1	5·1 ± 0·1	6·3 ± 0·1
Cation exchange capacity (meq/100 g)	50·5	81·2 ± 8·3	2·9 ± 0·2	2·1 ± 0·7	1·7 ± 0·4
Organic matter (mg/kg)	0·073	—	0·007	0·004	0·003
Moisture content at 10 kPc† (% dry wt)	55·6 ± 2·1	103·4 ± 7·2	4·1 ± 0·4	2·1 ± 0·1	1·6 ± 0·1
Moisture content at 33 kPc† (% dry wt)	41·9 ± 1.0	78·3 ± 1·7	2·1 ± 0·1	1·3 ± 0·0	0·8 ± 0·0
Moisture content at 50 kPc† (% dry wt)	37·6 ± 0·4	67·2 ± 2·0	1·8 ± 0·2	1·1 ± 0·1	0·6 ± 0·1
Moisture content at 100 kPc† (% dry wt)	34·1 ± 0·6	59·2 ± 6·0	1·7 ± 0·1	1·1 ± 0·0	0·7 ± 0·0
Moisture content at 300 kPc† (% dry wt)	28·2 ± 0·2	37·3 ± 2·0	1·4 ± 0·0	1·0 ± 0·0	0·5 ± 0·0
Saturated hydraulic conductivity (cm/d)	75·2		whole core ~ 350		

†kPc = kilopascal.

mol/litre $CaCl_2$-extractable Tc and 0·5 mol/litre $NaHCO_3$ (pH = 8·5)-extractable Tc.

The soil columns developed a redox gradient, which was detected by the chelate extraction of Fe and Mn, done at the end of the experiment. More soluble, and therefore reduced, forms of Fe and Mn occurred below 16 cm from the soil surface. Preliminary studies suggested that this condition developed during the first 14–21 days of the experiment.

The plant shoots contained Tc within 20 days of planting, as determined on plants thinned from the experiment. At collection, the plants in the fixed water-table systems had accumulated 78, 30 and 17 μg Tc/lysimeter from the shallow, middle and deep placements of Tc, respectively. The corresponding results for the fluctuating water-table systems were 46, 11 and 15 μg Tc/lysimeter. The plant roots were not deeper than 16–18 cm, suggesting that migration through the soil occurred to supply Tc to the plants from the deep placements.

Soil analyses indicated that Tc migrated toward the surface in all cases (see Fig. 1) and accumulated on the surface. The flux of water to supply evapotranspirational demand was almost entirely from the water table. The results clearly indicate a lower mobility of Tc in reduced soils. The

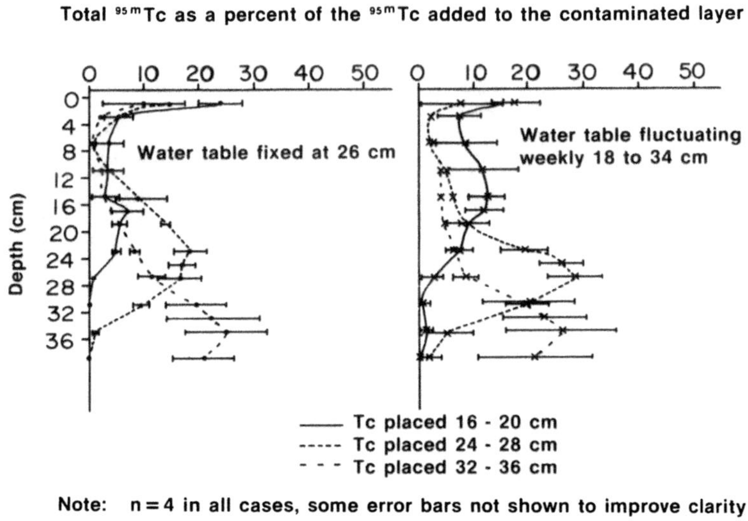

FIG. 1. Distribution of ^{95m}Tc in the field lysimeters at final harvest for each placement depth and for both the fixed and fluctuating water-table systems.

water-table interface was not a barrier to the upward migration of Tc, but it did cause a discontinuity in the degree of sorption of Tc. No Tc was detected in any of the soil extractions.

2.2. Second experiment

The second experiment consists of 80 undisturbed soil cores, 5·1 cm in diameter and averaging 52 cm in length, of a sandy gleyed Dystric Brunisolic soil. The soil properties by horizon are given in Table I and the cores extend from the surface to the lower B horizon. One-half of the 80 cores will be exhumed and sectioned to determine nuclide (Tc, I, Np, U, Th, Cs, Cr, Mo and Pb) distributions in the profile after one year, and the remaining one-half after four years. The one-year cores were spiked with $7{\cdot}4 \times 10^5$ Bq ^{99}Tc, as NH_4TcO_4, labelled with $7{\cdot}1 \times 10^5$ Bq ^{95m}Tc, and the four-year cores were spiked with ^{99}Tc only. One-half of each group were spiked just below the litter layer, but above the eluviated Ae horizon (leachate columns), and the other half were spiked in the lower B horizon, about 45 cm from the soil surface (groundwater columns). The cores were fitted with Perspex cups, which were sealed to the cores, and these units were placed in sleeves in the ground to maintain natural soil temperature and rainfall conditions (see Fig. 2).

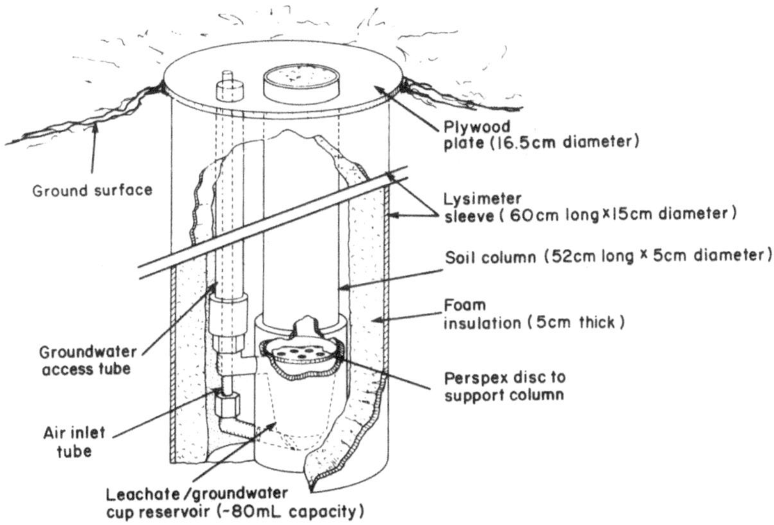

FIG. 2. Cut-away diagram of the soil core and leachate/groundwater cup assembly below the ground surface.

These cores have now experienced one season of freezing conditions and the soil temperatures in the cores, as well as at the coring location, are monitored (see Fig. 3). The soil temperatures in the columns are generally warmer and show greater temperature fluctuations than the canopy-covered coring location. Moisture contents are monitored routinely at the coring location and in three leachate and three groundwater columns in the experiment. Moisture contents were quantitatively similar at both sites before the first winter, but following snowmelt the moisture contents of the leachate column were substantially greater than those at the coring location (see Fig. 4). The trends of the natural system are present in the columns; however, the effects of transpiration are absent.

In the cores spiked near the surface, leachate is sampled after each significant rainfall (see Fig. 5). Preliminary results show that an average of 6·2% of the ^{95m}Tc placed below the litter layer flushed out of the cores with the spring snowmelt. Eight collectable leachate events have occurred since 17 April 1984, each removing from 0·1% to 7·8% of the initial ^{95m}Tc. Approximately 71% of the original ^{95m}Tc remains in the cores (see Fig. 5). The eight leachate events resulted from a total of 307 mm of rainfall, received from 17 April 1984 to 17 July 1984, representing approximately 19 field capacity volumes passing through the columns in this season.

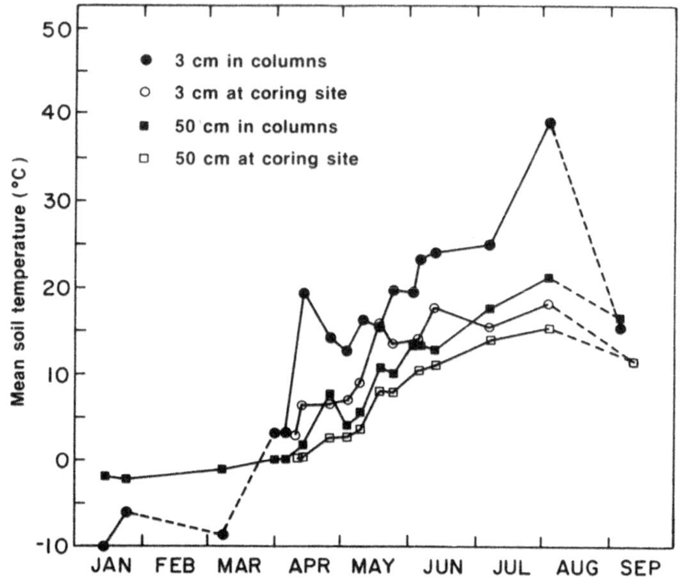

FIG. 3. Soil temperature versus time at two depths in the soil columns and at the coring site.

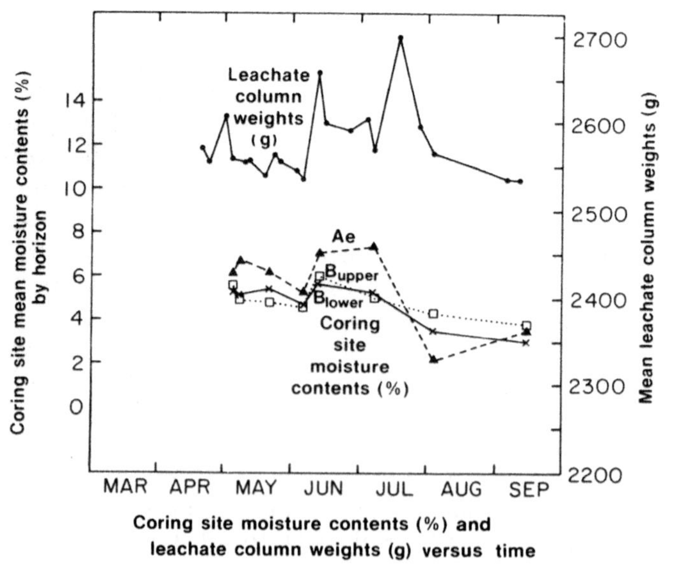

FIG. 4. Coring site moisture contents (% dry weight) and leachate column weights (g) versus time (2550 g column weight is about 11% moisture content).

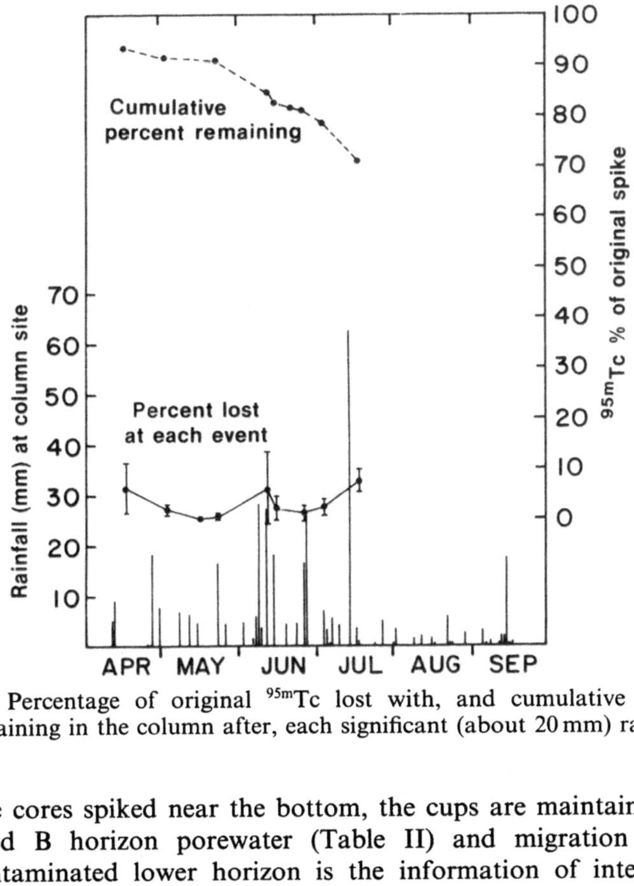

FIG. 5. Percentage of original ^{95m}Tc lost with, and cumulative percentage remaining in the column after, each significant (about 20 mm) rainfall.

In the cores spiked near the bottom, the cups are maintained full of simulated B horizon porewater (Table II) and migration up from this contaminated lower horizon is the information of interest. This information will be obtained upon exhumation.

A Tc sorption experiment carried out with soil from each of the Ae, B, C and litter horizons of this brunisol showed the mineral horizons had K_d values $<0{\cdot}05$ ml/g, which is not significantly different from zero. The litter horizon K_d value for Tc was $1{\cdot}1 \pm 0{\cdot}2$ ml/g.

2.3. Third experiment

In this experiment, three types of peat cores—a living sphagnum core (0–40 cm), a dead sphagnum core (40–80 cm) and a reed/sedge peat core (0–40 cm)—were confined in large round plastic containers (41 cm long × 28·5 cm in diameter) and spiked at the bottom with Tc, I, Np and U. These cores were then placed back in their natural habitat, and the

Table II
Brunisolic porewater analysis at about 45 cm depth—lower B horizon

Element	*Concentration (mg/litre)*
Na	15·6 ± 0·3
K	8·7 ± 0·3
Ca	15·9 ± 0·5
Mg	2·7 ± 0·3
Fe	0·6 ± 0·1
F	1·50 ± 0·04
Cl	18·2 ± 1·3
NO_3	<0·16
SO_4	27 ± 1
HCO_3	7 ± 2
pH	6·7 ± 0·1

water table was maintained via access tubes in the cores at the same level as that of the source bog, using the bog groundwater (see Table III). Water samples were withdrawn from the core surfaces for analysis, to pinpoint the time to surface breakthrough of the Tc. Similar cores were spiked with tritiated water, to follow the migration of water.

Table III
Surface groundwater composition of the sphagnum and reed/sedge peat bogs after spring thaw

Element	*Concentration*	
	Sphagnum (mg/litre)	*Reed/sedge (mg/litre)*
Na	1·7	4·8
K	2·4	2·6
Ca	8·9	39
Mg	3·8	38
Fe	<0·1	0·8
F	0·09	0·14
Cl	0·67	1·55
NO_3	<0·2	2·01
SO_4	1·2	20·2
pH	7·3	7·7

Preliminary results show that tritium was detectable at the surface of these cores two weeks after spiking and that tritium activities peaked about six weeks after spiking. No ^{95m}Tc has been detected yet at the peat core surfaces.

E_h measurements show these systems to be moderately reducing. A Tc sorption experiment showed that K_d values were two- to three-fold greater under anoxic conditions than under oxic conditions, in both living and dead sphagnum and in reed/sedge peat. The anoxic K_d values for Tc were $2{\cdot}7 \pm 1{\cdot}2$ and $3{\cdot}1 \pm 1{\cdot}5$ ml/g for living and dead sphagnum peat, respectively. The reed/sedge peat showed two distinct horizons, and the anoxic K_d values for Tc were 340 ± 100 and 100 ± 11 ml/g for the top layer and the second, more highly decomposed, horizon, respectively. The anoxic K_d value for Tc in the clay underlying this reed/sedge peat was $0{\cdot}3 \pm 0{\cdot}2$ ml/g. These peat cores will be exhumed and sectioned to determine the Tc distributions and the field sorption values.

3. CONCLUSIONS

The first experiment clearly showed that Tc mobility is lower in soil under reducing conditions, but that the water-table interface itself is not a barrier to the upward migration of Tc. The second experiment showed that, despite the very low K_d values ($<0{\cdot}05$ ml/g) measured in the laboratory, an undisturbed sandy soil profile retained 71% of its original Tc spike, placed just below the litter, after one year. In the third experiment, no Tc has reached the surface after three months, from placement 40 cm deep in either a natural sphagnum or a reed/sedge peat core. Under anoxic conditions, K_d values for Tc varied from about 3 ml/g in the sphagnum to 100 and 340 ml/g for the reed/sedge peat, indicating that the floristic composition of organic soils may strongly affect the migration of Tc in a swamp or bog. These cores will be sectioned to determine Tc distribution profiles and in situ K_d values.

The brunisol leachate–groundwater column study will continue for another three years, pending the outcome of the one-year exhumation results. Our plans are to continue describing the parameters necessary to predict migration of radionuclides through organic and highly organic mineral soils under both oxic and anoxic conditions. More types of organic soils, based on the floristic composition but predominantly on the degree of humification, will be incorporated into the programme. Further emphasis will be placed on sorption mechanisms, and the effects and mechanisms of Tc aging in soils, with some investigation into the release of radionuclides from organic soil when it is used as a soil amendment.

REFERENCES

1. Wuschke, D. M., Mehta, K. K., Dormuth, K. W., Andres, T., Sherman, G. R., Rosinger, E. L. J., Goodwin, B. W., Reid, J. A. K. and Lyon, R. B. (1981). Environmental and safety assessment studies for nuclear fuel waste management: Post-closure assessment. Atomic Energy of Canada Limited Report, TR-127-3, unrestricted unpublished report, available from SDDO, Atomic Energy of Canada Limited Research Company, Chalk River, Ontario K0J 1J0.
2. Sheppard, M. I., Vandergraaf, T. T., Thibault, D. H. and Reid, J. A. K. (1983). Technetium and uranium: Sorption by and plant uptake from peat and sand. *Health Phys.*, **44**, 635–43.
3. Hoffman, F. O., Garten, C. T. Jr, Huckabee, J. W. and Lucas, D. M. (1982). Interception and retention of technetium by vegetation and soil. *J. Environ. Qual.*, **11**, 134–41.
4. Cataldo, D. A., Wildung, R. E. and Garland, T. R. (1978). Technetium accumulation, fate, and behaviour in plants. In: *Proc. Environmental Chemistry and Cycling Processes*, Augusta, Georgia, April 28–May 1, 1976 (Ed. D. C. Adriano and I. L. Brisbin Jr), DOE Symposium Series, Vol. 45, pp. 538–49.
5. Wildung, R. E., McFadden, K. M. and Garland, T. R. (1979). Technetium sources and behavior in the environment. *J. Environ. Qual.*, **8**, 156–61.
6. Desmet, G. M. and Dirkse, W. G. (1984). Action of technetium on the growth, the manganese, zinc and iron concentrations and the fluorescence induction in young and old leaves of spinach plants. *Environ. Exper. Bot.*, **24**, 45–52.
7. Gast, R. G., Landa, E. R. and Thorvig, L. J. (1976). The behavior of technetium-99 in soils and plants. Progress report for the US Energy Research and Development Administration under Contract No. Ey-76-5-02-2447, COO-2447-5.
8. Berlyn, G. P., Dhillon, S. S. and Koslow, E. E. (1980). Technetium: A toxic waste product of the nuclear fuel cycle: Effects on soybean growth and development. *Envir. Managemt*, **4**, 149–56.
9. Mousny, J. M., Roucoux, P. and Myttenaere, C. (1979). Absorption and translocation of technetium in pea plant. *Environ. Exper. Bot.*, **19**, 263–8.
10. Myttenaere, C., Daoust, C. and Roucoux, P. (1980). Leaching of technetium from foliage by simulated rain. *Environ. Exper. Bot.*, **20**, 415–19.
11. Mousny, J. M. and Myttenaere, C. (1981). Absorption of technetium by plants in relation to soil type, contamination level and time. *Plant and Soil*, **61**, 403–12.
12. Franz, J. A., Martin, L. Y. and Wiggins, D. J. (1982). Behavior of reduced ^{99}Tc and ^{99}Tc organic complexes on Hanford soil, Pacific Northwest Laboratory Report, PNL-4178.

13. VANDERGRAAF, T. T., TICKNOR, K. V. and GEORGE, I. M. (1983). Reactions between technetium in solution and iron-containing minerals under oxic and anoxic conditions, Atomic Energy of Canada Limited Report, AECL-7957.
14. ROUTSON, R. C., JANSEN, G. and ROBINSON, A. V. (1975). Sorption of ^{99}Tc, ^{237}Np and ^{241}Am on two subsoils from differing weathering intensity areas, Battelle Pacific Northwest Laboratory Report, Richland, Washington, BNWL-1889.
15. SISSON, D. H., MACLEAN, S. C., SCHULZ, R. K. and BORG, R. J. (1979). A preliminary study of the migration of technetium in soil under hydrous conditions, Lawrence Livermore Laboratory Report, UCRL-52832.

12

Soil–Humic Acid Complexes of Technetium: Synthesis and Characterisation

L. VAN LOON, M. STALMANS, A. MAES, A. CREMERS and M. COGNEAU

Laboratorium voor Colloïdale Scheikunde, Fakulteit der Landbouwwetenschappen, KU Leuven, Kardinaal Mercierlaan 92, B-3030 Leuven, Belgium

1. INTRODUCTION

It has become well established in recent years that the soil-to-plant transfer factors of technetium decrease with time [1] and that such reduction in uptake is accompanied by a decrease in the water-soluble Tc fraction. These effects can be understood in terms of a reduction of Tc(VII) to less soluble oxidation states (IV) such as (hydrated) TcO_2 and to the formation of non-water soluble soil–organic matter complexes. These processes can be inferred from the generally higher extraction yields, using either peroxide or alkaline extractants.

The purpose of this paper is two-fold. Firstly, to give a survey of current knowledge regarding Tc–organic complexes, particularly in regard to synthesis conditions and the nature and combination of the functional groups, involved in coordinative binding in Tc-complexes. The survey is not intended to be exhaustive and is limited to those aspects which are deemed relevant to the Tc-complex chemistry in soil organic matter. More general reviews on the chemistry of Tc-complexes are to be found in the recent literature [2–4]. The second purpose is to present some examples of procedures to prepare Tc–humic acid complexes.

2. TECHNETIUM–ORGANIC COMPLEXES

The study of Tc-organics has received considerable impetus in the recent past, due to the widespread use of Tc-complexes as diagnostic imaging agents in nuclear medicine. A wide variety of complexes has been synthesised but an examination of the literature shows that the products may vary widely in properties depending on experimental procedure, as evidenced by spectroscopic studies and some pronounced differences in behaviour in vivo. Although in some cases oxidation state and coordinative structures appear to be fairly well established, these characteristics remain uncertain in most complexes formed and are to a large extent speculative. Moreover, data on thermodynamic stabilities are practically non-existent. Basically, complex formation is achieved by one of the following methods:

—Ligand exchange with appropriate complexing agents, using a halide technetate such as the hexabromotechnetate, K_2TcBr_6, as starting material.
—Reduction of TcO_4^-, using an appropriate reducing agent in the presence of the complexing agent of interest (which in some cases may also be the reducing agent, e.g. cysteine).

Bearing in mind the objectives stated in the introduction, complex formation generally involves moieties carrying a combination of two of the following functional groups: amino, carboxyl, hydroxyl and sulphydryl. For purposes of convenience, it is useful to subdivide these complexing agents into four main groups:

1. Hydroxycarboxylates: such as citrate, malate, tartrate, gluconate, β-hydroxyisobutyrate [5–11].
2. Aminocarboxylates: such as EDTA, DTPA, NTA, diaminocyclohexanetetra-acetate (DATA), ethylene diaminodiacetate (ENDA) [9, 12–16].
3. Polyamines: such as ethylenediamine, tri-, tetra- and pentamines, cyclam (1,5,8,12-tetra-azocyclotetradecane) [17–20].
4. Sulphydryl amino acids: such as methionine, cysteine and derivatives, e.g. N-acetylcysteine, penicillamine, cysteine-ethylester [21–25].

With the exception of the case of polyamines (which form quite stable Tc-complexes), it appears that the more stable configurations are achieved when coordinative binding involves different functional groups. In this context, attention is drawn to the fact that di- and tricarboxylic acids are unable to stabilise reduced Tc, unless at least one adjacent hydroxyl group is present.

Various reducing agents may be used such as stannous chloride, the most popular one in so-called radiopharmaceutical kits, an ascorbic acid–

ferric ion combination [26], hydrazine [27], sulphydryl amino acids such as cysteine or methionine, or electrolytic methods [28]. In general, the synthesis procedure is as follows: a reducing agent such as Sn^{2+} is mixed with the appropriate complexing agent, preferably under nitrogen atmosphere and, if appropriate, in acetate buffer. TcO_4^- is then introduced and the reaction is generally completed within minutes. However, when using cysteine as a reducer, considerably longer reaction times are necessary [24]. The same is true when synthesising Tc-complexes by ligand exchange procedures. Overall labelling conditions are specified in terms of pH and the relative concentrations of complexing agent, reducing agent and TcO_4^-. In general, it is found that, in order to obtain a 100% reduction yield, the Sn^{2+}/TcO_4^- ratio in the system should be two at least [13]. Possible loss of reduced but hydrolysed Tc may be overcome by operating at much higher (± 100) Sn^{2+}/TcO_4^- ratios, giving due attention to Sn^{4+} impurities in the reducer [13].

One of the disturbing features in the synthesis of Tc-complexes is that the nature of the reaction products may be quite variable, depending on experimental procedure. Not only are 'identical' complexes prepared by the ligand exchange or the reduction-complexation procedures quite different [29], but even in the case of using the same procedure, pronounced differences in the resulting complexes may be encountered, depending on labelling procedure. Depending on the excess of Sn^{2+} or ligand concentrations, either monomeric, dimeric or polynuclear complexes may be obtained [10, 25]. These effects are particularly striking in the case of citrate complexes which vary from monomeric at high citrate to polymeric at low citrate concentration.

A particularly interesting variant of the ligand exchange method is the synthesis of Tc–polyamine complexes through the transfer of Tc from aminocarboxylate complexes to the amine [19]. Such transfer, which is facilitated by opposite charge effects of the moieties involved, appears not to be feasible between aminocarboxylate entities, due to the repulsive nature of ligand–complex interactions [19].

Structural information on these Tc-complexes is relatively scarce. In general, it is acknowledged that the metal centre remains strongly bound to oxygen, forming a Tc=O core embedded within the multidentate ligand. Quantitative data on the state of oxidation of Tc are also quite scarce, and the general consensus is that Tc is in the IV state. However, oxidation states of III [9, 25] and V [9, 24] have been claimed, the lower oxidation state being more likely in strongly stabilising ligands and high Sn^{2+}/Sn^{4+} ratios [30].

Data on overall charge remain also rather scarce and are not in general agreement. It was reported that the monomeric aminocarboxylate complexes of EDTA and DTPA exhibit an overall charge of −2 at a pH of 7 [15]. In contrast, other reports would indicate a charge of −1 for the EDTA complex (with Tc in the V state) [3] or a zero charge in a dinuclear EDTA complex (with Tc in the IV state) [2].

In view of the foregoing description of the nature and combination of the functional groups required for the formation of stable Tc-complexes, one may expect humic acids, or soil organic matter in general, to be liable to stabilise Tc in reduced forms. In fact, all functional groups, and combinations thereof, as discussed above, are known to be present abundantly in soil organic matter, often in sterically favourable configurations.

3. EXPERIMENTAL SECTION

3.1. Technetium in soil extracts

A TcO_4^- contaminated podzol soil sample was incubated at 10% water content (or about 50% of field capacity) in the absence and presence of decomposable organic matter (10 mg glucose/g soil; nitrate amendment: C/N ratio of 20) and aqueous Tc extraction yields were monitored as a function of time. The results are shown in Figs 1 and 2 in terms of relative extraction yields. It is apparent that the rate of immobilisation is highly activated in the presence of glucose. In the absence of glucose, extraction yield appears to stabilise at about 0·5 after six months. Such levels are

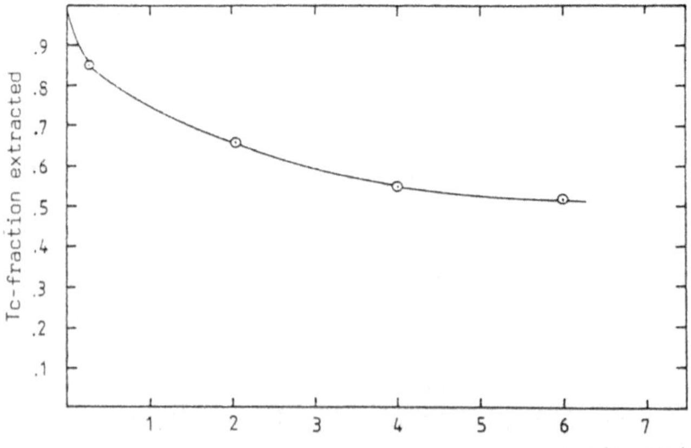

FIG. 1. Time dependence of aqueous extraction yield of Tc in a podzol soil.

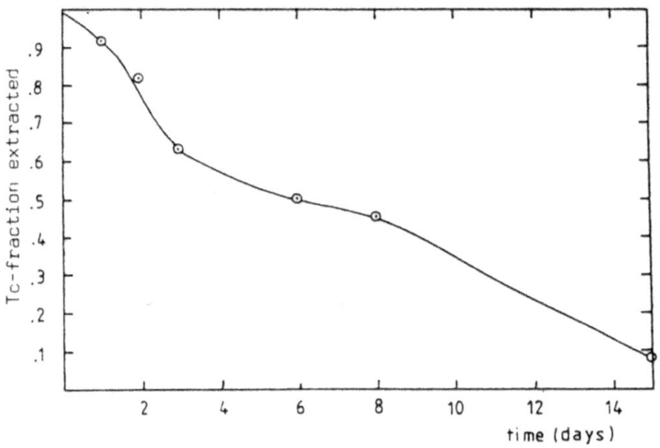

FIG. 2. Time dependence of aqueous extraction yield of Tc in a podzol soil in the presence of glucose.

already achieved within a week when glucose is present, the level of immobilisation increasing nearly linearly with time. Undoubtedly, this effect results in part from oxygen depletion in small water-filled soil pores, following proliferation of the soil microflora, and subsequently leading to reducing conditions.

Figure 3 shows an example of a gel chromatogram of a CO_3/HCO_3 soil extract (pH = 8·8) at some intermediate stage in the process. The extract was prepared by a 24 h extraction under N_2 atmosphere, using a 1:2 soil–water ratio. The extract was centrifuged at high speed (15 000 rpm, 1·5 h) and filtered (0·8 μm). It is sufficiently apparent that a considerable fraction (about $\frac{2}{3}$) of Tc is eluted as an organic complex. The extraction yield, obtained by a 0·005 M $CaCl_2$ treatment, was about 20% of the HCO_3 yield value.

The question obviously arises concerning the re-oxidation of Tc–humic acid complexes upon re-exposure to air. In Table I are shown some

Table I

Time (days)	*% Tc as a complex*
0	51·1
8	34·7
14	31·4
27	23·4
51	16·4

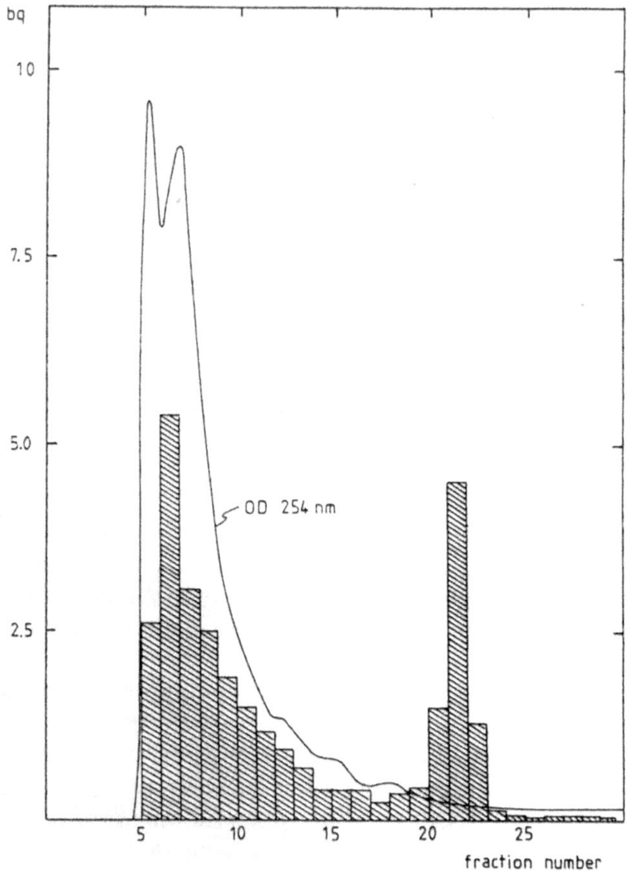

FIG. 3. Gel chromatogram (G-25) of a HCO_3/CO_3 (pH = 8·8) extract of a podzol soil.

preliminary data on the rate of re-oxidation of the Tc–organic fraction in a HCO_3/CO_3 (pH = 8·8) extract of a podzol soil upon air exposure. The data were obtained by gel chromatography.

In dealing with conditions of constant level of the oxidant (i.e. dissolved oxygen in equilibrium with atmospheric air) which is in any case present in large excess, as compared to the electron donor (Tc), it is logical to postulate a first-order rate behaviour for the decay process, i.e.

$$\frac{d(\text{Tc–OM})}{dt} = -k(\text{Tc–OM})$$

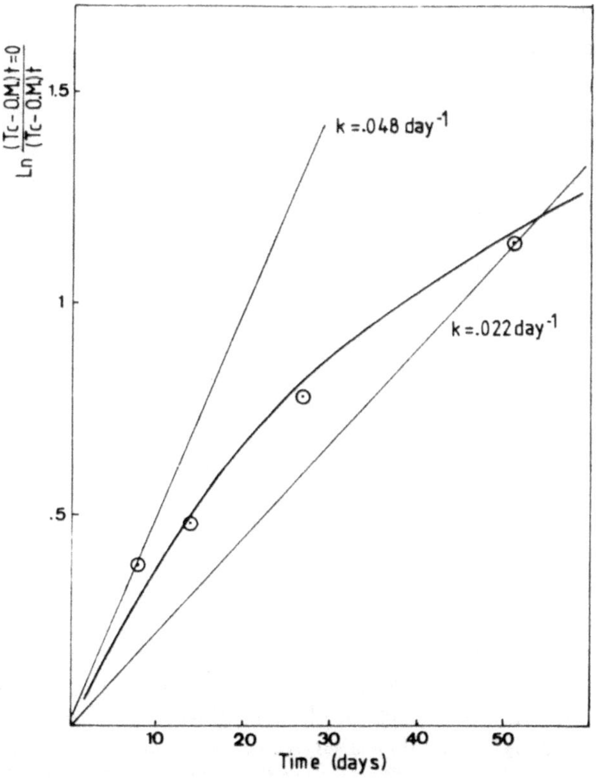

FIG. 4. First-order rate plot for the oxidative decay of Tc–HA complexes, extracted (HCO_3/CO_3; pH = 8·2) from a podzol soil. The curve is calculated for a 1/1 mixture of two complexes with rate constants of 0·08 day^{-1} ($t_{1/2}$ = 8·7 days) and 0·01 day^{-1} ($t_{1/2}$ = 70 days).

Yet, when the data are arranged in a first-order rate plot, i.e.

$$\ln\{(\text{Tc–OM})_{t=0}/(\text{Tc–OM})_t\}$$

versus time, the behaviour appears to be non-linear (see Fig. 4). However, it is likely that a variety of Tc-complexes occur in humic acids, depending on the particular combination of functional groups involved in coordinative binding. Therefore, the decay process is more likely to be describable in terms of a mixture of complexes, characterised by a range of first-order rate constants. Figure 4 shows a rather simple analysis of the re-oxidation rate in terms of two rate constants of 0·01 and 0·08 day^{-1} (or about 10^{-7} and 10^{-6} s^{-1}) and a 1/1 mixture of two groups of Tc-complexes, corresponding to characteristic 'half-life' values of 70 or 8·7 days. In view of the limited data available, it is obvious that, at this point, a more elaborate analysis is certainly not warranted. However, it would appear that, in studying re-oxidation rates of Tc–humic acid complexes, which may involve all sorts of combinations of the functional groups discussed above, one should be prepared to expect different 'kinds' of Tc-forms which may well differ by an order of magnitude in their re-oxidation rates. The consequences of this particular aspect of Tc-chemistry in soils will be assessed in the concluding section.

3.2. Synthesis of Tc–humic acid complexes by chemical reduction

Our early attempts at Tc–humic acid (HA) complex synthesis were based on the use of Sn^{2+} concentrations of 5×10^{-4} M in combination with a 10^{-4} M TcO_4 in the presence of HA. Complexation yields were generally rather poor, covering a range of 15–40%, depending on pH. The current procedure is routinely based on the following combination: Sn^{2+}, 10^{-3} M; TcO_4, 3×10^{-6} M; HA, 200–400 ppm. The procedure is as follows: appropriate amounts of Sn^{2+} (in HCl at pH $\approx$ 1·5) are added to a thoroughly deoxygenated (by nitrogen flow) humic acid preparation and the pH is adjusted to about 7. TcO_4^- is then introduced and the solution is shaken overnight. Finally, the system is centrifuged at high speed and the supernatant is separated from the small amount of sediment, containing the bulk of Sn, as a hydroxide precipitate, and minor fractions of reduced Tc as $TcO(OH)_2$ (generally less than 5%). This procedure leads to quantitative reduction, as evidenced by the total absence of TcO_4^- in the gel chromatogram. A typical example of a chromatogram on such a preparation is shown in Fig. 5.

3.3. Synthesis of Tc–humic acid complexes by microbial action

Tc–HA complexes can also be synthesised by making use of the reducing conditions which are being generated by the action of soil microflora in anoxic conditions. Eventually the process can be activated by inoculating the soils with *bacillus* or *Pseudomonas* strains. The procedure which is now being applied on a routine basis is as follows: about 50 g podzol (B_1) is mixed with 2 ml 1% glucose and 1 ml NH_4NO_3 10^{-2} M, and thoroughly homogenised. The system is incubated in an exsiccator over a saturated water atmosphere for some 3–5 days, with intermittent (twice daily) air exposures and thorough mixing. Thereafter 2 ml of TcO_4^- (Tc-99, labelled with Tc-95m) is added, raising the water content to some 10%. The mixture is then thoroughly homogenised and immediately transferred to

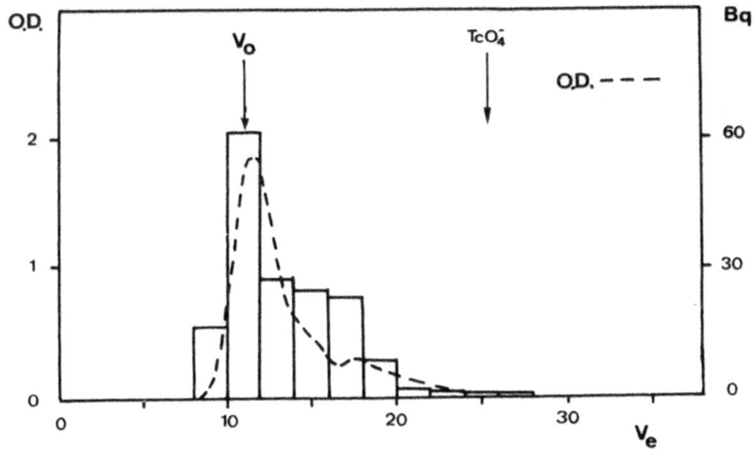

FIG. 5. Gel chromatogram (G-25) of a Tc–HA (Fluka) complex prepared by Sn^{2+} (10^{-3} M) reduction (eluent: NaOH 10^{-2} M; NaCl: 2×10^{-2} M).

an inert atmosphere. After 3–5 days, Tc is quantitatively reduced, as evidenced by the total absence of TcO_4^- in an aqueous extract (using solid/liquid ratios of 20 and deoxygenated distilled water) at pH = 6 (<0·1% TcO_4^-). Usually, minor quantities are present in such extracts but are associated with organic matter in the extract. A typical gel chromatogram on a HCO_3/CO_3 extract of such a soil is represented in Fig. 6, showing Tc to be quantitatively associated with the organic matter.

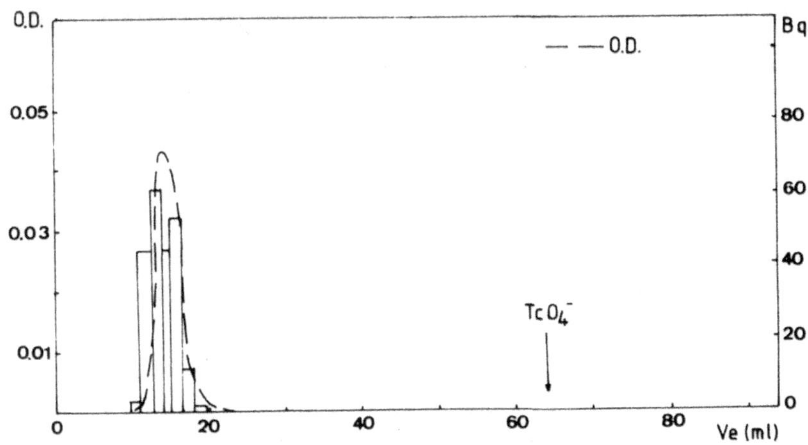

FIG. 6. Gel chromatogram (G-25) of a soil Tc–HA complex prepared by microbial action (eluent: NaCl 2×10^{-2} M).

4. CONCLUDING COMMENTS

Evidence is presented that, following local reducing conditions in soils, generated by microbial growth, soil organic matter is a most likely candidate for stabilising reduced technetium in chelating complexes, possibly in a variety of combinations of coordinating groups (such as amine, carboxyl, sulphydryl, hydroxyl). Such complexes can be readily synthesised, either directly by chemical reduction in the presence of HA or indirectly by microbial action.

From some preliminary data on the re-oxidation rate of Tc–HA complexes, it would appear that re-odixations may not be describable in terms of a single first-order rate process but by a combination of rate processes relating to the various combinations of functional groups involved in Tc-binding. The corresponding rate constants may well cover an order of magnitude ranging from 10^{-6} to $10^{-7}\,s^{-1}$ (or half-life values in the range of 10 to about 100 days). This may have a direct bearing on possible ageing effects of technetium in soils. If in fact soil Tc is subject to alternating conditions of redox regimes, leading to re-oxidation of a fraction of the more readily oxidisable Tc, then it is to be expected that in the long run Tc will be locked increasingly in less oxidisable forms. This discussion did not include reduced and uncomplexed $TcO(OH)_2$, the re-oxidation rate of which remains to be assessed and compared to rates of complexed Tc-forms. The question whether the 'unavailable' and immobilised fraction of Tc in soils will evolve towards either precipitated or humic acid-associated forms is therefore likely to be linked to their respective resistances towards re-oxidation and, in future research, such an approach may well be worth pursuing.

ACKNOWLEDGEMENTS

This work has been carried out under contract with the Commission of the European Communities, Radiation Protection Program, Contract No. BIO-B-531-83-B. One of us (M.S.) acknowledges the receipt of a doctoral fellowship from the IWONL.

REFERENCES

1. Hoffman, F. O., Garten, C. T., Lucas, D. M. and Huckabee, J. W. (1982). *Envir. Sci. Technol.*, **16**, 214.

2. JONES, A. G. and DAVISON, A. (1982). *Int. J. appl. Radiat. Isot.*, **33**, 867.
3. DAVISON, A. and JONES, A. G. (1982). *Int. J. appl. Radiat. Isot.*, **33**, 875.
4. SCHWOCHAU, K. (1983). *Radiochim. Acta*, **32**, 139.
5. SCHNEIDER, B. P. (1973). *J. Nucl. Med.*, **14**, 843.
6. CHARAMZA, O. and BUDCHIKOVA, M. (1969). *Nucl. Med.*, **8**, 301.
7. ARNOLD, R. W., SUBRAMANIAN, G., MCAFEE, J. G., ROSENSTREICH, M. E. and GEBHART, M. E. (1974). *J. Nucl. Med.*, **15**, 574.
8. USCHIDA, T., MIKI, M., UEDA, M., TANAKA, A. and VEDA, I. (1977). *Nucl. Med.*, **16**, 36.
9. STEIGMAN, J., MEINKEN, G. and RICHARDS, P. (1975). *Int. J. appl. Radiat. Isot.*, **26**, 601.
10. MUENZE, R. and GROSSMAN, B. (1977). *Radiochem. Radioanal. Lett.*, **31**, 95.
11. GROSSMAN, B. and MUENZE, R. (1982). *Int. J. appl. Radiat. Isot.*, **33**, 189.
12. GORSKI, B. and KOCH, H. (1970). *J. Inorg. Nucl. Chem.*, **32**, 3831.
13. SRIVASTAVA, S. C., MEINKEN, G., SMITH, T. D. and RICHARDS, P. (1977). *Int. J. appl. Radiat. Isot.*, **28**, 83.
14. BAKER, R. J., DIAMANTI, C. I., GOODWIN, D. A. and MEARES, C. F. (1981). *Int. J. Nucl. Med. Biol.*, **8**, 159.
15. RUSSELL, C. D., CRITTENDEN, R. C. and CASH, A. G. (1980). *J. Nucl. Med.*, **21**, 354.
16. NOLL, B., SEIFFERT, S. and MUENZE, R. (1983). *Int. J. appl. Radiat. Isot.*, **34**, 581.
17. SIMON, J., KITRING, A., TROUTNER, D. E., VOLKERT, W. A. and HOLMES, R. A. (1979). *Radiochem. Radioanal. Lett.*, **38**, 133.
18. TROUTNER, D. E., SIMON, J., KETRING, A. R., VOLKERT, W. A. and HOLMES, R. A. (1980). *J. Nucl. Med.*, **21**, 443.
19. VOLKERT, W. A., TROUTNER, D. E. and HOLMES, R. H. (1982). *Int. J. appl. Radiat. Isot.*, **33**, 891.
20. STALMANS, M. (1983). Technetium Komplexchemie en Gedrag in Bodemsystemen, Thesis, K.U. Leuven.
21. TUBIS, M. and ENDOW, J. S. (1968). *Int. J. appl. Radiat. Isot.*, **19**, 835.
22. TUBIS, M., KRISHNAMURTHY, G. T. and ENDOW, I. S. (1972). *J. Nucl. Med.*, **13**, 652.
23. YOKOYAMA, A., SAJI, H., TANAKA, H., ODORI, T., MORITA, R., MORI, T. and TORIZUKA, K. (1976). *J. Nucl. Med.*, **17**, 810.
24. JOHANSON, B., SYHRE, R., SPIES, H. and MUENZE, R. (1978). *J. Nucl. Med.*, **19**, 816.
25. YOKOYAMA, A. and HORIUCHI, K. (1982). *Int. J. appl. Radiat. Isot.*, **33**, 929.
26. STERN, H. S., ZOLLE, I. and MCAFEE, Z. G. (1965). *Int. J. appl. Radiat. Isot.*, **16**, 283.
27. GALATEANU, I., BRATU, C. and PETRIDE, A. (1977). *Radiochem. Radioanal. Lett.*, **28**, 95.
28. STEIGMAN, J., ECKELMAN, W. C., MEINKEN, G., ISAACS, H. S. and RICHARDS, P. (1974). *J. Nucl. Med.*, **15**, 75.
29. MUENZE, R. (1980). *Radiochem. Radioanal. Lett.*, **43**, 219.
30. SUNDREHAGEN, E. (1979). *Int. J. appl. Radiat. Isot.*, **30**, 739.
31. STEVENSON, F. J. (1982). *Humus Chemistry*, New York, Wiley-Interscience.

13

Measurement of the Complexation Capacity of Organic Matter in Dilute Extracts of Soils and Sediments

M. Stalmans, S. De Keijzer, A. Maes and A. Cremers

Laboratorium voor Colloïdale Scheïkunde, Fakulteit der Landbouwwetenschappen, KU Leuven, Kardinaal Mercierlaan 92, B-3030 Leuven, Belgium

1. INTRODUCTION

It is well known that the solid/liquid partition coefficient (K_D) of many (complex forming) radionuclides is dependent on the presence of liquid-phase complexing agents which may either have been added intentionally or dissolved from the sediment [1–5]. These effects, which result from the competition of solute complexing agents with solid phase sorption sites or functional groups, are a direct indication of radionuclide organic matter (OM) associations in the extract. They are however not directly related to radionuclide speciation in the sediment or soil.

The K_D dependence on OM content in the extract was discussed in some detail recently [6] and shown to be related to the solid/liquid ratio, used in batch equilibria. It was furthermore shown that the radionuclide to organic matter ratio in the extract may form the basis for a quantitative method for assessing the relative contribution of the OM geochemical sink to radionuclide binding in sediments. The method is discussed in detail in another paper at this seminar [7].

The key data to be obtained in the method are the OM concentrations in dilute extracts. In view of pronounced variations in molecular weight, which would make comparisons of humic acids (HA) of different sources quite difficult, the molarity scale for expressing HA concentration is quite impractical. Therefore, HA concentration is preferably expressed in equivalent terms, i.e. in terms of functional group capacity. Current methods for determining functional group number (total acidity, COOH,

OH, titration) are suited for relatively large samples [8] and can hardly be used for very dilute extracts. Moreover, total organic carbon is also impractical, in being a poor measure of binding capacity. Eventually optical measurements could be used but are complicated by the pronounced dependence of extinction coefficients on molecular weight or degree of aromaticity and conjugation [9, 10] (fulvics versus humics), the effect being particularly pronounced in the 250–280 nm range.

We are therefore proposing a simple and expedient method, based on total metal binding capacity [8], using the cationic silver–thiourea complex as an index cation. This paper presents some preliminary data obtained on dilute OM extracts of different sources.

2. THE SILVER–THIOUREA METHOD

2.1. Principle of the method

In earlier papers from this laboratory, it was shown that the Ag–(thiourea)$_n$ (Ag^+-TU) complex shows an exceedingly high selectivity for clays [11, 12] and resinous ion exchangers [13], making it a suitable candidate for a one-step method in the measurement of the Cation Exchange Capacity (CEC) of soils [14]. It was shown that CEC values of soils obtained on the basis of Ag^+-TU agreed quite well with those obtained by the traditional ammonium acetate method, irrespective of OM content. In fact, the increasingly higher pH dependence of the Ag^+-Tu CEC with increasing OM content shows that, in effect, the OM contribution is included in this method.

In short, the method is based upon the addition of a buffered Ag^+-TU solution containing a four to fivefold excess of silver with respect to the ion capacity of the adsorbent. The pronounced coagulating effect of the Ag^+-TU complex allows easy solid–liquid separation by centrifugation and the CEC is simply obtained by difference of the silver levels in the saturating and supernatant fluid.

2.2. Experimental procedure

The sample, a dilute OM extract, is mixed with an Ag^+-TU solution of known Ag^+ concentration (labelled with ^{110m}Ag), containing a 'free' TU concentration in the range 10^{-2}–10^{-1} (ligand number of silver at such conditions is about 3). Lower TU concentrations lead to 'unstable' solutions; preferably, solutions are prepared freshly for use and then stored in the dark. In view of the pronounced pH-dependence of the

CEC, measurements are standardised at some given reference pH, using a buffer system. Two buffers are tested below: ammonium acetate (0·1–1 M) and a boric acid–triethanolamine mixture (0·2 M or 4×10^{-3} M), a buffer which has been used in CEC measurements of calcareous and gypsiferous soils [15]; moreover, triethanolamine is often used in CEC measurements of soils (Ba-triethanolamine at pH = 8) [16].

The exchange of the Ag^+-TU complex is relatively fast but, as a matter of routine, systems are usually shaken overnight, the coagulated samples are centrifuged and the supernatant is assayed for ^{110m}Ag activity. The capacity is obtained by difference.

3. RESULTS AND DISCUSSION

3.1. Exchange capacity of humic acid systems

A series of examples of Ag-TU adsorption in HA is shown in Figs 1–3. Figure 1 shows the data, plotted as amount adsorbed versus added Ag^+-TU in a commercial humic acid (Fluka) in NH_4Ac buffer (ranging from 0·01 to 0·09 M). The total amount of humic acid in the samples was either 2 or 10 mg in a total volume of 40 ml (including the Ag^+-TU solution

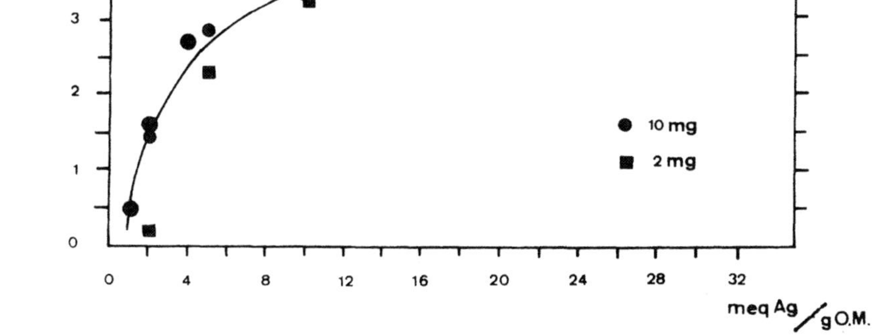

FIG. 1. Adsorption data of Ag^+-TU in a commercial humic acid, using NH_4OAc as buffer (pH = 6·7 ± 0·2).

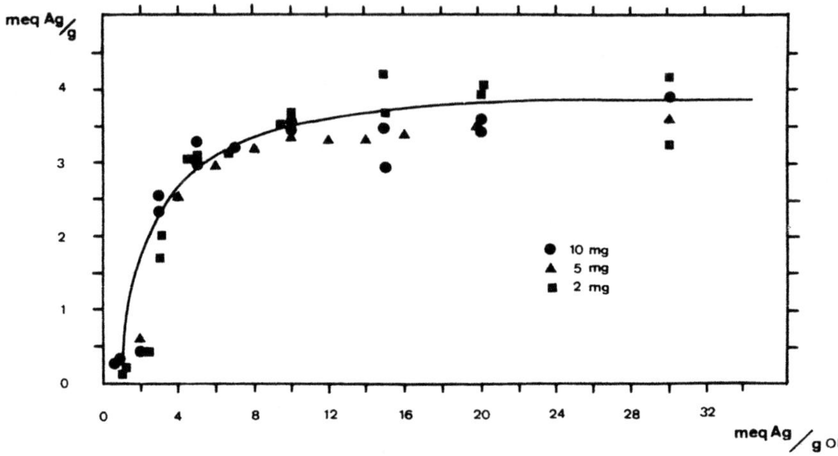

FIG. 2. Adsorption data of Ag^+-TU in a commercial humic acid, using boric acid–triethanolamine as buffer (pH = 6·7 ± 0·2).

added) corresponding to 50 or 250 ppm. It appears that about a five-fold excess is required to ensure complete saturation of the HA. The reproducibility is reasonably good, and similar CEC values are obtained in the 2 and 10 mg samples, about 3·8 meq/g.

Figure 2 shows the adsorption data on the same material using the boric acid–triethanolamine buffer under otherwise identical conditions

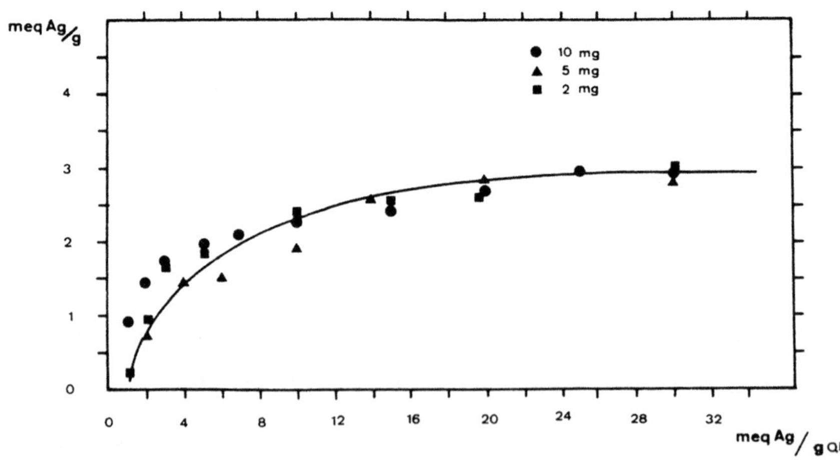

FIG. 3. Adsorption data of Ag^+-TU in humic acid, extracted from a podzol, using boric acid–triethanolamine as buffer (pH = 6·7 ± 0·2).

and on sample quantities of 2, 5 and 10 mg (or 50, 125 and 250 ppm in the reaction mixture). Taking into account the minute quantities of humic material involved, the reproducibility is again reasonably good, the CEC value being identical to the one obtained in the presence of the ammonium acetate buffer.

A third example of Ag^+-TU adsorption is represented in Fig. 3 showing the data on a humic acid extract from a podzol soil. The buffer used is boric acid–triethanolamine, the sample quantities being again 2, 5 and 10 mg. The CEC is about 3 meq/g. Repeated efforts were made to further extend the method to smaller quantities ($\sim$2 mg). Such measurements can be made but are both less reproducible and reliable in that they systematically lead to overestimates of functional group capacity ($\sim$20%).

The question obviously arises about the nature of the functional groups which are being measured by Ag^+-TU. It appears that the numerical values of the CEC, reported above, cover the range 3–3·8 meq/g. These values are well below the total acidity (i.e. COOH and 'acidic' OH or phenolic) reported for humic acids: 6–8 meq/g [17]. They are however within the range commonly reported for the number of COOH groups, 'average' values being around 3·6 meq/g [8]. It would therefore appear that, besides COOH, no other functional groups are being measured by the Ag-TU method. In forming the strongest complexes with lanthanides and actinides, these COO^- groups are of course the functional groups which are liable to be most specifically involved in environmental conditions, prevailing in sediments and water bodies.

The reliability of the Ag^+-TU capacity as a quantitative measure of metal binding capacity is demonstated by a comparison with titration data and Eu-binding capacity of an HA extract of a podzol. Organic matter was extracted with 0·5 M NaOH and HA precipitated at pH = 1. The sediment was dialysed against 10^{-2} M HNO_3 and finally washed 'electrolyte free', and diluted with distilled water ($\sim$300 ppm). The capacity, as determined by NaOH titration in 0·1 M Na^+ to pH = 7, generally considered to reflect the COOH capacity [18], was 1·35 meq/litre. The binding capacity for Eu, using 5×10^{-3} M $Eu(NO_3)_3$ (labelled with ^{152}Eu) at pH = 7 in 10^{-2} M NH_4Ac and 0·1 M NaCl, was 1·32 meq/litre. Finally, the Ag^+-TU capacity (0·015 M TU, 0·07 M NH_4Ac) was 1·51 meq/litre.

3.2. Effects of thiourea and ammonium acetate concentration

It was shown earlier that the TU concentration may have an effect on silver adsorption, particularly at TU levels in a range corresponding to

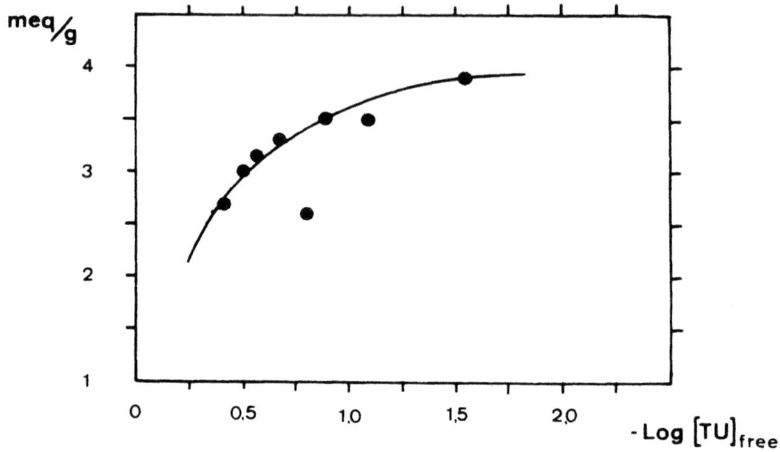

FIG. 4. Effect of TU concentration on the Ag^+-TU CEC of a commercial humic acid.

the addition of a third ligand in the metal coordination sphere [12, 13]. This effect was investigated in the Fluka HA sample by measuring the CEC values on 10 mg batches. Constant amounts of Ag were added (15 meq/g HA) at varying TU concentrations keeping the NH_4Ac concentration constant (0·05 M). The data are summarised in Fig. 4 showing the CEC dependence on TU concentration covering a range of 0·025–0·4 M. It appears that the CEC variation is quite small, between 0·025 and 0·1 M TU, but that a relatively rapid decrease occurs above TU concentrations of 0·1 M.

The decreasing selectivity of the silver complex with increasing TU concentration is similar to what was found in the case of clays [12] and resinous exchangers [13], a phenomenon which was interpreted as resulting from a relatively unfavourable (with respect to the liquid phase) coordination of the third ligand in the ion exchanger. No efforts were made in this case to establish the ligand number of silver in the HA but the TU dependence clearly indicates that the species adsorbed is the silver complex. In view of the strong dependence of CEC on TU at high concentrations, it is recommended to operate in the range 0·01–0·1 M, preferably at 0·1 M to ensure good stability of the silver–thiourea solutions.

The effect of NH_4Ac concentration as a buffer was studied on 10 mg batches of humic acid (Fluka) varying the buffer concentration in the range 0·01 to 1 M, keeping the TU concentration constant at 0·1 M in the reaction mixture. In each case, the amount of silver added corresponds

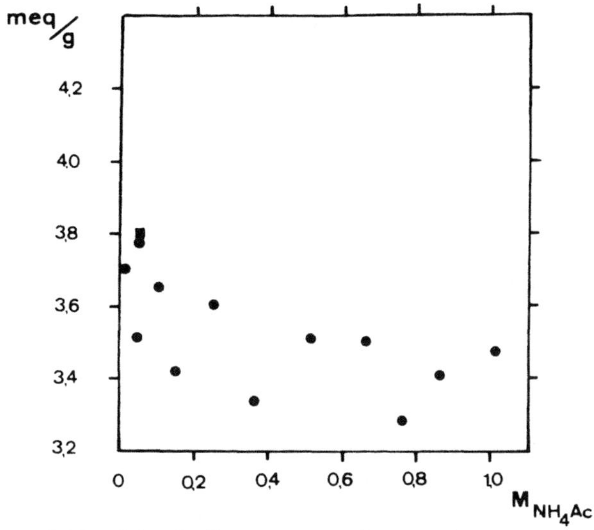

FIG. 5. Effect of NH_4OAc buffer concentration on the Ag^+-TU CEC of a commercial humic acid.

to about a five-fold excess with respect to the adsorbent capacity. The results are summarised in Fig. 5. The equilibrium pH in the systems covers the range 7·10–7·35. Although some scatter is apparent, there is a noticeable trend towards lower CEC values at high NH_4Ac concentration: about 3·6–3·8 meq/g at low and 3·3–3·5 meq/g high buffer concentrations. This effect is most likely due to some weak competition by the ammonium ion for the exchange sites. When working with dilute OM extracts, a buffer concentration in the range of 0·05 to 0·1 M appears to be adequate in routine measurements.

3.3. Capacity measurements on soil extracts

In the method relating to the quantitative assessment of the organic matter sink to radionuclide binding [7] in soil sediments, an extraction step is required for the purpose of isolating a measurable fraction of organic matter (OM). In view of the known effect of alkali extraction on the breakdown and autoxidation of OM (as evidenced by oxygen consumption) during extraction [8], it is recommended that such manipulations be carried out under conditions as mild as possible, preferably in an inert atmosphere.

Some preliminary results, obtained on the B_1 horizon of a podzol (Kalmthout, Belgium) are reported below. The characteristics of the soil

Table I
Characteristics of the podzol soil (B_1 horizon)

CEC (NH_4Ac, pH = 7)	8·5 meq/100 g
% organic carbon	2·53%
Organic matter (ashing)	5·33%
Humine (with respect to OC)	29·80%
Humic acids (with respect to OC)	57·80%
Fulvic acids	12·3%
% clay	0·014%
pH (H_2O)	3·72
pH (KCl, 1 M)	3·19

[19] as sampled are listed in Table I. It is apparent that only traces of clay are present and that the cation capacity is exclusively of an organic nature, i.e. an ideal sample for the purpose in mind. The soil is extremely acidic and was amended with $CaCO_3$ to raise its pH to 6, prior to use. Figure 6 shows the adsorption isotherm of the Ag^+-TU complex, as obtained in the presence of 0·1 M NH_4Ac, buffered at pH = 6. It is seen that the exclusively organic exchange complex is saturated with Ag^+-TU at concentrations of the order of 10^{-4} M. Strictly speaking, this is not an

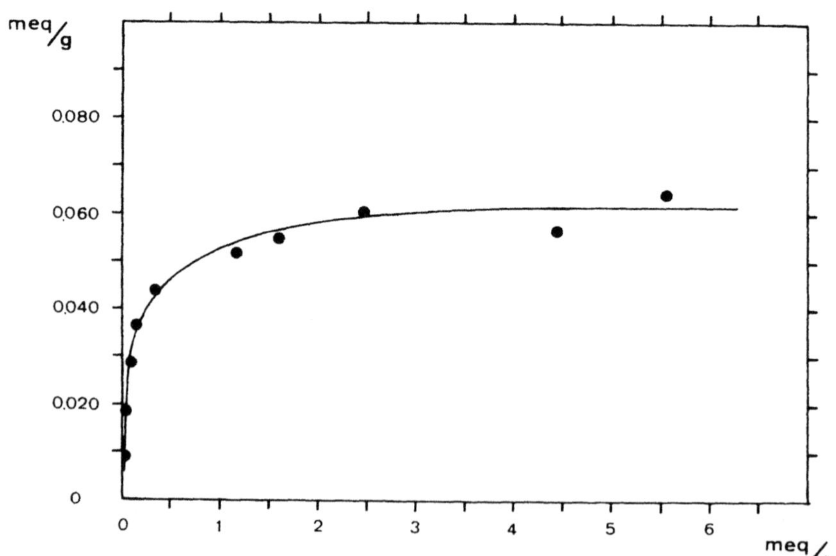

FIG. 6. Adsorption isotherm (25°C) of Ag^+-TU in a podzol soil, using NH_4OAc as buffer (pH = 6).

adsorption isotherm as such; the equilibrium concentrations of silver are obtained from measurements on the supernatant in which fractions of Ag^+-TU may be bound to the OM phase (i.e. not all Ag^+-TU is present as 'free' ions at low levels of saturation). At saturation however, a colourless supernatant is obtained and the OM is quantitatively coagulated. The statistical average of the plateau values is about 6 meq/100 g. The value obtained at pH = 7 under otherwise identical conditions is 7·6, a result which compares favourably with the one obtained by NH_4Ac, given in Table I.

It is obvious that these numbers are well below the value that can be expected on the basis of a 2·53% organic carbon content. The total number of carboxyl groups, taking a CEC value of 3·5–4 meq/g, would be expected to be in the range of 16–20 meq/100 g. Evidently, this is due to the fact that a CEC measurement, based on the use of NH_4^+ or Ag^+-TU, fails to detect the so-called 'potential' CEC, connected with the 'block' sites which are not exposed as a result of coiling of the HA polymer [8, 20]. Such sites will however be exposed and measured when dispersing the soil in alkaline solutions. In fact, the number of blocked sites may equal or even exceed the value, as obtained in a conventional CEC measurement [8]. It was also shown that the liming of a soil may lead to the release of blocked sites [21, 22].

In expressing radionuclide activity per unit amount of functional groups in the extract and the sediment residue (or alternatively obtaining meaningful quantitative distribution coefficients of OM), it is imperative that such values be calculated with respect to the effective number of functional groups, available at the particular conditions of extraction. Evidently, it can be foreseen that, at increasing pH, numbers of groups will be exposed but it is immaterial that the newly exposed groups were not available for complex formation at 'in situ' conditions (provided of course that the relative effect of 'decoiling' is the same on the OM in both phases).

The effect of extraction conditions on the functional group capacity was tested in the soil, both in buffered and unbuffered conditions, covering a pH range of 6·5–9. The entire procedure was carried out under inert atmosphere in a glove box. In the first case, soil samples were dispersed at liquid/solid ratios of 10 (2 g in 20 ml) in a 0·05 M NH_4Ac solution containing increasing levels of NaOH, covering a range of 50–250 μM per g soil. In the second case (unbuffered) extractants were dilute NaOH solutions corresponding to a range of 20–100 μM per g soil (L/S = 10). All systems were shaken for 3 days, centrifuged at high speed and

Table II
Effect of extraction pH on organic matter extraction yields (in equivalent terms) and overall function group capacity

Initial pH (extractant)	*Equilibrium pH*	**OM conc. in extract (meq/litre)*	*Total CEC (µeq/g)*	*Extraction yield (%)*	K_D *(OM)*
Buffered systems					
8·49	6·46	5·13	122	42	13·8
8·87	7·69	5·86	139	42	13·7
9·13	8·46	6·47	148	44	12·9
9·40	8·91	7·44	157	47	11·1
9·54	9·15	8·63	148	58	7·2
Unbuffered systems					
11·3	6·59	5·62	118	48	10·9
11·6	7·05	4·60	96	48	10·8
11·8	7·61	6·84	125	55	8·2
11·9	8·53	8·56	127	67	4·87
12·0	8·96	10·30	139	74	3·55

* Total OM extracted (in µeq/g) corresponds to column value × 10.

decanted. OM concentrations are measured in the extract and the sediment residue (after prior neutralisation to pH 7) using Ag^+-TU in 0·1 M NH_4Ac. The 'net' amount of functional groups in the sediment residue is obtained after due correlation of the OM content in the interstitial liquid phase, remaining in the residue after centrifugation (on the basis of weight and OM levels in the supernatant). The results are summarised in Table II.

Examination of the data shows clearly that, both in buffered and unbuffered systems, a distinct increase occurs in the overall CEC, maximum values (at pH $\approx$ 9) being respectively 0·157 and 0·139 meq/g, i.e. about twice as high as the original CEC at pH = 7 (0·076 meq/g). This shows that the release of blocked sites is nearly identical to the number of sites measured by a conventional CEC measurement. The maximum value found may well be a good approximation of the total number of COOH groups in the soil since it is consistent with a CEC of 3·45 meq/per g OM (i.e. 15·7/(2·53 × 1·8)).

It appears quite unlikely that the extraction treatment has resulted in any significant hydrolytic degradation, as evidenced by the fact that the total CEC, after extraction in the most alkaline conditions, is similar in both buffered and unbuffered systems (a small trend is seen towards lower values in unbuffered conditions). In fact, during the initial stages of

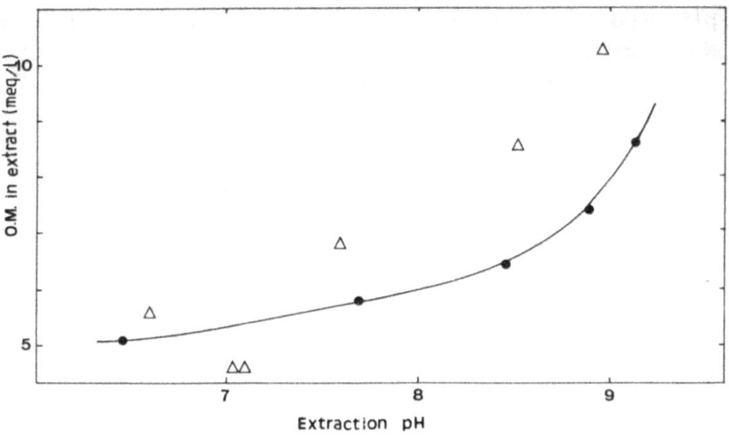

FIG. 7. Effect of extraction pH on the organic matter extraction yield in buffered (●) and unbuffered (△) conditions (L/S ratio: 10).

extraction the unbuffered systems have been exposed to more alkaline conditions, which may perhaps be responsible for the slightly higher extraction yields. The extent of this effect was tested in the most alkaline system, i.e. initial pH of 12. Immediately after dispersion, the pH drops to 11, decreasing further to a value of 9·35 within 30 min. If, in fact, hydrolytic degradation had occurred, then this would be reflected in higher CEC values. Since this is not the case, we are confident that the procedure is sufficiently mild to exclude the possibility of OM degradation.

The extraction yields of OM (also expressed in terms of K_D) are somewhat higher in unbuffered conditions and may appear relatively high, in view of the mild conditions used. Of course, these data may not be compared with OM extraction yields based on OC content since, in the present case, yields are expressed in terms of equivalents extracted, the HA extracted being precisely those with high functional group density. Moreover, the high yield may further be facilitated by the absence of adsorption on aluminosilicates. The higher extraction yields in unbuffered conditions may well be related to the higher pH values during the initial stage of the extraction process. It would appear that extraction in buffered conditions is to be preferred in that extraction yields show a rather smooth dependence on extraction pH. In standardising extraction methods, it would therefore be advantageous to have some predictable pattern in extraction yields. The pattern in the case of unbuffered conditions is somewhat erratic, as illustrated in Fig. 7. This may well be due to rather large pH variations during extraction. In contrast, in the buffered systems, pH differences between initial and final conditions are quite small and not exceeding one pH unit in the pH range of 7–9.

4. CONCLUSIONS

A method is presented which allows reproducible measurements of the COOH functional group capacity in dilute extracts of soils, based on the use of the Ag^{+}-TU complex as an index cation. The method is shown to be applicable to a podzol soil and leads to reliable values of the OM distribution coefficient, necessary in the evaluation of the importance of the OM sink of radionuclide binding [7]. It would be of interest to extend these methods to mixtures of clays, oxides and OM such as marine or estuarine sediments. In such systems however, the situation is complicated by the difficulty of obtaining the fraction of the exchange capacity to be ascribed to the organic matter fraction. The study of appropriate methods, possibly by using other index cations such as the cobalt–hexamine complex, is now in progress in this laboratory. In view of the fact that complex formation of lanthanides and actinides with HA is quite insensitive to ionic strength, the present approach may well be extended to marine and estuarine sediments; the in situ pH values of 7·5–8 are liable to afford measurable extraction yields, using low ionic strength extractants as described in this paper.

ACKNOWLEDGEMENTS

This work has been carried out under contract with the Commission of the European Communities, Radiation Protection Program, contract BIO-B-531-83-B. One of us (M.S.) acknowledges the receipt of a doctoral fellowship from the IWONL.

REFERENCES

1. Means, J. L., Crerar, D. A. and Duguid, J. O. (1978). *Science*, **200**, 1477.
2. Nelson, D. M., Okartunnen, J., Orlandini, K. O. and Larson, R. P. (1981). Influence of dissolved organic carbon on the sorption of plutonium to natural sediments. Radiol. Environ. Res. Div. Ann. Rep., ANL-80-115. Part III. Argonne Nat. Lab., Argonne, Illinois.

3. SANCHEZ, A. L., SCHELL, W. R. and SIBLEY, T. H. (1982). In: *Environmental Migration of Long-lived Radionuclides*, Vienna, IAEA, p. 188.
4. PIMPL, M. and SCHUTTELKOPF, H. (1983). C.E.C. Seminar, The Transfer of Radioactive Materials in the Terrestrial Environment Subsequent to an Accidental Release to the Atmosphere, Dublin, p. 251.
5. SIBLEY, T. H. and CLAYTON, J. R. JR. (1984). In: Proceedings Int. Seminar on the Behaviour of Radionuclides in Estuaries, Renesse, in press.
6. CREMERS, A. and HENRION, P. N. (1984). In: Proceedings Int. Seminar on the Behaviour of Radionuclides in Estuaries, Renesse, in press.
7. STALMANS, M., MAES, A. and CREMERS, A. Role of organic matter as geochemical sink for technetium in soils and sediments (this volume, pp. 91–113).
8. STEVENSON, F. J. (1982). *Humus Chemistry*, New York, Wiley-Interscience.
9. LAUD, J. N. (1969). *Soil Sci.*, **107**, 303.
10. SWIFT, P. S., THORNTON, B. K. and POSNER, A. M. (1970). *Soil Sci.*, **110**, 93.
11. CREMERS, A. and PLEYSIER, J. (1973). *Nature Phys. Sci.*, **243**, 86.
12. PLEYSIER, J. and CREMERS, A. (1975). *J. Chem. Soc. Faraday, Trans. I*, **71**, 256.
13. MAES, A. and CREMERS, A. (1978). *J. Chem. Soc. Faraday, Trans. I*, **74**, 2470.
14. CHHABRA, R., PLEYSIER, J. and CREMERS, A. (1975). In: *Proc. of the Int. Clay Conf., Mexico City*, Wilmette, Illinois, Applied Publ. Ltd., p. 439.
15. VAN ROSMALEN, H. A. (1980). Internal Report BO 80-3, Amsterdam, Royal Tropical Institute.
16. BLACK, C. E. (Ed.) (1965). *Methods of Soil Analysis*, Madison, Wisconsin, American Society of Agronomy.
17. SCHNITZER, M. In: *Soil Organic Matter* (Ed. M. Schnitzer and S. U. Khan), Amsterdam, Elsevier, Ch 1.
18. TORRES, R. A. (1982). Humic acid complexation of europium, americium and plutonium, Ph.D. thesis, Florida State University.
19. LOON, L. VAN. (1982). Fixatie van zink in humusrijke podzol bodems, Thesis, K. U. Leuven.
20. SCHNITZER, M. (1965). *Nature*, **207**, 667.
21. MCLEAN E. O. and OWEN, E. J. (1969). *Soil Sci. Soc. Amer. Proc.*, **33**, 855.
22. MCLEAN, E. O., REICOSKY, D. C. and LAKSHMANAN, C. (1965). *Soil Sci. Soc. Amer. Proc.*, **29**, 374.

14

^{99}Tc in Environmental Waters

M. GARCÍA-LEÓN,[a] C. PIAZZA[b] and G. MADURGA[a]

[a] *Facultad de Fisica, Universidad de Sevilla, Apdo. 1065, 41080 Sevilla, Spain*

[b] *Facultad de Química, Universidad de Sevilla, Sevilla, Spain*

1. ^{99}Tc IN WATERS

^{99}Tc is produced at a rate of ~6% by fission of ^{235}U and ^{239}Pu. This, together with other sources of production, such as activation and the progressive use of ^{99m}Tc in nuclear medicine, is leading to a presumably considerable presence of ^{99}Tc in the environment, since the birth of the nuclear era.

Being a radioisotope of long half-life ($2{\cdot}14 \times 10^5$ y), the importance of its contribution, therefore, to the future collective dose to the population becomes apparent.

However, there exists a rather scanty amount of data of its concentration in nature. And this is due, probably, to the difficulties encountered in its chemical behaviour and its radioactive counting (it emits only a relatively soft β^- radiation, $E_{max} = 292$ keV).

Recently, the studies on ^{99}Tc in the environment have received much attention, as this seminar evidences.

Since 1982 our department has devoted part of its research work to the determination of ^{99}Tc in the environment, specially, although not only, in waters. In this work we present some of our results in this field.

A basic problem in the treatment of water samples is the volume involved. Indeed, due to the low specific activity of this radionuclide, it must be preconcentrated from as high a volume as possible. Several methods have been proposed for this purpose. Evaporation, by Golchert and Sedlet [1], the use of ionic interchange resins by Riley and Siddiqui [2], the reduction-coprecipitation method by Holm *et al.* [3], which in part is described in Section 2, are the most prominent. None of them, however, can deal with as high a volume of water as is needed to preconcentrate ^{99}Tc from environments where a sizeable contamination of ^{99}Tc is not 'a priori' expected. We carefully describe a possible alternative to this problem in Section 2, as well as the analysed samples. Their activities and some comments on the identification of them are presented in Section 3. Some conclusions and prospects of the method form Section 4 and close the paper.

2. SEPARATION METHODS

Two separation methods will be reported here. A new one takes advantage of a high concentration of radionuclides found in the flocs from water treatment plants as described later. The other one was basically developed in Lund and recently published [3]. We are applying this method to samples of rainwater collected in Seville since 1982. Experiments are in progress and we hope to be able to publish results in the near future.

Briefly, the method consists of the following steps: a given volume of water is spiked with about 1000 Bq of ^{99m}Tc. $FeSO_4$ is added to reduce Tc in the sample that is precipitated upon adding $FeCl_3$ and drops of $NH_4(OH)$ until a basic pH is reached. The precipitate is dissolved in 3 M H_2SO_4, together with $(NH_4)_2S_2O_8$ and H_2O_2. The solution is boiled to oxidise Tc to its +VII state. Filtration of the solution is carried out and Tc is extracted from the liquid phase with tributylphosphate (TBP), previously equilibrated with 4·5 M H_2SO_4, in the presence of HF. After washing with 3 M H_2SO_4 and HF, Tc is back-extracted with 2 M NaOH from the organic phase that was diluted in xylene. Electrolysis is carried out over 2 h at 0·35 A in the NaOH solution. The anode is a platinum wire and the cathode a stainless steel planchet that is dried and prepared for counting.

The radiochemical yield is determined by measuring in a γ-spectrometer the γ emission ($E_\gamma = 140$ keV) from ^{99m}Tc that remains in the final sample.

With this method, of course, large volumes of water can be analysed, since many litres of water are reduced to a few grams of precipitate. Nevertheless, in some cases the concentration of ^{99}Tc in waters is so low that unworkable volumes of water need to be analysed.

In 1979, Roy *et al.* [4] proposed the measurement of γ-emitting radionuclides in flocs collected from water treatment plants. They showed that the floc formed by the coagulation and precipitation produced in the

Table I
Results on the different subsamples of floc of a water treatment plant from Valencia, Spain

Subsample	*Weight (g)*	*Yield (%)*	*Activity (mBq/g)*	*Date of measurement*	Δ
L1	20	32·0 ± 0·8	1·52 ± 0·63	23 February 1983	
			1·30 ± 0·59	19 April 1983	0·39 ± 0·20
			1·22 ± 0·63	21 June 1983	
			1·30 ± 0·63	27 June 1983	0·46 ± 0·18
L2	100	7·0 ± 0·4	0·56 ± 0·59		1·2 ± 0·8
L3	20·5	8·8 ± 0·1			
L4	20·7	43·9 ± 0·5	1·00 ± 0·37		0·40 ± 0·11
L5	20	62·9 ± 1·0	0·56 ± 0·26	22 July 1983	
			0·52 ± 0·26	6 September 1983	0·55 ± 0·15
			0·48 ± 0·30	4 October 1983	

process of purification of waters by adding $Al_2(SO_4)_3.14\,H_2O$ contains many radionuclides. The sensitivity of the method becomes obvious since a few grams of these flocs can be related to hundreds of litres of waters.

In this work we report on the achievements so far reached in our laboratory in trying to look for ^{99}Tc in the flocs.

We have used several samples. Firstly, a 200 g floc sample from a purification plant in Valencia (no nuclear installation is working in its vicinity at the moment), Spain, provided by M. L. Sánchez-Mayoral, was analysed. It was divided into several subsamples to test the chemical method and to study the counting and identification procedures. They will be referred to, hereafter, by L1–L5, as is shown in Table I, column 1. In column 2 of the same table we give the analysed masses of each one. The yields obtained, column 3, show a progressive improvement of the chemical method.

After some trials, a final form of the radiochemical method was reached. Some 20 g of floc are spiked with 200 Bq of ^{99m}Tc, to measure the yield. After a calcination of 1–2 h at 600°C, the ashes are suspended in about 200 ml of distilled water, together with 10 g of $(NH_4)_2S_2O_8$ and 2 ml of H_2O_2. Some H_2SO_4 is poured on the suspension until an acid pH is reached, and then the volume of H_2SO_4 to make it 3 M is slowly added. The mixture is carefully boiled for 20 min, while maintaining the suspension with a magnetic stirrer. The mixture is filtered and the solid phase washed with 25 ml of 3 M H_2SO_4. Tc is extracted with 15 ml of TBP, previously equilibrated with the same volume of 4·5 M H_2SO_4, and 2 ml of HF. The aqueous phase is neglected, and a washing of the organic phase with 25 ml of 3 M H_2SO_4 and 0·5 ml of HF done. Tc is back-extracted and the organic phase diluted in 30 ml of xylene, with 15 ml of NaOH. The aqueous phase is directly transferred to an electrolytic cell. The electrolysis is carried out for 2 h at 0·35 A. The yield is obtained as previously explained.

Samples L4 and L5 were analysed following this method and show relatively good chemical recoveries. But in samples L2 and L3 a far lower yield was obtained. Neither of them were previously calcinated. This suggests that possibly Tc in this kind of sample is mostly bound to organic matter. In the case of sample L1, in which the first analysis was carried out, a different pretreatment of the floc was done. Indeed, after the calcination, direct dissolution of the ashes in 80 ml of 3 M H_2SO_4 was attempted. During the boiling some spattering may have occurred, producing some losses. In fact, losses in the electrolysis were of only 9%, which does not explain the low yield obtained.

Table II
Characteristics of the Canadian samples

Sample	*Town*	*River*	*Date of sampling*	*Mass analysed (g)*
L6	Hull	Ottawa	7 July 1983	20·7
L7	Ste Foy	St Lawrence	18 March 1983	20·1
L8	Ste Foy	St Lawrence	10 June 1983	20·9
L9	Hull	Ottawa	7 July 1983	20·5
L10	Trois-Rivieres	St Maurice	22 April 1983	20·1
L11	Ste Foy	St Lawrence	10 June 1983	20·0
L12	Trois Rivieres	St Maurice	22 April 1983	20·0

Recently, Dr J.-C. Roy, from Quebec, provided us with several floc samples from Canada. Their characteristics are listed in Table II. They were sampled at three different stations: one at Hull on the Ottawa river, one at Trois-Rivieres on the St Maurice river and two at Ste Foy, a town near Quebec, on the St Lawrence river. Hull is located 200 km downstream from Chalk River, where a nuclear research reactor, NRX (200 km west of Ottawa), exists. In the same way, Ste Foy is located 750 km from it. As for Trois-Rivieres, no nuclear installation is working in its vicinity.

The date of sampling is given in column 4 of Table II, where in column 1 its identification name (L6–L12) is presented.

The mass analysed was in all cases about 20 g (column 5 of Table II) and replicate analyses were carried out for three of them: the one from

Table III
Results on the different Canadian samples

Sample	*Yield (%)*	*Activity (mBq/g)*	Δ
L6	22·2 ± 0·3	2·01 ± 0·59	
L9	50·4 ± 0·8	2·37 ± 0·30	0·24 ± 0·09
L7	62·3 ± 0·8	1·16 ± 0·22	0·42 ± 0·16
L8	47·5 ± 0·6	0·47 ± 0·24	
L11	65·0 ± 0·9	0·73 ± 0·15	0·35 ± 0·14
L10	43·1 ± 0·7	1·00 ± 0·31	0·28 ± 0·24
L12	41·5 ± 0·7	1·19 ± 0·22	

Hull, L6–L9; that from Ste Foy taken on 10 June 1983, L8–L11; and the sample from Trois-Rivieres, L10–L12.

In Table III, column 2, the yields obtained for each one are given. The radiochemical method was that previously explained in detail. In the case of sample L6 a lower yield is obtained, probably related to a bad calcination. This was improved for sample L9, and a higher yield was obtained. Similar remarks can be made for samples L8 and L11.

3. ACTIVITIES OBTAINED AND THEIR IDENTIFICATION

The activities have been measured with a low-background P-10 gas-flow proportional counter, which is described elsewhere [5]. It was calibrated for ^{99}Tc β radiation with 18 standard samples of activities ranging from 0·63 to 66 150 dpm, prepared by J. Rioseco from the University of Lund.

At the best working conditions for ^{99}Tc, the background reaches a value of about 1 cpm and the efficiency 37%. Details on the calibration work can be found in ref. 6.

^{99}Tc in the floc samples was identified using our Δ-test method [5, 6]. Basically, it takes advantage of the fact that the ratio between the counting rates at two different values of the HV, say $\Delta \equiv$ (net counts with 1200 V)/(net counts with 1450 V), critically depends in our detector system on the energy of the β spectrum. This dependence is well defined for energies below $E_{max} = 600$ keV, while Δ is practically the same for higher energies. And the eventual existence of counts above an upper level discriminator in the detector single-channel analyser reveals the existence of an α contamination.

In the present working conditions we expect a $\Delta \simeq 0{\cdot}38$ for ^{99}Tc, to be compared with 0·5 for ^{147}Pm ($E_{max} = 228$ keV) or 0·15 for ^{36}Cl ($E_{max} = 710$ keV).

So with this method we can simultaneously detect spurious β and α activities in the sample which could affect the ^{99}Tc counting. The details of the method can be seen in refs 5 and 6.

The results of the activities are given in column 4 of Table I for the Valencia samples. They are presented in mBq per gram of floc mass. The errors are assigned by conventional propagation of those produced by the calibration and the chemical yield. The purity of the ^{99m}Tc added to the sample was checked in all cases. When impurities were found, they were assumed, and sometimes verified, to be ^{99}Tc produced. Then their contributions were subtracted from the total number of counts obtained with the floc sample. In these cases, the error produced in the operation was added to those from the other sources.

In general, if we consider the experimental errors, the activities of the subsamples are in good agreement with each other. However, we cannot exclude a possible inhomogeneous distribution of ^{99}Tc in the whole sample. Due to the low yield of subsample L2, the activity measured has a large error which makes it compatible both with the activities measured for other subsamples and with absence of ^{99}Tc. We have recorded this measurement following the recommendations of Collé *et al.* [7] and omitted the measurement of subsample L3 which has an equally low yield.

As for the samples from Canada, the results are given in column 3 of Table III.

There appears to be good agreement between the activities for each pair of samples of the same floc, although the comments about inhomogeneity can be made, specially for the samples of lower activities. The samples L6–L9 show higher activities than the others, as expected. In fact, their γ-spectrum [8] presents a wide variety of activation and fission products (^{103}Ru, ^{106}Ru, ^{137}Cs, ^{95}Zn, ^{95}Nb, ^{54}Mn, ^{46}Sc, ^{59}Fe and ^{60}Co), probably due to the fact that the sampling station is not far away from Chalk River (200 km).

A similar effect is noted for samples L7 and L8–L11. Indeed, they were taken at the same station (750 km from Chalk River) but on different dates. Only ^{137}Cs appears in the γ-spectrum for the samples taken on 10 June 1983 (L8–L11). However, for that taken on 18 March 1983 (L7), ^{131}I, ^{103}Ru, ^{106}Ru, ^{137}Cs, ^{95}Zn, ^{95}Nb, ^{46}Sc and ^{60}Co appear and a higher ^{99}Tc activity is obtained.

Finally, for samples L10–L12 a slightly higher activity is seen than for samples L8–L10, although no nuclear installation is operating in the proximity of the sampling station. But if we observe its γ-spectrum, a slight amount of ^{106}Ru is present, together with a typical fallout ^{137}Cs peak.

For both the Valencia and Canadian samples very low ^{99}Tc activities are obtained. This makes difficult the analysis of the results because of the errors involved. So we feel that a larger amount of floc should be analysed, and an improvement of the yield achieved. This can be specially important in applying the Δ-test or any other identification test.

As already mentioned, we have identified ^{99}Tc by measuring Δ. From its definition high errors are expected for low activities.

In Table I, column 6, Δ is given for the Valencia samples and in column 4 of Table III for the Canadian samples. For the latter, Δ was measured only for the sample of higher yield of each pair.

In general, Δ reached a reasonable value to make it plausible that all the measured activities are essentially ^{99}Tc (see ref. 6). This can be said even for the somewhat ambiguous cases, L2 and L10, in view of their very low activities and, therefore, the relative significance of the numbers obtained. For sample L9 a slight high-energy β contamination is suspected, and more measurements are being made. In any case, these results reveal the need for an accurate knowledge of the background at the time of measurement.

As an additional test, subsamples L1 and L5 were measured several times throughout the year 1983 (column 5 of Table I). From the results a long half-life activity in the subsamples is confirmed. For subsample L1 Δ was measured twice on different dates. The values obtained were in agreement.

4. CONCLUSIONS AND PROSPECTS OF THE METHOD

Our results confirm the possibility of extending the method proposed by Roy *et al.* [4] to preconcentrated ^{99}Tc from environmental waters. Since a few grams of the flocs from water treatment plants can be related to hundreds of litres of waters, the sensitivity of the method is obvious.

At the moment, we cannot translate the ^{99}Tc activity per gram of floc into activity per litre of water, because we do not know so far the degree of retention of ^{99}Tc in these flocs, or the amount of water (litres) to be related to grams of dried floc.

There are several methods for finding this last quantity. The one most often used is based on the ratio of K in the dried floc and in the suspended matter in raw water [9, 10]. This will be applied to our samples in the near future.

As for the degree of retention of ^{99}Tc in the flocs, experiments are in progress. A value of about 50% is suspected [8], although this may well be an optimistic guess. Indeed, very preliminary laboratory experiments reveal that Tc(VII) is not retained in the precipitation of $Al_2(SO_4)_3 . 14\,H_2O$ but Tc(IV) is. However, the actual environmental conditions are very different from the laboratory ones, and this result can be modified, specially due to the presence of suspended matter in the water. Tc is commonly present in nature in its VII state, and hence it is possible that the measurement of ^{99}Tc in flocs gives information on the amount of reduced Tc in waters. And this would be an interesting datum to learn about its behaviour in the aqueous environment.

Finally, of course, the continuous analysis of floc samples in the same environment could tell us about ^{99}Tc evolution. And, since in some countries these flocs are used to fertilise agricultural soils, it would be possible to study the transfer of ^{99}Tc to man by the additional analysis of vegetable samples.

ACKNOWLEDGEMENTS

We thank Dr J.-C. Roy and M. L. Sánchez-Mayoral, who provided us with the floc samples.

Special thanks are due to Dr E. Holm and J. Rioseco, who gave to one of us (M.G.-L.) the opportunity of a valuable stay at Lund, and for the ^{99}Tc calibration samples.

Financial help from the Instituto de Estudios Nucleares, Madrid, is acknowledged.

REFERENCES

1. GOLCHERT, N. W. and SEDLET, J. (1969). *Anal. Chem.*, **41**, 669.
2. RILEY, J. P. and SIDDIQUI, S. A. (1982). *Anal. Chim. Acta*, **139**, 167.
3. HOLM, E., RIOSECO, J. and GARCÍA-LEÓN, M. (1984). In: Proc. Semin. Alpha Particle Spectrometry and Low-level Measurements, Harwell, England, 10–13 May 1983. *Nucl. Instrum. Methods*, **223**, 204.
4. ROY, J.-C., BARBEAU, C., COTÉ, J.-E. and TURCOTTE, J. (1979). *Nucl. Instrum. Methods*, **160**, 187.

5. GARCÍA-LEÓN, M., GARCÍA-MONTAÑO, E. and MADURGA, G. (1984). *Int. J. appl. Radiat. Isot.*, **35**, 195.
6. GARCÍA-LEÓN, M., PIAZZA, C. and MADURGA, G. *Int. J. appl. Radiat. Isot.* (in press).
7. COLLÉ, R. *et al.* (1980). Upgrading of environmental radiation data, Health Physics Society Committee Report, HPRS-1.
8. ROY, J.-C. (1984). Private communication.
9. ROY, J.-C., LUPIEN, M., COTÉ, J.-E. and TURCOTTE, J. (1982). *Nucl. Instrum. Methods*, **206**, 553.
10. TURCOTTE, J., DODDRIDGE, P., BEAULIEU, P., ROY, J.-C. and COTÉ, J.-E. To be published in *Analysis*.

15

Properties and Solubility of Technetium Dioxide

CH. HANKE, B. JÄHRLING and K. H. LIESER*
Fachbereich Anorganische Chemie und Kernchemie, Technische Hochschule Darmstadt, D-6100 Darmstadt, Federal Republic of Germany

1. INTRODUCTION

One of the elements which significantly contributes to the long term biological hazard is the fission product technetium.

The solubility behaviour of technetium depends mainly on its chemical state and on the chemical environment. The valencies of technetium range from $-$I to $+$VII, but in aqueous solutions the oxidation states $+$IV and $+$VII predominate.

The redox potential–pH diagram (Fig. 1) shows that in an oxidising environment (such as air-saturated water) the lower oxidation states are converted to the highly soluble pertechnetate. Under reducing conditions (absence of oxygen) pertechnetate is transformed into the very sparingly hydrated TcO_2.

It is not known whether in nuclear waste Tc is present as Tc^{VII} or Tc^{IV}.

Due to the oxidising atmosphere during the production of the glass, Tc^{VII} will predominate. On the other hand high temperatures and the possibility of radiochemical reduction will favour Tc^{IV}. It is also expected that in the reducing environment of the repository Tc^{VII} is transformed to Tc^{IV}. Due to this fact it is important to investigate the chemical properties and the solubility of TcO_2.

*To whom correspondence should be addressed.

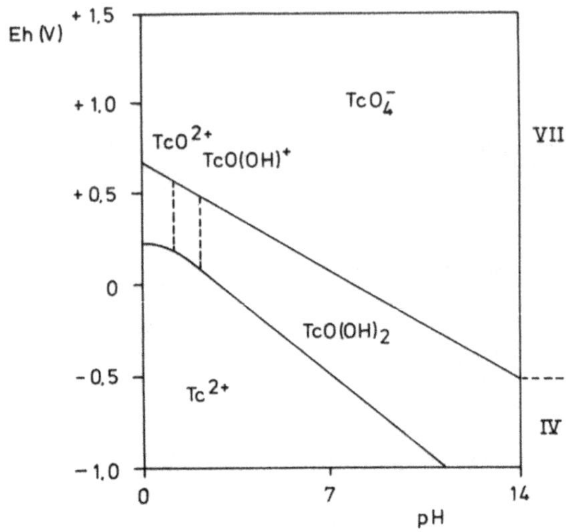

FIG. 1. E_h–pH diagram of Tc [1].

2. PREPARATION OF TcO_2

In the literature several procedures for the preparation of TcO_2 are described. The most frequently used methods are:

1. Decomposition of NH_4TcO_4 in a nitrogen stream at 700 to 800 °C or at 450 °C in vacuum [2]:

$$NH_4TcO_4 \rightarrow TcO_2 + 2H_2O + \tfrac{1}{2}N_2$$

2. Hydrolysis of K_2TcCl_6 or K_2TcBr_6 in neutral medium [3]
3. Reduction of a pertechnetate solution with

   Zn/HCl [4]
   Sn^{2+} [5]
   hydrazine [6]
   or electrochemical reduction [7]

4. Hydrolysis of a ternary oxide of Tc^{VI} with subsequent disproportionation of the intermediate oxidation state in solution to give $HTcO_4$ (aq) and $TcO_2\,xH_2O$ (s) [8]

$$2H_2O + 4HTcO_4 \rightarrow 4H_2TcO_4 + O_2$$

$$3H_2TcO_4\ (\text{red}) \rightarrow 2HTcO_4 + 2H_2O + TcO_2$$

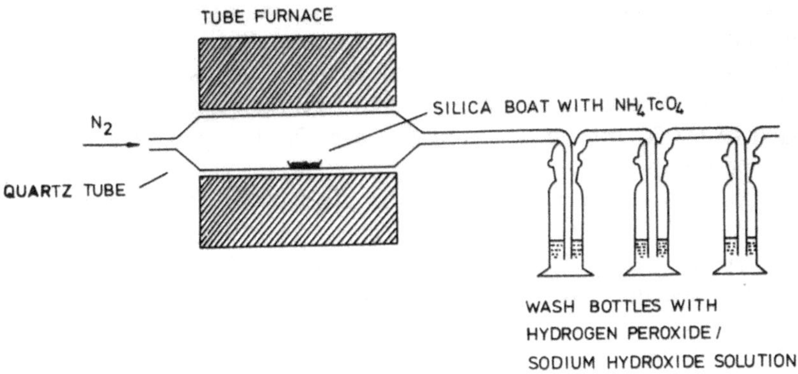

FIG. 2. Decomposition of NH_4TcO_4 at 800°C for 2 h—product A.

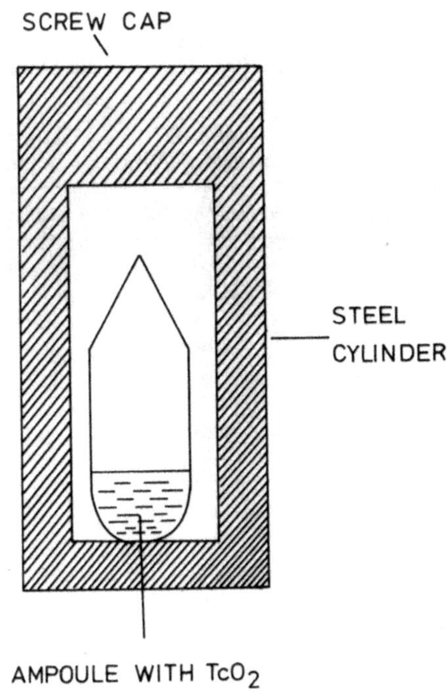

FIG. 3. Heating of TcO_2 at 860°C for 15 h—product B.

In our experiments method (1) was preferred, because method (3) gave colloidal products and high values for the solubility.

The $NH_4{}^{99}TcO_4$ was placed in a silica boat, which was placed in a quartz tube (Fig. 2). In a stream of pure nitrogen the quartz tube was heated to 800 °C for 2 h in a tube furnace. Product A was obtained by this procedure.

To decompose traces of NH_4TcO_4 in product A, it was placed into a quartz ampoule which was evacuated, closed by melting and placed into a steel cylinder which was closed by a screw cap (Fig. 3). In this device, the TcO_2 was reheated to 860 °C for 14 h and product B was obtained.

Table I
Cell dimensions of products A and B (Debye–Scherrer diagram)
Specific surface area: 2·63 m^2/g

	Literature [2]	*Product A*	*Product B*
a [Å]	5·55 ± 0·01	5·54 ± 0·02	5·52 ± 0·02
b [Å]	4·85 ± 0·01	5·02 ± 0·02	4·99 ± 0·05
c [Å]	5·62 ± 0·01	5·58 ± 0·02	5·57 ± 0·03
β [°]	121·90 ± 0·10	121·80 ± 0·10	121·93 ± 0·39

Table II
Grain size distribution of Products A and B

Grain size (μm)	*Product A (%)*	*Product B (%)*
⩽ 10	63·0	64·4
10–20	23·5	23·9
20–30	7·3	5·8
30–40	2·4	1·6
40–50	1·6	1·3
50–60	0·9	1·3
60–70	0·3	0·2
70–80	—	0·1
80–90	0·3	0·4
90–100	—	0·3
100–110	0·3	—
110–120	—	0·4
120–130	0·1	0·1
130–140	0·1	—
140–150	—	—
150–160	—	—
160–170	0·1	—

3. CHARACTERISATION OF TcO_2

TcO_2 was characterised by Debye–Scherrer diagrams, grain size distribution and specific surface area. The results are shown in Tables I and II.

In Table I the dimensions of the unit cell of products A and B are given. The values are in good agreement with literature data.

The specific surface area was determined by methylene blue adsorption. For the calculation of the surface area it was assumed that one molecule of methylene blue covers an area of $1{\cdot}3\,nm^2$. For product B a specific surface area of $2{\cdot}63\,m^2/g$ was obtained.

In order to determine the the grain size distribution, products A and B were photographed through a microscope. It was found that the majority of all particles have a grain size less than $10\,\mu m$.

4. SOLUBILITY EXPERIMENTS

In Fig. 4 a flow sheet of the solubility experiments is shown. TcO_2 was equilibrated with pure water. The suspension was shaken by a rotation machine which rotated at time intervals to give grinding effects. The concentration of Tc in solution was measured as a function of time. At the end of the experiment the phases were separated by centrifugation, filtration ($0{\cdot}45\,\mu m$) and ultrafiltration ($0{\cdot}001\,\mu m$). In the aqueous solution the activity of ^{99}Tc was measured by liquid scintillation, and the redox potentials were also determined.

Speciation was done by spectrophotometry and by extraction method based on the fact that under these conditions only TcO_4 is extracted in the organic phase.

In Fig. 5 a UV-spectrum shows that TcO_4^- absorbs strongly at 248 and 288 nm.

5. RESULTS AND DISCUSSION

In Figs. 6–8 the results of some solubility experiments in presence of air are shown. It can be seen that the solubility depends on the mass of TcO_2 and that it increases linearly with time. It was also found that the amount of Tc in solution does not depend on the pore size filters used for the separation of the phases ($0{\cdot}45$ or $0{\cdot}001\,\mu m$). As can be seen from Fig. 7, the results are very similar which means that colloids are not present in

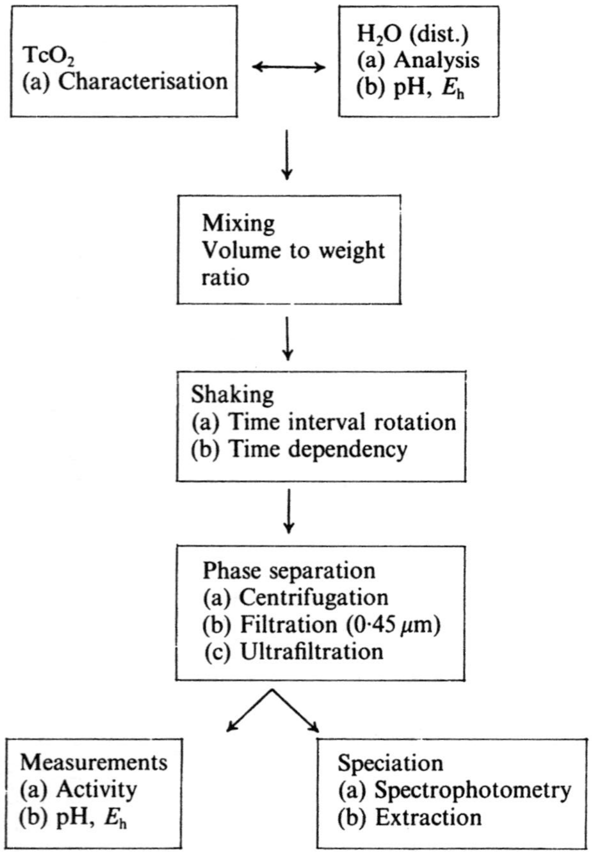

FIG. 4. Flow sheet of the solubility experiments.

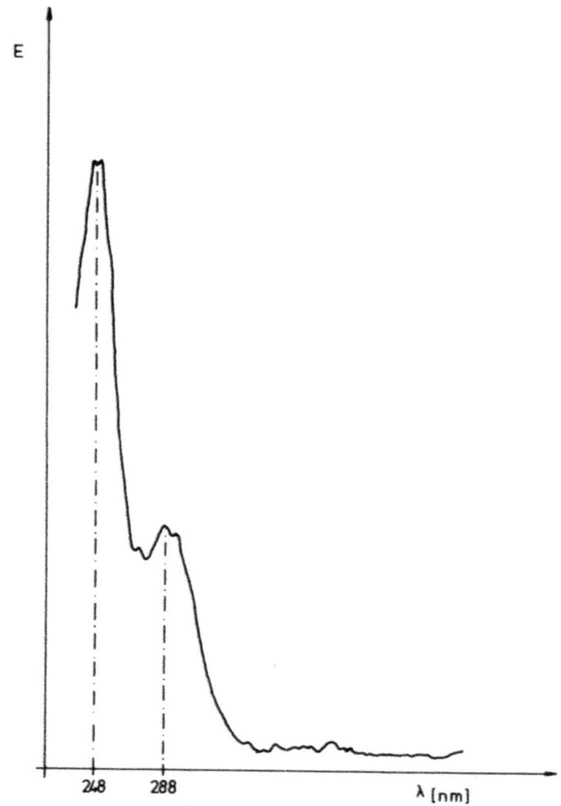

FIG. 5. UV spectrum of TcO_4.

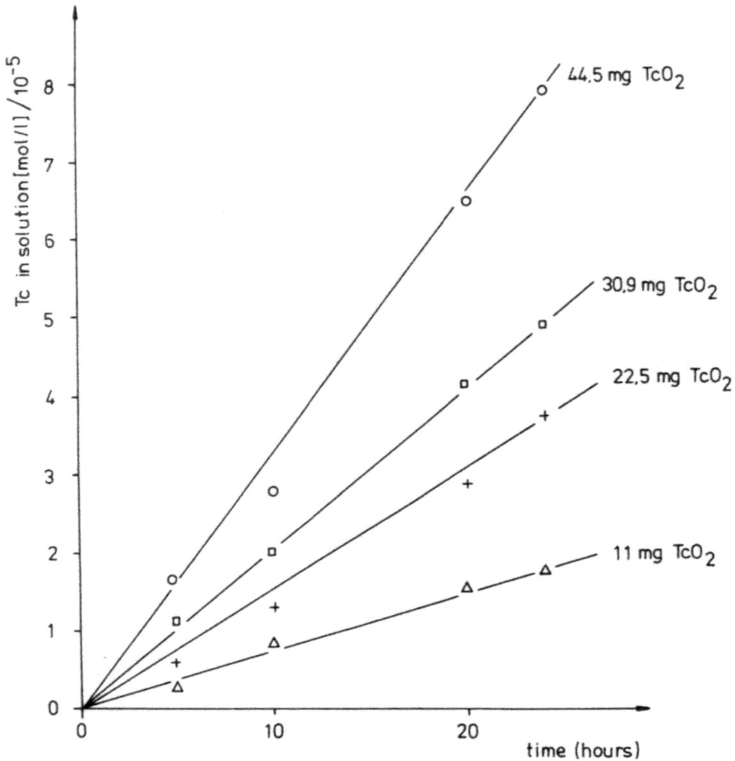

FIG. 6. Product A/H_2O dist. as f(t) (aerobic) (volume = 5 ml; pH = 5·2; E_h = 213 mV).

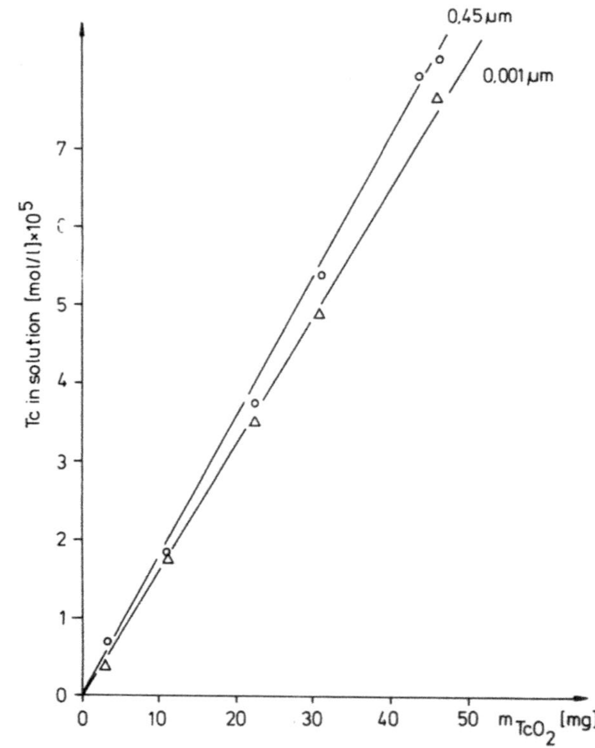

FIG. 7. Product A/H_2O (aerobic) (shaking time = 1 day).

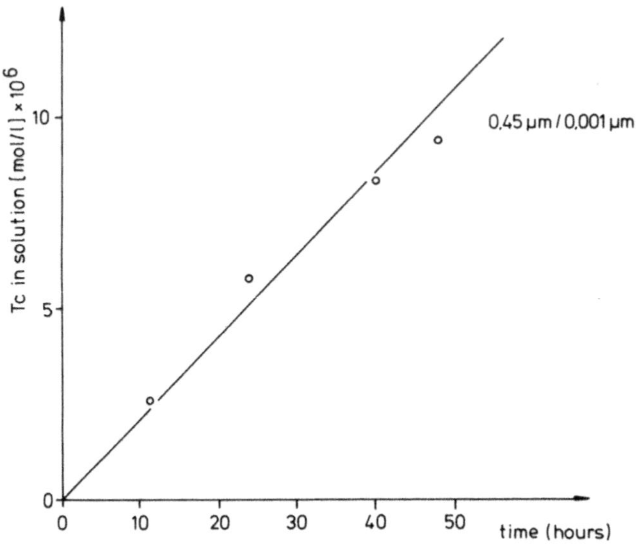

FIG. 8. Product B/H_2O as f(t) (aerobic) (pH = 5·2; E_h = 213 mV; V = 5 ml; m_{TcO_2} = 4 mg).

measurable concentrations. The same was found for product B (Fig. 8): filtration through a filter with a pore size of 0·45 μm and 0·001 μm gave the same values.

6. CONCLUSIONS

1. Under the given conditions Tc is dissolved as TcO_4^-. This was proved by extraction and by spectrophotometry.
2. The rate of dissolution is apparently equal to the rate of oxidation and depends on the surface area and the mass.
3. Colloidal Tc species are not formed under the conditions of the experiments, as was proved by ultrafiltration.

Solubility experiments and speciation studies under anaerobic conditions are in progress.

REFERENCES

1. TORSTENFELDT, B. *et al.* (1981). *Technetium in the Geologic Environment—a Literature Survey*, Programraadet foer Radioaktivt Axfall, Stockholm (Sweden), Sul, 29 pp.

2. MÜLLER, O., WHITE, W. B. and ROY, R. (1964). *J. Inorg. Nucl. Chem.*, **26**, 2075.
3. SCHWOCHAU, K. (1964). *Angew. Chem.*, **76**, 9.
4. NELSON, C. M., BOYD, G. E. and SMITH, W. T. (1954). *J. Am. Chem. Soc.*, **76**, 348.
5. SUNDERHAGEN, E. (1979). *Int. J. appl. Radiat. Isot.*, **30**, 739.
6. STEIGMANN, J., MEINKEN, G. and RICHARDS, P. (1975). *Int. J. appl. Radiat. Isot.*, **26**, 601.
7. SPITZYN, V. I., KUZINA, A. F., OBLOVA, A. A., GLINKINA, M. T. and STEPOVAYA, L. I. (1976). *J. Radiat. Chem.*, **30**, 561.
8. KELLER, C. and UANELLAHOPULAS, B. (1965). *J. Inorg. Nucl. Chem.*, **27**, 787.

16

Relevance of the Study of Technetium Accumulation and Speciation in Plants

G. M. DESMET

Commission of the European Communities, Radiation Protection Programme, Rue de la Loi 200, B-1049 Brussels, Belgium

Radiation protectionists dealing with the accumulation and behaviour of a defined radionuclide in one compartment of the biosphere should not forbear to examine and to take into account what is going on in its preceding and its next ecological compartment.

One could wonder whether it is necessary and even indispensable to study in depth, for example, the chemical transformations which technetium might undergo when having been absorbed and subjected to the biochemical machinery of a growing plant.

When involved in the estimations of the potential danger for whole populations, one could merely be worried by what happens from a quantitative point of view once technetium has been transferred from one link of the food chain to another. When, however, one is trying to assess what might happen to an individual, a more subtle approach on a more qualitative basis should be followed.

Basically, the sources of technetium released from nuclear power installations are TcF_6 and TcO_4^- eventually [1]. When pertechnetate enters the biosphere and possibly the human food chain, one of its first steps will be the penetration of soils of all kinds, where it is reduced to various degrees [2, 3]. Plants are the next step in the terrestrial ecosystem and its concomitant food chain. One could wonder whether, how and how much technetium will become absorbed, and equally one wants to know what happens to this technetium when it is getting involved in plant metabolism. Why, on the other hand, should one get so excited about all this; can one not just be satisfied with straightforward measurements of concentration factors? To answer this it is necessary to investigate the very complex

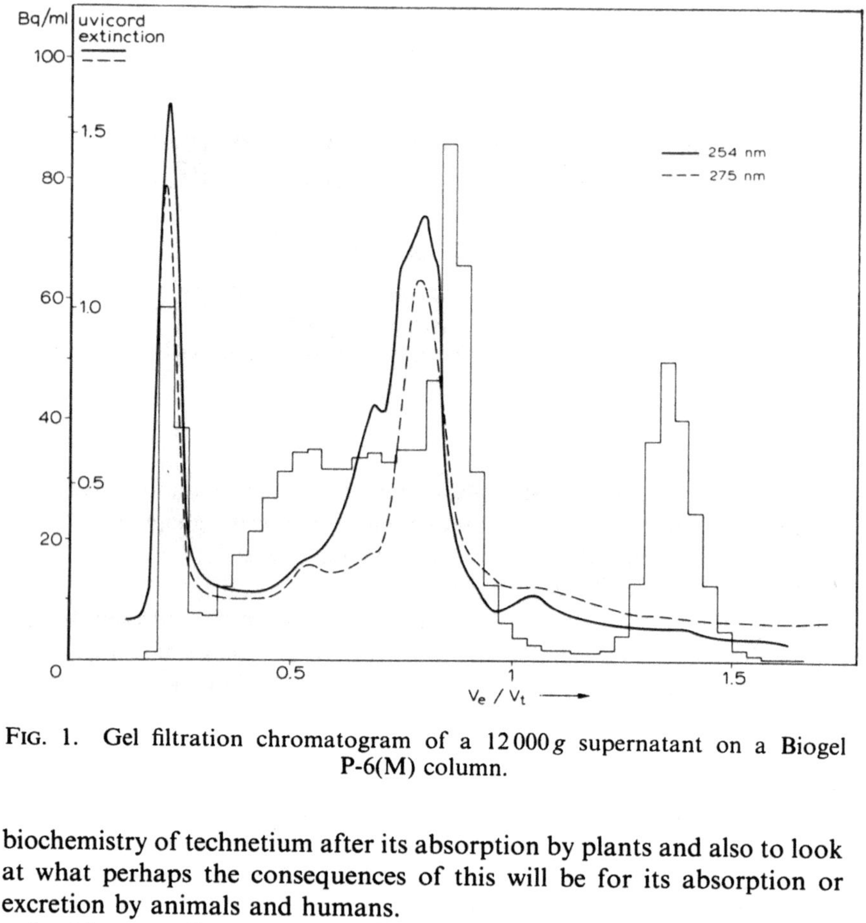

FIG. 1. Gel filtration chromatogram of a 12 000 *g* supernatant on a Biogel P-6(M) column.

biochemistry of technetium after its absorption by plants and also to look at what perhaps the consequences of this will be for its absorption or excretion by animals and humans.

In this paper, I want to deal primarily with cases where TcO_4^- was the chemical species which was given to spinach plants growing on nutrient solutions of well-known composition [4]. Besides a few initial measurements of concentration factors, attention was paid particularly to its chemical form in the plants. The first problem was to find out whether or not the initial pertechnetate was metabolised. Therefore leaves of plants grown during periods of various lengths were harvested and extracted with a mild buffer and treated further, as described elsewhere [5]. The supernatant of a 12 000 *g* centrifugate was placed on top of columns which contained different gels, preferably of high mechanical strength [5]. Four groups of labelled compounds could be distinguished on the Biogel

P-6 chromatogram (Fig. 1). The first one, of more than 6000 g mol^{-1}, moving with the void volume ($\sim$12%) clearly separated from another set of compounds ($\sim$42%). A third group with a low molecular mass eluted at a volume close to the V_{total} (about 27%). The remainder (about 19%) eluted behind the V_t. A Fractogel TSK-HW 40 column is particularly suitable for the fractionation of low molecular weight substances as it has a fractionation range of between 10^4 and 10^2 g mol^{-1}. The last compound, which was already revealed on the Biogel P-6 column, was even more retained, with an elution ratio of 8·3 ($V_e/V_t = 8{\cdot}3$) (Fig. 2). Suspecting that this could possibly be TcO_4^- , a number of co-chromatographic separations were run on different gels. The rationale behind this is to use columns containing different gels and of different lengths, and to define whether the elution volume of spike pertechnetate-99 always coincided with the last eluted peak of the plant, which is truly confirmed in

Table I
Relative elution volume (V_e/V_t) for TcO_4^- and for the last compound of the radiochromatogram of a 12 000 *g* supernatant

Column	*φ (cm)*	*L (cm)*	V_e/V_t
Biogel P-6(M)	2·6	94·5	1·6
	1·6	61·5	1·4
Biogel P-2(F)	2·6	90·5	3·3
	1·6	61·0	2·7
	1·0	6·0	2·4
TSK-HW 40(F)	1·0	38·5	8·3
	1·0	4·0	4·1

It is concluded from the table that TcO_4^- is present in homogenised spinach leaves.

Table I. Hence, combining what has been said hitherto, the pertechnetate clearly is reduced into a number, previously uncounted, of Tc–bio-organic complexes.

It is important to know how the reduction occurs since we are interested both in the chemical and physical characteristics of these compounds for reasons which I shall discuss now. Therefore, a brief outline of the use of technetium-99m as a radiopharmacon is given. Technetium-99m has been utilised in medical diagnostics to trace tumours or other illnesses. Two main classes of technetium-labelled radiopharmaca exist: firstly,

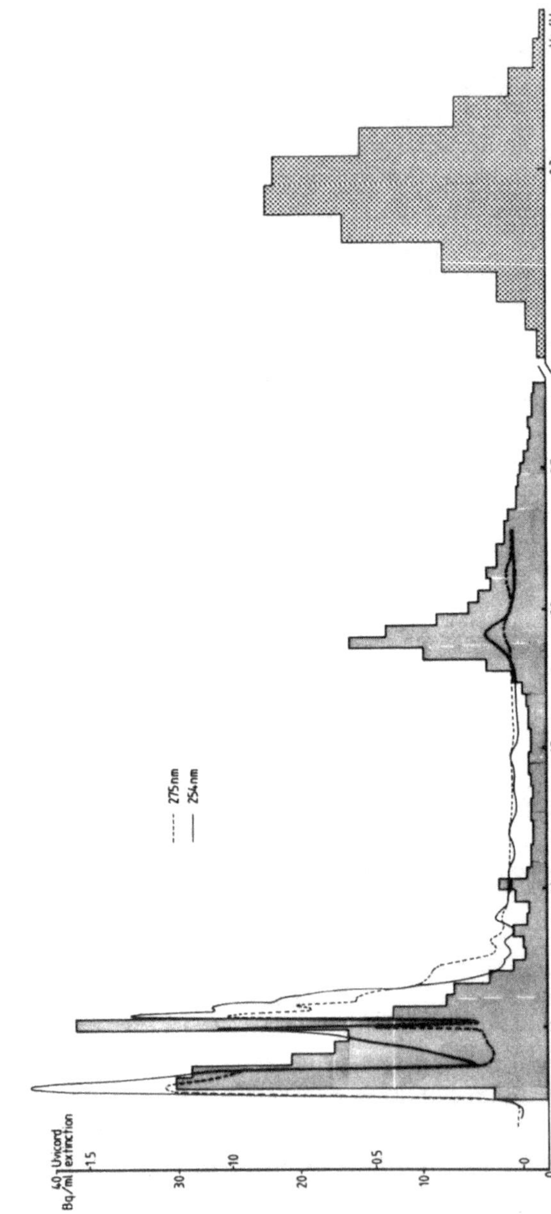

FIG. 2. Gel filtration chromatogram of a 12 000 *g* supernatant on a Fractogel TSK-HW 40(F) column (*φ*, 1·0 cm; *L*, 38·5 cm).

technetium-tagged radiopharmaceuticals where the organic part of the molecule determines the actual spot to be traced, and secondly, technetium-essential radiopharmaceuticals where the entire molecule is necessary to reach its tracing spot [6]. There is a definite preference for certain organs, dependent on the chemical nature of the Tc compounds, e.g. there are bone tracing agents, kidney function agents, kidney structure agents, infarct avid agents and some are hepatobiliary agents. Looking into even more detail, to a biochemist or a plant physiologist a few appear very familiar, like, for example, Tc-citrate, Tc-gluconate, Tc-pyrophosphate and especially Tc-dihydrothioctic acid. Dihydrothioctic acid is the reduced form of lipoic acid which contains a reactive disulphide group and which plays a key role in the production of coenzyme A, one of the most important enzymatic cofactors in living organisms [7].

Two main questions after all remain to be solved. First, what are the mechanisms laying behind the reduction of TcO_4^- in plants and what kind of compounds will be formed? In other words, what is the nature of the electron donor or donors and what is the nature of the ligands in vivo? Secondly, do these compounds have a physical nature that will make them behave like those radiopharmaca and, based on the knowledge of their physical nature, will their behaviour be predictable? As stated by McAfee [8], four of the most important factors affecting the biodistribution of drugs are

(1) molecular size,
(2) lipid solubility,
(3) plasma protein binding,
(4) specific binding sites.

To assess the first problem, I shall make a quick run-down of only a few of these biochemical processes, which presumably could play a role in TcO_4^- reduction. For the in vitro reduction of TcO_4^- into a fairly stable complex, a few conditions mostly are to be fulfilled. A strong reductant, such as, for example, tin, with a standard reduction potential of -136 mV is necessary [9]. Also, the presence of a ligand containing sulphur, oxygen or nitrogen is highly recommended, at least when the reduced Tc is to be kept in solution. These reactions are much favoured by an acidic pH [10]. The cysteine reaction with TcO_4^- is rather exceptional since it occurs spontaneously even at neutral pH, resulting in some three Tc-cysteine complexes of most probably different oxidation states [11]. The question is whether similar situations exist in plants. A few possibilities may be given. In illuminated chloroplasts a very low reduction potential of

-600 mV arises. Also there are ferredoxin molecules containing iron–sulphur clusters in their photosynthetic chains. Close to this, a ferredoxin–thioredoxin system with active dithiol groups regulates light-mediated enzymes [7]. This ferredoxin–thioredoxin has also been described as a precursor of ribonucleotide reductase, which converts ribose moieties of ribonucleoside into desoxyribonucleoside [12, 13], being the building blocks of ATP and DNA. Its mechanism demonstrates the way in which an iron II/III transition is involved, together with possible redox changes of dithiol groups. Such biochemical configurations, with their appropriate redox potentials and chemical ligands, may be liable to reduce pertechnetate in plants. All enzymatic candidates for the reduction of TcO_4^- could possibly be among the high molecular mass compounds we find on our gel chromatograms. Nevertheless, they are surely not the only candidates for reducing and binding Tc. Chromatograms run on Fractogel TSK-HW 40 columns revealed the existence of several very small molecules of by now undefined nature. Emphasis lays now on the identification of these small molecules and on the understanding of how they are produced.

Conclusively, a few statements can be made:

—Pertechnetate absorbed by plants is only partially reduced.
—The number of Tc binding compounds is not very sure yet, nor are their physical and chemical properties. The bound Tc sticks also to molecules with a molecular mass below 6000 g mol^{-1}.

For radiation protection purposes it is particularly important to get as much as possible acquainted with their chemical and physical characteristics in order to satisfy the above criteria of McAfee, indispensable for making reliable predictions of their biodistribution.

REFERENCES

1. Wildung, R. E., McFadden, K. M. and Garland, T. R. (1979). Technetium sources and behaviour in the environment. *J. Env. Qual.*, **8**, 156.
2. Saas, A., Grauby, A., Denardi, J. L. and Quinault, J. M. (1979). Comportement du technetium-99 et du molybdène stable dans un sol brun calcaire. *Radioprot.*, **14**, 241.
3. Landa, R. E., Thorvig, L. H. and Gast, R. G. (1979). Effect of selective dissolution, electrolyte aeration and sterilisation on technetium-99 sorption by soils. *J. Env. Qual.*, **6**, 181.
4. Desmet, G. M. and Dirkse, W. G. (1984). Action of technetium on the growth, the manganese, zinc and iron concentrations and the fluorescence induction in young and old leaves of spinach plants. *Envir. Exp. Bot.*, **24**, 45.

5. Lembrechts, J. F. and Desmet, G. M. Accumulation of Tc–bio-organic complexes in spinach plants in relation to growth (this volume, pp. 295–300).
6. Burns, H. D., Worley, P., Wagner, H. N. Jr, Marsilli, L. and Risch, V. (1979). Design of technetium radiopharmaceuticals. In: *The Chemistry of Radiopharmaceuticals* (Ed. N. D. Heindel *et al.*), Masson Publishing USA Inc.
7. Goodwin, T. W. and Mercer, E. I. (1983). *Introduction to Plant Biochemistry*, Oxford, Pergamon Press.
8. McAfee, J. G. (1975). Radioactive diagnostic agents: current problems and limitations. In: *Radiopharmaceuticals* (Ed. G. Subramanian *et al.*), New York, The Society of Nuclear Medicine Inc.
9. Steigman, J., Meinken, G. and Richards, P. (1975). The reduction of pertechnetate by stannous chloride. *Int. J. appl. Radiat. Isot.*, **26**, 601–9.
10. Münse, R., Bechstein, J., Grossmann, B., Hartman, E., Hoffman, J., Johannsen, B., Mocker, D., Noll, B., Seifert, S., Spies, H. and Syhre, R. (1983). Untersuchungen zum chemischen Zustand von Modellverbindungen für Technetium-Radiopharmaka. Fartschrittbericht Zentralinstitut für Kernforschung Rossendorf bei Dresden, Akademie der Wissenschaften der DDR, pp. 1–153.
11. Spies, H., Johannsen, B., Münse, R. and Unverferth, K. (1978). Die Reaktion von Pertechnetat mit Cystein. *Z. Anorg. allg. Chem.*, **447**, 215–20.
12. Thelander, L. (1973). Physico-chemical characterisation of ribonucleoside diphosphate reductase from *Escherichia coli*. *J. Biol. Chem.*, **248**, 4591.
13. Thelander, L., Larsson, B., Hobbs, J. and Echstein, F. (1976). Active site of ribonucleoside diphosphate reductase from *Escherichia coli*. *J. Biol. Chem.*, **352**, 1398.

17

A Review of the Biological and Geochemical Behaviour of Technetium in the Marine Environment

T. M. BEASLEY* and H. V. LORZ

College of Oceanography, Oregon State University, Corvallis, Oregon 97331, USA

1. INTRODUCTION

In 1971, the National Academy of Sciences published the review *Radioactivity in the Marine Environment* [1] which summarised current understanding of the behaviour of radionuclides in the ocean. Notably absent, in the tabulations of organism concentration factors and seawater activity levels, was any reference to ^{99}Tc. The paucity of data, at that time, can most reasonably be attributed to the low total activities expected to be present in the ocean as a consequence of the testing of nuclear devices. For example, Beasley *et al.* [2] estimated that 160 terabecquerels (TBq) of ^{99}Tc ($\sim$4300 curies) would have been released into the environment between 1945 and 1963 based on a total of 194-megaton equivalents of fission. By contrast, some 78×10^4 TBq (21 megacuries) of ^{90}Sr and $1{\cdot}1 \times 10^4$ TBq (0·3 megacuries) of $^{239,240}Pu$ were produced in this same time period.

In the years directly following maximum fallout input (1964–67), ^{90}Sr activities in the mixed surface layer of ocean waters in the 40–60°N latitude band showed maximum values ranging between 1·7 and 3·4 Bq/100 kg seawater ($\sim$100–200 dpm/100 kg seawater) [3]. Assuming that ^{99}Tc behaved similarly to ^{90}Sr, i.e. essentially conservatively, expected ^{99}Tc activities in these same waters would have been only 0·35–0·70 mBq/100 kg seawater (0·02–0·04 dpm/100 kg seawater). Coupled with the relatively low energy of the β^- particle emitted in the decay of ^{99}Tc ($E_{max} = 0{\cdot}292$

* Present address: Env. Res. Div., Argonne National Laboratory, 9700 S. Cass Ave., Argonne, Illinois 60439, USA.

MeV), such activity levels present formidable analytical challenges for their precise determination.

However, in the mid-1970s, interest concerning Tc in the environment was stimulated by several developments. First, Denham *et al.* [4] estimated the potential dose to man which could result following the release of high-level wastes from a hypothetical arid storage site. Five radionuclides were shown to be of particular significance. They were ^{99}Tc, ^{3}H, ^{129}I, ^{237}Np and ^{79}Se. Technetium was assumed to be poorly sorbed to oxygenated soils, an assumption which was confirmed by Routsen *et al.* [5] and discussed in more detail recently by Bondietti and Francis [6]. As such, Tc might be expected to readily migrate through oxygenated soils to groundwaters and into streams and rivers. Water from these various sources which are used for agricultural purposes would ultimately bring Tc to man by ingestion of food crops or through potable water supplies. Subsequently, Wildung *et al.* [7] demonstrated that accumulation of Tc by plants was an efficient process, lending further support to the potential radiological relevance of Tc in the environment.

Secondly, two important papers confirmed the present-day dissemination of ^{99}Tc into aquatic ecosystems as part of the nuclear fuel cycle. Prompted by an inability to balance gross β^- measurements in selected marine samples collected from the Irish Sea with specific radiochemical isolation techniques for known β^--emitting radionuclides, Dutton and Ibbett [8] verified the presence of ^{99}Tc in these materials; the isotope was shown to concentrate to surprisingly high levels in the seaweed *Fucus vesiculosus* (16 kBq/kg wet wt). Subsequently, Blaylock and Witherspoon [9] demonstrated that ^{99}Tc was released in waste effluents associated with uranium enrichment facilities in which irradiated uranium fuels that had undergone preliminary treatment to remove fission products are further processed to recover unspent ^{235}U. Apparently, fission product removal is not completely effective in removing ^{99}Tc which is isolated during the gaseous diffusion process and subsequently appears in waste treatment facilities and in discharged effluents.

Thirdly, Spies [10] reported on the uptake, retention, distribution and loss of ^{95m}Tc in the red abalone *Haliotis rufescens*, in which equilibrium whole-body concentration factors ranged between 135 and 205 and which evidenced biological elimination half-lives of 60 days when calculated from uptake curves. Shortly thereafter, Gromov [11] published the first information on the bioavailability of Tc to phytoplankton; concentration factors of $7 \pm 3 \times 10^4$ were observed. These data, and those from the Irish Sea measurements, implied that uptake of Tc by marine biota might possibly be of such magnitude as to substantially influence the vertical transport of Tc within the ocean and perhaps constitute an important transport pathway back to man following its introduction to marine systems.

Finally, from the mid-1970s to today, studies have been conducted to determine the technical feasibility of using sites in selected areas of the ocean seafloor as a repository for high-level radioactive wastes. Understanding the biogeochemical behaviour of the very long-lived radionuclides contained in such wastes is central to predicting the consequences of their migration from implanted canisters to the surrounding sediment and conceivably back to the sediment–water interface. Technetium-99 was recognised early on as an important component of such waste [12]. It is not surprising, then, that more information has accrued on the environmental behaviour of Tc in the last half decade than in the preceding two. Here, we summarise the general progress which has been made in understanding Tc biogeochemistry in the marine environment. The review is not exhaustive; rather it is intended to point out the 'first-order' conclusions which have been reached to date at the expense of describing individual investigations in depth. Bibliographies within the references cited provide an extensive source of such information which can be consulted when specific details are required.

2. GEOCHEMICAL BEHAVIOUR OF Tc IN MARINE SYSTEMS

2.1. Chemical forms

Figure 1 illustrates the calculated stability relationships expected for the various chemical forms of Tc at different environmental values of E_h and pH. In oxygenated seawater where E_h and pH are assigned values of 0·40 V and 8·0, respectively [13], the most stable form of Tc would be as pertechnetate, the negatively charged oxyanion. As E_h decreases, reduced species can be formed which could include $TcO_{2(s)}$. It is also possible that in seawater intermediate species such as $TcO(OH)_2$ and $TcCl_6^{-2}$ would be present [14]. It should be emphasised, however, that stability diagrams such as Fig. 1 are derived from thermodynamic data and provide no information on the rates of transformation between species.

Fortunately, a simple diagnostic test does exist for distinguishing TcO_4^- from other reduced forms in aqueous solutions and the rates at which

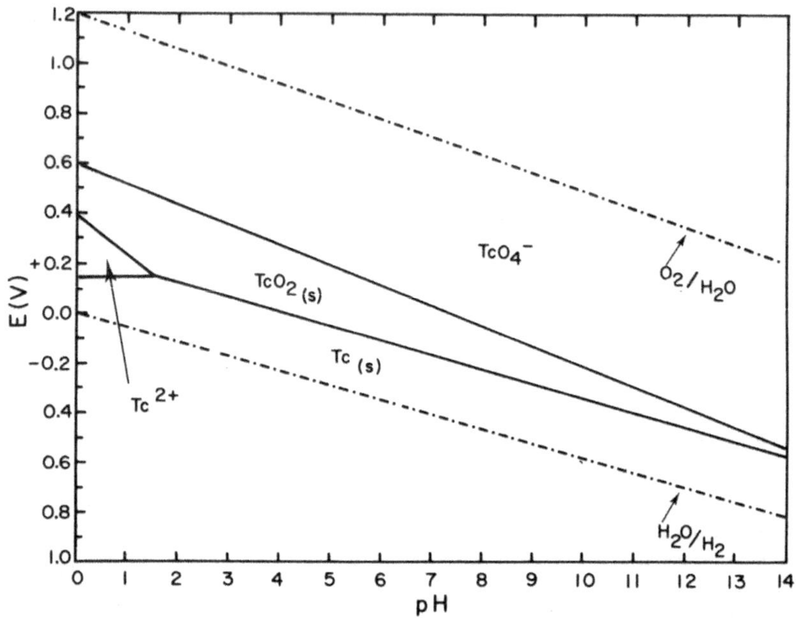

FIG. 1. Stability diagram of the various chemical forms of Tc under different environmental conditions of E_h and pH. For well-oxygenated seawater, $E_h = 0{\cdot}4$ V and pH = 8; predicted form of Tc under these conditions is TcO_4^-.

oxidation–reduction reactions occur. The test is based on the fact that, in alkaline solutions, reduced forms of Tc are carried on $Fe(OH)_3$ precipitates while TcO_4^- is not [14]. Figure 2 shows the rate at which pertechnetate, reduced to lower valence states by hydroxylamine hydrochloride, is reoxidised in fresh seawater at room temperature (25°C). More than 90% of the reduced species reappears as pertechnetate within 24 h. Similar results are observed when pertechnetate is reduced with hydrazine sulphate [15]. A fraction of the reduced species (~10%) can remain in lower oxidation states for appreciable periods of time (~14 days [15]), but these data do confirm that the principal form of Tc in oxygenated seawater, as expected, is TcO_4^-. The chemical form of Tc expected to be present in low-level liquid effluents entering the environment from fuel reprocessing plants would be TcO_4^- [16] and indirect evidence from measurements made in the Irish Sea support this contention [12]; ^{99}Tc appears to be far less susceptible to removal from the water column than does ^{137}Cs.

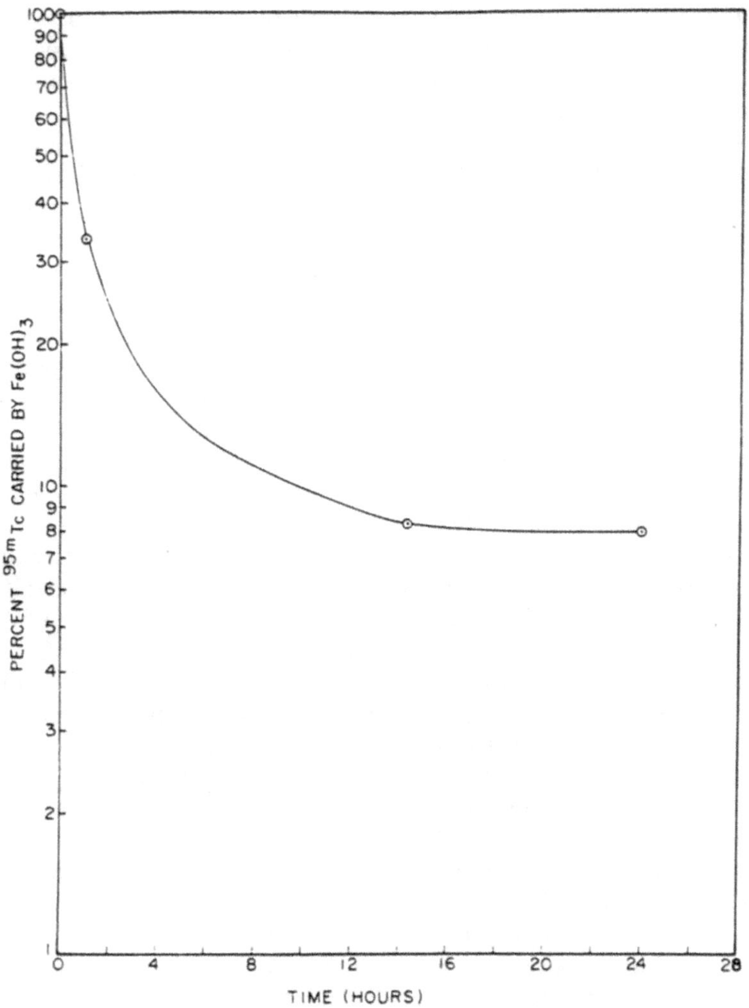

FIG. 2. Percentage of reduced ^{95m}Tc remaining in oxygenated seawater as a function of time. Reduced ^{95m}Tc added to seawater following reduction with hydroxylamine hydrochloride. As reduced forms become oxidised to $^{95m}TcO_4^-$, less of the isotope is carried by ferric hydroxide precipitate.

2.2. Adsorption of Tc to inorganic particulate material

Most of the experimental studies to date dealing with the interaction of different chemical forms of Tc with inorganic particulate matter have employed the classical approach of determining a distribution coefficient, K_D, for these forms between the aqueous and solid phases where this distribution coefficient is expressed as $K_D = \text{activity g}^{-1}\ \text{solid} \div \text{activity g}^{-1}$ solution. As might be expected from its behaviour with $Fe(OH)_3$ in alkaline media, TcO_4^- is not strongly sorbed to the mixed mineral facies of marine sediments low in organic matter. Masson *et al.* [17] conducted a series of experiments in which both ^{99}Tc and ^{95m}Tc, as pertechnetate, were amended to marine sediments with differing concentrations of organic matter, including petroleum hydrocarbon residues. Distribution coefficients for sediments low in organic matter content were essentially unmeasurable. Concurrent studies by Fowler *et al.* [18] and unpublished data obtained in our laboratory have confirmed that with such sediments, K_Ds for TcO_4^- rarely exceed values of 1–4. Reduced forms of ^{95m}Tc amended to these sediment types which are then exposed to oxygenated seawater display rapid loss of Tc, arguing again for the efficient oxidation of the reduced species to TcO_4^- [18].

In contrast, K_Ds for TcO_4^- on sediments high in organic matter can be appreciable. Masson *et al.* [17] have recorded K_Ds as high as 1500 in sediments containing 7% organic matter following 50 days of equilibration in seawater. Interestingly, over the first 8 days of contact, K_Ds were extremely low but tended to increase sharply with time. The authors postulated that bacteria present in the sediments (as a consequence of the organic matter substrate) most probably reduced the TcO_4^- to lower valence states which were then adsorbed to the sediments.

Using NE Atlantic deep-sea sediments of low (0·3%) organic matter content, Fowler *et al.* [18] also observed initial low K_Ds (~ 1) which increased in magnitude with time. The induction period for increase (~ 10 days) closely parallelled those observed by Masson *et al.* Of particular note, however, is that this behaviour was observed whether the Tc was amended to the sediments as TcO_4^- or as reduced (+4) Tc. Loss of both forms of Tc from the sediment when placed in unlabelled oxygenated seawater was rapid and quantitatively identical; transfer coefficients of Tc to organisms placed into the sediment were inconsequential. It would be very instructive to have data on the rates of Tc loss from high organic matter sediments initially labelled with pertechnetate to determine whether desorption rates are dependent upon the diagenetic alteration of the organic matter with time.

2.3. Diffusion of Tc within the sediment column

As yet no field data have been reported which relate to the post-depositional behaviour of Tc in sediment cores raised either from the open ocean or from coastal low-level waste discharge sites. This is not surprising considering the levels of activity to be anticipated in each regime as a consequence of the low sediment–water distribution coefficients discussed above. However, laboratory experiments [19] using smectite-rich deep-sea sediments (with low organic matter content) in contact with atmospheric oxygen show diffusion coefficients of the order of 3×10^{-6} cm^2/s, a value comparable to those predicted for the sulphate ion (temperature range = 0–25°C [20]). Under anoxic conditions in these sediments, TcO_4^- appears to undergo slow reduction with a concomitant decrease in the diffusion coefficient to a value near 1×10^{-8} cm^2/s. These data suggest that Tc in sediments found below the horizon at which sulphate reduction occurs would remain relatively immobile. The implications of these findings are obvious as they relate to the implacement of high-level wastes in the seabed at depths of tens of metres below the sediment–water interface.

Thus, evidence to date suggests that binding of pertechnetate with inorganic particulate matter in oxygenated seawater is slight. At the sediment–water interface of organic-rich sediments and particularly in anoxic basins some TcO_4^- removal to underlying sediments could occur, especially if bottom water transport is sufficient to produce intense nepheloid layers where particle populations are high. The quantitative aspects of such removal processes, however, cannot be generalised and will require site-specific studies to better clarify their importance in reducing water column inventories.

3. BIOLOGICAL BEHAVIOUR OF Tc IN MARINE ORGANISMS

3.1. Phytoplankton

As previously mentioned, Gromov [11] first reported on the uptake of Tc by marine phytoplankton. Using a mixed population collected from the upwelling region off the coast of Peru and seawater inoculated with pertechnetate, Gromov recorded wet weight concentration factors (CF = activity g^{-1} biota ÷ activity g^{-1} seawater) of $7 \pm 3 \times 10^4$; however, a monocellular culture of the green algae *Platymonas viridis* treated in an identical manner as the mixed phytoplankton culture showed virtually no

$^{99}TcO_4^-$ uptake. To date further experiments by other investigators have failed to confirm that phytoplankton show any marked affinity for Tc either as TcO_4^- or in reduced forms. The most comprehensive of these investigations is that recently reported by Fisher [21], in which monocultures of seven species of marine phytoplankton were incubated with Tc in the +7 and +4 oxidation states for 4 days. Only one species evidenced a concentration factor near 20 (*Heterocapsa pygmaea*; CF = 17); the remainder displayed concentration factors <10. Fowler *et al.* [15] observed virtually no TcO_4^- uptake by a population of mixed flagellates cultured from the Mediterranean Sea and Beasley *et al.* [2] demonstrated no TcO_4^- uptake by the phytoflagellate *Pseudoisochrysis paradoxa* over a two-day labelling period with the cells in log phase growth. The preponderance of evidence suggests, therefore, that uptake of Tc by oceanic phytoplankton is not a significant process.

3.2. Macroalgae

In contrast to phytoplankton, certain marine macroalgae do concentrate Tc to a significant degree [8, 12, 17, 22–25]. However, this bioaccumulation has been observed to apply almost exclusively to brown algae where laboratory concentration factors range between 250 and 2500. The single exception to this generalisation to date is the low concentration factor measured in the brown algae *Colpomenia sinuosa* reported by Topcuoglu and Fowler [25] (CF = 14). All species of red and green algae investigated show little affinity for Tc either in the +7 or +4 valence state.

Fixation of Tc by the brown algae appears to be an active process rather than one of simple, passive adsorption [25]. Uptake is both temperature and light dependent, with rapidly growing parts of the plant showing the highest activities per unit weight. Elimination of Tc from labelled algae does differ markedly between species but is generally characterised by one or more initial rapid loss components ($T_{1/2(b)} = 1$–3 days) followed by a single, slower loss component of significantly longer duration ($T_{1/2(b)} = 16$–196 days). In general, however, these elimination kinetics are not unlike those observed in most biological species, as discussed below.

3.3. Zooplankton

There are few reported data on the biokinetic behaviour of Tc in either small or large zooplankton. However, the information in hand suggests that live zooplankton may not show a marked affinity for Tc either in the +7 or +4 valence state. Schulte *et al.* [26] showed that uptake of

$^{95m}TcO_4^-$ by the brine shrimp *Artemia salina* reaches an equilibrium concentration factors from water of only 2·7 after an exposure period of 2 weeks, and Fowler *et al.* [27] observed a Tc concentration factor (amended to water as TcO_4^-) near 1 for the euphausiid *Meganyctiphanes norvegica* after an exposure period of 6 days. Interestingly, however, Fowler *et al.* [27] did observe a pronounced increase in the uptake of TcO_4^- by dead euphausiids (CF = 50) over a 7-day period. The authors suggest that bacteria on the surfaces of the carcasses either directly accumulate TcO_4^- or reduce the higher valence state Tc to lower oxidation states which is then bound to the exoskeleton material. Bacterial mediation was confirmed by 'poisoning' seawater containing dead euphausiids and $^{95m}TcO_4^-$ with sodium azide, with the result that no activity of consequence was transferred from the seawater to the carcasses. However, the magnitude of the concentration factor observed in this experiment is not sufficiently high to suggest that cast moults and carcasses would constitute an appreciable pathway for moving Tc from surface waters to depth. Nevertheless, additional information would be desirable in this area so that approximate removal rates could be calculated based on organism abundance, moulting rate, moult sinking rates and the degree to which Tc is retained by this material over time.

3.4. Other crustaceans

Pronounced differences have been observed in the biokinetic behaviour of Tc in crustaceans, i.e. shrimp, crab, isopods, amphipods and lobsters. With the exception of lobster, all the species mentioned have whole-body concentration factors which range in value between 2 and 12 [2]; for lobster, values between 1000 and 1400 have been recorded. However, within all members of this group Tc appears to concentrate, principally, in the hepatopancreas and visceral organs of the animals. Particularly high concentration factors occur in the lobster hepatopancreas (CF = $\sim 10^4$ [17, 28]).

Pentreath [22] observed significantly different biological elimination rates of ^{95m}Tc in the lobster (*Homarus gammarus*) depending upon whether the animal was labelled by water uptake or from ingestion of a single ration of labelled shrimp. In the first case, biological half-lives were long (211 and 339 days for two animals), while excretions from the ingestion experiment were significantly shorter ($T_{1/2(b)} = 66 \pm 8$ days; mean of five animals). The longer retention times from the water labelling experiment may result from the relatively high concentration factors observed by Pentreath [28] for the shell of the lobster (CF = $\sim 10^3$), which might be expected to display slow turnover rates. If correct, moulting of these animals would constitute a significant mechanism for reducing their body burdens of Tc where uptake is by the water pathway.

The benthic isopod *Anilocra mediterranea* studied by Fowler *et al.* [15] displayed biphasic loss of ^{95m}Tc following uptake from water (as TcO_4^-), with approximately 20% of the initial label being lost in the first 24 h. Thereafter, turnover was slow; the biological elimination half-life approached 300 days. Moulting of the animals did not occur during the experiment, nor were dissections performed to determine tissue distributions of the ^{95m}Tc. In a subsequent experiment using gammarid amphipods collected from the Mediterranean [27], ^{95m}Tc was amended to seawater in both the +7 and +4 valence states, and at equilibrium significant differences were observed in the uptake for these two chemical forms. For Tc +4 concentration factors reached values near 30, while for Tc +7 concentration factors were ~4 (similar to the value observed for *A. mediterranea*). However, during this experiment several animals did moult. In two amphipods exposed to Tc +7, approximately 30% of their ^{95m}Tc body burden was shown to be associated with the cast moult. In the case of one amphipod exposed to Tc +4, only 8% of the ^{95m}Tc was found in the moult.

Loss of ^{95m}Tc in the shrimp *Palaemon elegans* following uptake from labelled seawater (TcO_4^-) has been shown to be rapid and temperature dependent [29]. Depuration measured at both 10°C and 20°C followed biphasic kinetics. At the higher temperature, biological half-lives of 2·6 and 5·9 days were computed for the fast and slow components of loss, respectively. After 21 days in loss, more than 97% of the incorporated label had been lost. At 10°C, the biological half-lives for elimination for the same loss components were 8 and 25 days. Interestingly, moults contained only traces of ^{95m}Tc activity and were not a significant factor in the elimination rates observed.

The excretion of ^{95m}Tc by the shrimp *Lysmata seticaudata* following a single ingestion of brine shrimp, labelled by both Tc +7 and Tc +4 valence states, has been reported by Fowler *et al.* [15]. In both cases, elimination was again observed to be biphasic. In the case of Tc +7, 85% of the ingested label was excreted within the first 6 days, with the remainder being lost with a biological half-life near 40 days. For the Tc +4 experiment, only 30% of the initial label was excreted in the first 6 days following ingestion but the biological half-life of the slow loss component was reasonably fast ($T_{1/2(b)} = \sim 20$ days).

Assimilation of Tc by crustacea following ingestion of labelled food is reasonably high. In the case of lobster (*H. gammarus*) and crab (*Cancer pagarus*), approximately 80% of the initial activity is retained following a single feeding of labelled shrimp [28]. In the case of the shrimp themselves [15], assimilation has been shown to depend strongly on the valence state of Tc used to label the food. For *L. seticaudata*, approximately 26% assimilation was observed following the ingestion of brine shrimp labelled with $^{95m}TcO_4^-$. When fed brine shrimp labelled with reduced forms of Tc, the assimilation reached 80%. However, for euphausiids which were fed brine shrimp labelled with Tc +7 and +4, assimilation was virtually identical (40%), as was elimination [30]. The sharp differences observed between the shrimp and euphausiid in these experiments are as yet unexplained.

3.5. Molluscs

One of the more interesting findings of the biological behaviour of Tc in marine organisms has been the very low uptake of pertechnetate by bivalve molluscs. For three different species of mussel (*Mytilus edulis, Mytilus galloprovincialis* and *Mytilus californianus*), steady-state concentration factors for $^{95m}TcO_4^-$ (following uptake from labelled seawater) never exceeded a value of 2 [2, 15, 17]. The same low value applies for identical experiments conducted with the oyster *Crassostrea gigas* [2]. Technetium accumulation is highest in the visceral mass and lowest in the mantle and adductor tissue for both mussel and oyster. Shell concentration factors are comparable to whole-body concentration factors and dead mussel shell accumulates $^{95m}TcO_4^-$ to levels comparable to that of live animals [2].

Elimination of Tc by mussels and oysters is again biphasic; approximately 50% of the Tc accumulated by both species is eliminated within the first 5 days of depuration [2, 15]. The remaining 50% is eliminated more slowly, with biological elimination half-lives of between 100 and 150 days being common. No substantial difference is observed in the depuration rates under laboratory versus field conditions.

The low uptake of pertechnetate by bivalve molluscs appears to be unrelated to any ability of these invertebrates to regulate their Tc content. Figure 3 shows the relationship between ^{99}Tc uptake in *M. californianus* and Tc concentration in seawater [31]. Technetium accumulation in mussel soft parts is directly proportional to the Tc concentration in seawater over a ten-fold change in Tc concentration. If active regulation were invoked as a possible explanation for the low uptake of Tc observed in molluscs, non-linearity would most certainly have been observed in the

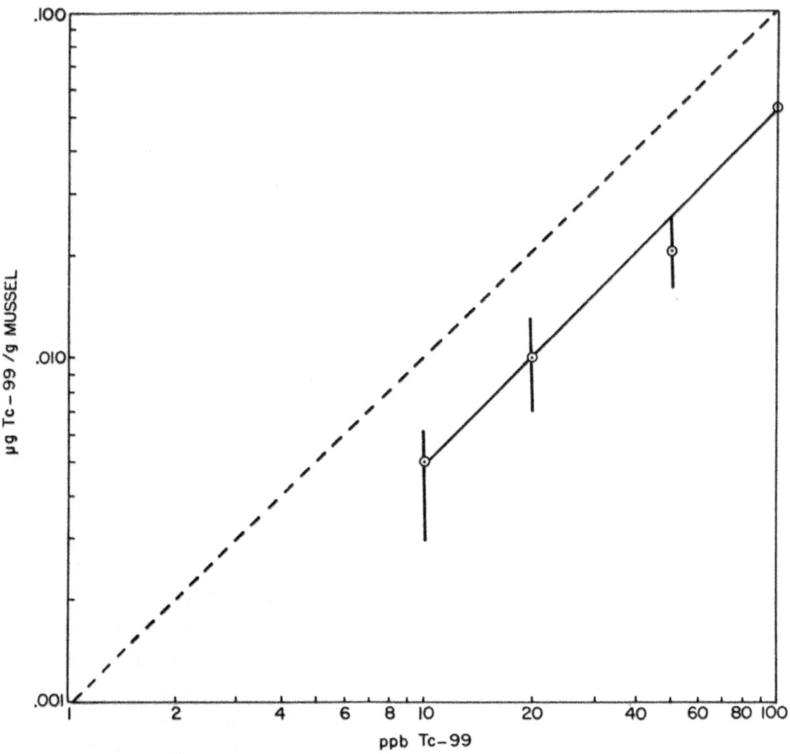

FIG. 3. Uptake of ^{99}Tc by the soft parts of *M. californianus*. Dashed line represents the theoretical slope of soft tissue concentrations *v.* seawater concentrations if mussels do not regulate their ^{99}Tc intake. Circles represent measured concentrations of ^{99}Tc in mussels as a function of ^{99}Tc in seawater. Exposure period: 8 days.

uptake experiments where absolute concentrations of Tc were very much higher. It should also be emphasised that the range of Tc concentrations used in this experiment greatly exceeds the atom concentrations of ^{95m}Tc used in any of the bivalve uptake experiments discussed above where this electron-capture decay radionuclide was used to label the molluscs.

Other members of the phylum Mollusca do not show the same low steady-state concentration factors as do mussels and oysters, a fact demonstrated by Spies' data [10] and that of Beasley *et al.* [24] for the archeogastropod *H. rufescens*. Here, steady-state concentration factors for water uptake (TcO_4^-) exceed 10^2. At equilibrium, highest activities were again observed for the visceral mass and the total radioactivity in the animals was partitioned nearly equally between soft tissue and shell.

In juvenile abalone, depuration of Tc following water uptake has been observed to be biphasic with the initial fast component showing a $T_{1/2(b)}$ equal to 7 days, while the $T_{1/2(b)}$ for the slower loss component equalled 58 days [24].

Uptake and loss kinetics of pertechnetate following a labelling by a food pathway have been done for juvenile *H. rufescens* by labelling the brown algae *Nereocystis luetkeana* with $^{95m}TcO_4^-$ and then feeding the algae to the abalone [24]. Technetium assimilation was high ($\sim$45%) and the biphasic loss observed gave biological elimination half-lives which were quite comparable to those observed in the direct water uptake experiments.

The gastropod *Aporrhais pespelicani* has been shown to display uptake and loss kinetics for both Tc +7 and Tc +4 which are similar to those observed for mussels [30]. Steady-state concentration factors for both valence states following uptake from water never exceeded 3; loss was biphasic with approximately 30% of the initial label being lost within the first 6 days. The slow loss component exhibited a $T_{1/2(b)}$ between 70 and 75 days for both valence state experiments.

Thus, as was observed within the phylum Arthropoda, distinct differences exist in the biokinetic behaviour of Tc for animals within the phylum Mollusca. However, one clear conclusion from experiments with molluscs is that the use of bivalves as sentinel organisms for monitoring ^{99}Tc in the marine environment, as has been proposed for a wide range of entities in seawater [32], would not be advisable.

3.6. Annelids

Within the phylum Annelida, polychaetes have shown a variation in their biokinetic behaviour of Tc not unlike that observed for molluscs and arthropods. For example, Masson *et al.* [17] exposed three species of polychaetes to $^{95m}TcO_4^-$ in seawater for a period of 40 days. Concentration factors at the end of this period ranged between 10 (*Arenicola marina*) and 1100 (*Perinereis cultrifera*); *Nereis diversicolor* obtained intermediate values of 100. Pentreath [22] observed concentration factors >500 for *Nereis* sp. exposed to ^{95m}Tc as pertechnetate in a period of only 26 days. Fowler *et al.* [15] exposed *N. diversicolor* to both Tc +7 and Tc +4 for 14 days and found that Tc +7 was taken up in preference to Tc +4 (CF = 343 *v.* 131, respectively). During depuration, those worms labelled with Tc +7 lost approximately 40% of their Tc body burden within 5 days; thereafter, loss followed a biological elimination half-life near 245 days. For those worms labelled with Tc +4, approximately 60% of the initial body burden was excreted by day 5 and elimination thereafter was nearly twice as fast as was observed for Tc +7 ($T_{1/2(b)} = 138$ days). Technetium was shown in these experiments to be unassociated, to any large degree, with the mucus coating of the organisms, arguing for true systemic deposition of the various valence states of Tc during the uptake phase.

3.7. Fishes

Few experiments have been reported to date which deal with uptake, assimilation and loss of Tc in marine fish. Pentreath [22] measured the equilibrium concentration factor for the plaice *Pleuronectes platessa* following accumulation of $^{95m}TcO_4^-$ from seawater. A mean value near 9 was observed. Excretion of the ^{95m}Tc (for the slow component of loss) gave biological half-lives of 46 days. When fed polychaetes (*Nereis* sp.) labelled by uptake of pertechnetate from seawater, assimilation was not pronounced (3–10%) and excretion was rapid; only about 20% of the ingested ^{95m}Tc was retained 3–5 days after feeding. The long-term loss component of that ^{95m}Tc which was retained gave a biological elimination half-life near 36 days. Using a similar experimental design [28], the thornback ray *Raja clavata* attained equilibrium concentration factors from water near 8 and a preliminary estimate of the slow loss component for excretion was about 30 days. Assimilation of ^{95m}Tc from labelled food was greater than that observed for the plaice (10–30%) and the rates of elimination following labelling by feeding were somewhat slower.

The distributions of ^{95m}Tc at the organ/tissue level in the two species of fish were remarkably similar for both the direct water uptake and the labelled food experiments. Highest activities were found in the kidneys, stomach and gut, while the lowest activities were found in muscle tissue.

These data are to be compared with the concentration factor of ~ 2 observed in the blenny (*Blennius pholis*) by Masson *et al.* [17] following nearly 40 days of exposure to $^{95m}TcO_4^-$ labelled seawater. Highest activities observed in these animals occurred in the liver, digestive tract and gills. The lowest activities were again recorded for the muscle tissue.

Schulte *et al.* [33] labelled the polychaete *Marphysa bellii* using ^{95m}Tc-pertechnetate in seawater and subsequently fed these worms to the seabass *Dicentrarchus labrax* over a period of 18 days; retention averaged $22 \pm 3\%$ (5 fish). Loss of ^{95m}Tc from the fish was biphasic with a $T_{1/2(b)}$ for the fast loss component equal to 1·1 days. Approximately 48% of the incorporated isotope was lost during this rapid elimination. The remainder of the label

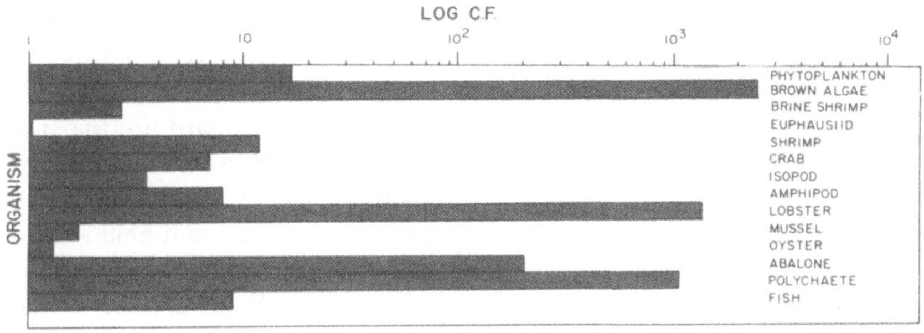

FIG. 4. Concentration factors for TcO_4^- in marine organisms as determined in laboratory experiments using $^{95m}TcO_4^-$. Only maximum values are shown where a range of values have been reported.

was eliminated with a $T_{1/2(b)}$ near 15 days. The majority of the residual radioactivity at dissection following 114 days in loss was found, once again, in the stomach, liver, kidney and gut.

4. SUMMARY OF Tc CONCENTRATION FACTORS IN MARINE ORGANISMS

Figure 4 is a graphical representation of the concentration factor data discussed above for the different marine organisms investigated to date. In those cases where a range of values has been observed for a particular species, only the maximum value has been plotted. All of the data in Fig. 4 relate to direct water uptake where the initial form of the Tc was pertechnetate. Moreover, all of these data were obtained from laboratory experiments. We are unaware of any concentration factor estimates which have been published for marine animals collected from contaminated sites where the activity levels of Tc in the water have been measured concurrently.

It is evident that likely sentinel organisms for monitoring changing levels of Tc in seawater would be relegated to brown macroalgae, certain species of polychaetes and perhaps lobster. Of these three target materials, brown macroalgae would seem the most logical candidate.

5. EXTRAPOLATION OF LABORATORY DATA TO CONTAMINATED FIELD CONDITIONS

As stated earlier, fallout levels of ^{99}Tc in the ocean are of such low magnitude as to preclude this source of Tc as a means of studying its environmental behaviour. Therefore, 'ground-truthing' of laboratory investigations to field measurements is most reasonably accomplished in areas where discharges of low-level radioactive wastes containing ^{99}Tc are presently occurring. The two obvious choices in this regard are the Irish Sea (Sellafield) and the NE Atlantic near La Hague. Support for this view comes by comparing the global fallout estimate of ^{99}Tc ($\sim$4300 Ci) with an annual discharge, in 1978, of some 5000 Ci of ^{99}Tc from Sellafield to the Irish Sea [12]. While such discharges are periodic, it would not be unreasonable to assume that inputs of ^{99}Tc from Sellafield alone may have exceeded global fallout inputs of this radionuclide to the ocean by as much as an order of magnitude over the last decade and a half.

Where such 'ground-truthing' has occurred, the laboratory studies and field measurements are in general agreement. Those species of organisms which have shown an affinity for $^{95m}TcO_4^-$ in seawater labelling experiments such as the brown algae, certain polychaetes and lobster exhibit easily measurable concentrations of ^{99}Tc, while those which have shown little ability to concentrate Tc from water (red and green algae, bivalve molluscs, etc.) have consistently lower concentrations of ^{99}Tc [12, 17, 22, 23]. Moreover, the distribution of ^{99}Tc in the organs and tissues of these animals follows quite closely the distributions observed in the laboratory experiments. It is, in fact, gratifying that such a close correspondence between laboratory and field studies has, so far, been demonstrated. As a logical next step, it will be important to characterise the chemical form of ^{99}Tc in the liquid wastes discharged to the ocean at sites such as Sellafield and La Hague and to observe any changes in these forms which occur in the water column during transport away from the discharge site. In addition, information on the turnover rate of ^{99}Tc in organisms collected from these areas would be invaluable for comparison with present laboratory data.

6. CONCLUSIONS

Substantial advances have been made in our understanding of Tc behaviour in marine systems over the last half decade. Inference from laboratory studies suggests that major sinks for Tc removal from the water column have yet to be identified. Pertechnetate does not display a pronounced affinity for inorganic sedimentary material, nor is it efficiently taken up by biota at trophic levels 1 and 11. Until otherwise demonstrated, a reasonable first assumption for describing TcO_4^- behaviour in seawater would be to classify this anion as conservative. If correct, the distribution of ^{99}Tc following its introduction into seawater (as pertechnetate) will be governed by the rates at which the receiving waters are mixed horizontally and vertically. Even though organic-rich sediments ($>7\%$ organic matter) might sequester Tc, they are quantitatively unimportant in an ocean-wide context, as are anoxic basins and trenches where pertechnetate might be reduced to lower valence states which are more readily adsorbed to inorganic particulate matter. As a first-order approximation, then, inventories of ^{99}Tc in the ocean should increase in proportion to input until a steady-state condition obtains. This would likely require times comparable to or greater than the physical half-life of ^{99}Tc itself [34].

It is now possible to identify major classes of marine organisms which actively concentrate Tc from seawater and those which show no such ability. Evidence is accumulating on the kinetics of Tc uptake and turnover in a variety of marine animals as well as the sites (organs and tissues) where Tc concentrates to high levels. Therefore, information now in hand can be used in preliminary assessments of the radiological importance to man of Tc discharges to coastal ocean ecosystems.

Finally, laboratory or field measurements have yet to identify a stable element analogue of Tc in seawater which could be used to further predict its behaviour in the ocean. Spies [10] noted that the Tc concentration factors he observed in abalone were substantially lower than those recorded for Mn in other molluscs even though both elements belong to group VIIB of the periodic table of the elements. He suggested that iodate might more closely mimic pertechnetate in this regard. However, Beasley *et al.* [2] observed no influence of either I^- or IO_3^- in altering concentration factors of pertechnetate in mussels exposed to labelled seawater. Stable rhenium would be a logical choice for comparison with technetium, however less is presently known about rhenium biogeochemistry in the ocean than is known for Tc! Identifying such a stable element analogue would permit its measurement in both deep-sea organisms and seawater to better predict the behaviour of Tc released to these abyssal environments. It is from such environments that information on Tc behaviour will be the most difficult to obtain.

ACKNOWLEDGEMENTS

Our research is sponsored by the Ecological Research Division, Health and Environmental Research, Office of Energy Research, US Department of Energy. We gratefully acknowledge this support.

REFERENCES

1. NAS (1971). *Radioactivity in the Marine Environment*, Washington, DC, National Academy of Sciences.
2. Beasley, T. M., Gonor, J. J. and Lorz, H. V. (1982). Technetium: uptake, organ distribution and loss in the mussel *Mytilus californianus* (Conrad) and the oyster *Crassostrea gigas* (Thunberg). *Mar. Envir. Res.*, **7**, 103–16.
3. Volchok, H. L., Bowen, V. T., Folsom, T. R., Broecker, W. S. and Bien, G. S. (1971). Oceanic distributions of radionuclides from nuclear explosions. In: *Radioactivity in the Marine Environment*, Washington, DC, National Academy of Sciences, pp. 42–89.
4. Denham, D. H., Baker, D. A., Soldat, J. K. and Corley, J. P. (1973). Radiological evaluations for advanced waste management studies. *Pacific Northwest Laboratory Report BNWL-1764*, Richland, WA, 79 pp.
5. Routsen, R. C., Jansen, G. and Robinson, A. V. (1977). Technetium-99, ^{237}Np and ^{241}Am sorption on two United States subsoils from differing weathering intensity areas. *Health Physics*, **33**, 311–17.
6. Bondietti, E. A. and Francis, C. W. (1979). Geological migration potentials of technetium-99 and neptunium-237. *Science*, **203**, 1337–40.
7. Wildung, R. E., Garland, R. T. and Cataldo, D. A. (1977). Accumulation of technetium by plants. *Health Physics*, **32**, 314–17.
8. Dutton, J. W. R. and Ibbett, R. D. (1973). The determination of technetium-99 in marine biological materials. In: *Symposium on the Determination of Radionuclides in Environmental and Biological Materials*, London, Central Electricity Generating Board, 7 pp.
9. Blaylock, B. G. and Witherspoon, J. P. (1975). Dose estimation and prediction of radiation effects on aquatic biota resulting from radioactive releases from the nuclear fuel cycle. In: *Impacts of Nuclear Releases into the Aquatic Environment*, Vienna, International Atomic Energy Agency, pp. 377–93.
10. Spies, R. B. (1975). Uptake of technetium from seawater by red abalone, *Haliotis rufescens. Health Physics*, **29**, 695–9.
11. Gromov, G. G. (1976). Uptake of plutonium and other nuclear wastes by plankton. *Mar. Sci. Comm.*, **2**, 227–47.
12. Pentreath, R. J., Jefferies, D. F., Lovett, M. D. and Nelson, D. M. (1980). The behaviour of transuranic and other long-lived radionuclides in the Irish Sea and its relevance to the deep sea disposal of radioactive wastes. In: *Marine Radioecology*, 3rd NEA Seminar, Paris, Organization for Economic Cooperation and Development, pp. 203–21.
13. Bass Becking, L. G. M., Kaplan, I. R. and Moore, D. (1960). Limits of the natural environment in terms of pH and oxidation–reduction potentials. *J. Geol.*, **68**, 243–84.
14. Anders, E. (1960). *The Radiochemistry of Technetium*. Nuclear Science Series Document NAS-NS-3021, Washington, DC, US Department of Commerce.
15. Fowler, S. W., Benayoun, G., Parsi, P., Essa, M. W. A. and Schulte, E. H. (1981). Experimental studies on the bioavailability of technetium in selected marine organisms. In: *Impacts of Radionuclide Releases into the Marine Environment*, Vienna, International Atomic Energy Agency, pp. 319–39.

16. Till, J. C., Hoffman, O. and Dunning, D. E. (1979). A new look at ^{99}Tc releases to the atmosphere. *Health Physics*, **36**, 21–30.
17. Masson, M., Aprosi, G., Laniece, A., Gueguenat, P. and Belot, Y. (1981). Approches experimentales de l'etude des transferts du technetium a des sediments et a des especes marine benthiques. In: *Impacts of Radionuclide Releases into the Marine Environment*, Vienna, International Atomic Energy Agency, pp. 341–59.
18. Fowler, S. W., Aston, S. R., Benayoun, G. and Parsi, P. (1983b). Bioavailability of technetium from artificially labeled northeast Atlantic deep-sea sediments. *Mar. Envir. Res.*, **8**, 87–100.
19. Schreiner, F., Fried, S. and Friedman, A. (1981). Diffusion of neptunyl (V) and pertechnetate ions in marine sediments. In: *The Scientific Basis for Nuclear Waste Management* (Ed. S. Topp), Amsterdam, Elsevier, pp. 273–7.
20. Li, Y. H. and Gregory, S. (1974). Diffusion of ions in seawater and in deep sea sediments. *Geochim. Cosmochim. Acta*, **38**, 703–14.
21. Fisher, N. (1982). Bioaccumulation of technetium by marine phytoplankton. *Envir. Sci. Technol.*, **16**, 579–81.
22. Pentreath, R. J. (1981). Biological availability to marine organisms of transuranium and other long-lived nuclides. In: *Impacts of Radionuclide Releases into the Marine Environment*, Vienna, International Atomic Energy Agency, pp. 241–72.
23. Jeanmarie, L., Masson, M., Patti, F., Germain, P. and Cappellini, L. (1981). Technetium-99 content in some marine organisms collected near La Hague, France. *Mar. Poll. Bull.*, **12**, 29–32.
24. Beasley, T. M., Lorz, H. V. and Gonor, J. J. (1982). Biokinetic behavior of Tc in the red abalone *Haliotis rufescens*; a reassessment. *Health Physics*, **43**, 501–7.
25. Topcuoglu, S. and Fowler, S. W. (1984). Factors affecting the biokinetics of technetium (^{95m}Tc) in marine macroalgae. *Mar. Envir. Res.*, **12**, 25–43.
26. Schulte, E. H., Secondini, A. and Scoppa, P. (1983). Trasferimento del tecnezio attraverso le catene alimentari marine. In: *Atti del XXII Congresso dell'Associazione Italiana di Protezione Contro le Radiazioni* (Ed. S. Belletti), Frascati, Poligrafica Laziale, SRL, pp. 563–70.
27. Fowler, S., Heussner, S., Larosa, J. and Topcuoglu, S. (1983). The biokinetic behaviour of technetium in marine organisms. In: Biennial Report, 1981–1982, IAEA TECDOC-297, Vienna, International Atomic Energy Agency, pp. 45–9.
28. Pentreath, R. J. (1981). Biogeochemical studies of technetium in marine and estuarine ecosystems. Report DOE/EV-10251-3, Washington, DC, US Department of Energy, 75 pp.
29. Schulte, E. H., Scoppa, P. and Secondini, A. (1982). Accumulo e rilascio del tecnezio da parte di alcuni organismi marini: (1) *Palaemon elegans*. *Boll. Soc. Ital. Biol. Sperim.*, **LVIII**, 1361–7.
30. Fowler, S. W. (1981). Biogeochemical studies of technetium in marine and estuarine ecosystems. Report DOE/EV-10251-3, Washington, DC, US Department of Energy, 75 pp.
31. Beasley, T. M. (1981). Biogeochemical studies of technetium in marine and estuarine ecosystems. Report DOE/EV-10251-3, Washington, DC, US Department of Energy, 75 pp.

32. International Mussel Watch (1980). Report of workshop sponsored by the Environmental Studies Board, Commission on Natural Resources, National Research Council, Washington, DC, National Academy of Sciences, 248 pp.
33. Schulte, E. H., Scoppa, P. and Secondini, A. (1982). Comportamento del tecnezio nell'ambiente marino: mobilita' dell'anione pertecnetato. In: *IV Convegno Nazionale sulla Attivita di Ricerca nei Settori della Radiochimica e della Chimica Nucleare, delle Radiazioni e dei Radioelementi* (Ed. U. Casellato and P. A. Vigato), Padova, Litografia La Photograph, pp. 185–94.
34. Aprosi, G. and Masson, M. (1984). Bilan des etudes experimentales de transferts de technetium a des sediments et a des especes marines benthiques et comparaison a des resultats *in situ*. *Radioprotection*, **19**, 89–103.

18

Accumulation and Loss of Technetium by Macrophytic Algae

C. Benco, S. Cannarsa, I. Ceppodomo and A. Zattera
ENEA Centro Ricerche Energia Ambiente, PO B 316, La Spezia, Italy

1. INTRODUCTION

Artificial radionuclides have been introduced into the natural environment for more than three decades and a large body of data relating to their effects and behaviour is now available. In assessing the detriment for the future, some interesting aspects will come from long-living radionuclides and among them from Tc.

Tc radioisotopes enter the environment through fall-out, enrichment plants, fuel reprocessing plants and medical uses [1,2]. Production of long-living ^{99}Tc will increase in the future. Burkholder *et al.* [3] estimate for the year 2000 an inventory of ^{99}Tc of some $1{\cdot}69 \times 10^5$ kg or 0·11 EBq in the United States. For the world production by the year 2000 about $5{\cdot}6 \times 10^6$ kg or 3·6 EBq will have been produced [4].

In recent years attention has been paid to the behaviour of Tc in the marine environment and data on its behaviour in marine organisms are now available [2, 5–14]. As a rule, marine organisms do not concentrate Tc to a great extent.

Among algae, it seems that they do not accumulate Tc to a high degree with the clear exception of macrophytes dealing with the class *Phaeophyta*. Fisher [9] tested different phytoplankon, dealing with *Diatomeae*, *Chlorophyceae* and *Coccolithophorideae*, and found concentration factors (CF) from 0 to 17 as maximum (CF = concentration factors = ratio of radioactivity per gram of wet algae divided by radioactivity per millilitre of seawater). Green and red macrophyte algae, both in experimental and field conditions, do not accumulate Tc to a great extent. On the contrary this element is highly accumuiated by brown algae [2, 15, 16]. Masson *et al.* [15] measured CF greater than 10^3 in *Fucus serratus*. An unpublished source cited by Birks [1], estimates for some brown algae concentration factors up to 10^4. It seems quite clear the potential role as bioindicators that this class of algae can play. In addition Beasley [17] reported also: 'As such these macroalgae serve as an important concentrating mechanism for TcO_4^- from seawater and consequently as a significant source of Tc to the benthic food chain. As detrital material, and as an important carbon source to detrital feeders, as well as bacteria, their potential role for focusing Tc into biota must be important. In the future, more attention must be given to this area of research'.

Despite the important role of these macroalgae, no studies (with the notable exception of the work by Topcuoglu and Fowler [14] have been carried out. The goal of the present work is to present preliminary results on the accumulation of Tc by some macroalgae.

2. MATERIAL AND METHODS

^{95m}Tc (a gamma–emitter with $T_{1/2} = 61$ days) was produced by the cyclotron of the JRC at Ispra (Italy) and kindly supplied to us. It was present in two physico-chemical (anionic) forms (see Results). All the samples were measured with a single channel analyser (Packard auto gamma analyser) coupled with a NaI (Tl) well crystal. The counting efficiency was about 50% and the propagates counting error was generally less than 5% at 1σ level.

Algae used in experiments were collected in the Fiascherino bay (Gulf of La Spezia, Italy), immediately aerated and then used for the experiments. The four species of brown algae that were sampled and tested are: *Sargassum vulgare*, *Cystoseira complexa*, *Dictyopteris membranacea*, *Dictyota dichotoma implexa*. One species of green algae (Chlorophyta) *Ulva rigida* was also examined.

Experiments were carried out on batches with aged Mediterranean seawater. A radioactive solution was added to reach a concentration of about 18·5 kBq/litre (0·5 μCi/litre). The different batches were immediately aerated and after 2 h the algae were added. In all experiments, the pH was between 8·04 and 8·06 and E_h between 460 and 475 mV. The experiments were conducted in daylight illumination conditions and at cooled room temperatures (20 ± 1°C). At intervals, portions of algae and volumes of seawater were sampled and measured. Before counting, the pieces of algae were rinsed for 1 min in non-contaminated seawater and the acccompanying water removed with filter paper. After accumulation, the algae were resuspended in non-contaminated seawater and sampled, as in the accumulation experiment, to study the elimination.

3. RESULTS

Chromatography (chromatography paper Whatman 3MM) carried out directly on the ^{95m}Tc source, as supplied by the JRC Ispra, shows (see Fig. 1(A)) two peaks ($R_f = 0{\cdot}74$ and $R_f = 0{\cdot}89$) indicating that Tc was present in two physico-chemical species. This was the result of the treatment of the irradiated Mo target which was dissolved in 7M HNO_3 and HDEHP (di-hexyl-phosphoric acid) 0·5 N in mesitylene. The two identified species were anionic since they were retained by a cationic resin (Dowex 50W-X8). However, by adding ^{95m}Tc source solution to aerated seawater only one form of ^{95m}Tc resulted (see Fig. 1(B)) with the typical peak of pertechnetate ($R_f = 0{\cdot}64$). Seawater plus source, treated as above, were employed for the accumulation experiment.

Attempts were also made to carry out experiments with reduced technetium. For this purpose ^{95m}Tc was conjugated with hydrazine sulphate and, as such, added to aerated seawater. Since the reduction peak disappears (see Fig. 2) it can be assumed that in seawater the reduced form of Tc is not a stable one.

Tc (pertechnetate) accumulation kinetics in various algae are reported in Figs. 3 and 4. One can see that with the exception of *Cystoseira complexa*, the accumulation is very rapid and CF decrease from *Sargassum vulgare* to *Ulva rigida*.

In the graphs for the various sampling times two points are displayed: they correspond to the pooled results of two distinct experiments. Considered as a whole, the results show different accumulations for brown seaweed and lower accumulations for *Ulva* (Chlorophyta).

All the results seem to be governed by a one-compartment kinetics of the type: $\mathrm{CF}(t) = \mathrm{CF}(1 - e^{-kt})$ in which $\mathrm{CF}(t) = \mathrm{CF}$ of the isotope at time t, CF is the concentration factor obviously at equilibrium; k = exchange rate ($k = 0{\cdot}693/tb_{1/2}$). Considering such kinetics, the values given in Table I can be calculated.

At the end of accumulation, aliquots of media were filtered using the double layer 0·45 μm filter technique and paper chromatography was carried out. The results show in all batches that 0·3% of ^{95m}Tc was in the

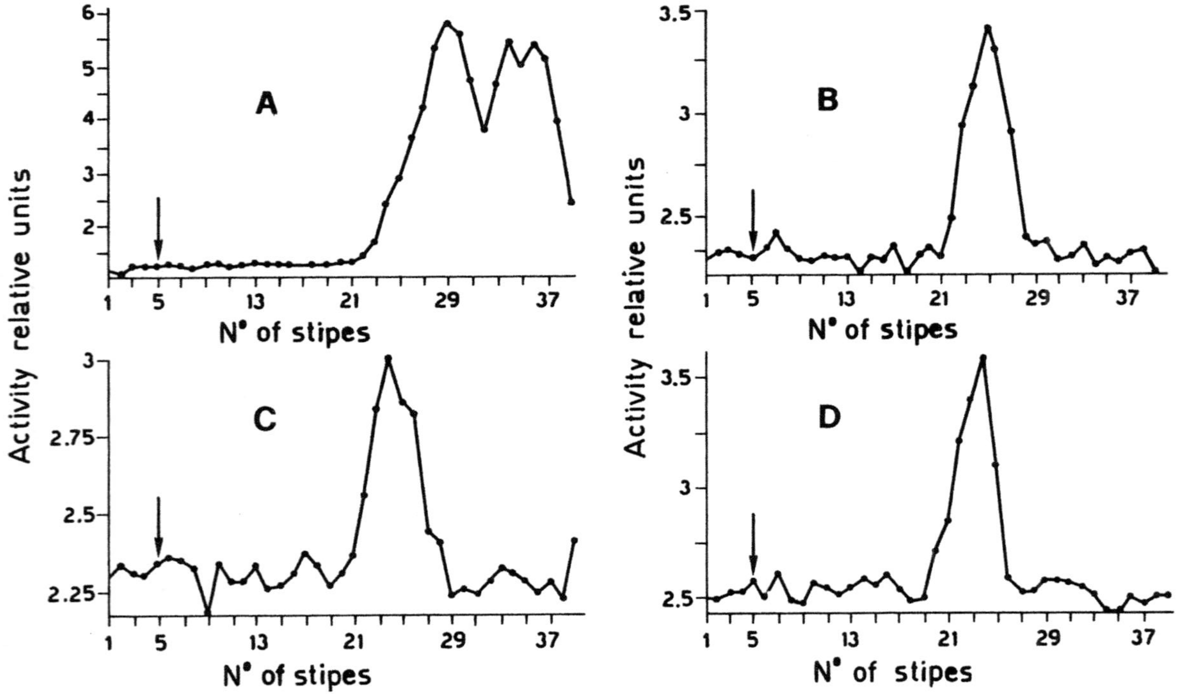

FIG. 1. Technetium speciation. (A) Species in the source. (B) Species in seawater after 2 h. (C) Species in seawater to the end of experiment in which there were *Dictyota dichotoma implexa*. (D) Species in seawater to the end of experiment in which there were *Ulva rigida*.

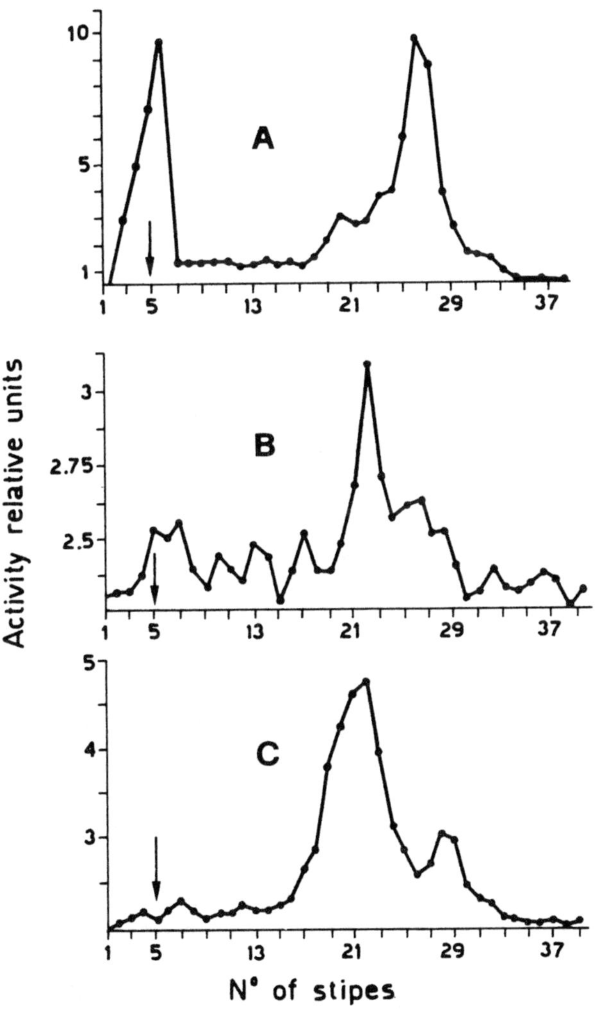

FIG. 2. Behaviour of reduced technetium. (A) Reduced Tc in seawater with hydrazine sulphate. (B) After 2 h of addition to seawater of reduced technetium. (C) After 24 h of addition to seawater of reduced technetium.

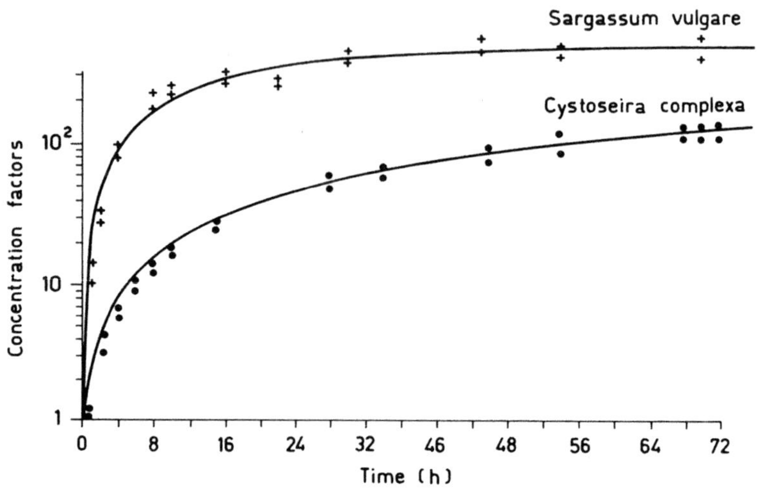

FIG. 3. Accumulation of ^{95m}Tc by brown algae.

particulate phase and that the physico-chemical species did not change with time (see for example Fig. 2(A, B, C)).

Dissection of *Cystoseira* indicates different contents of ^{95m}Tc at the end of the experiment. The relative distribution of radioactivity in the parts of dissected *Cystoseira* is: main cylindrical axis 23%, lateral axis 26% and stipes 51%, indicating a greater affinity of stipes for Tc.

The elimination of Tc from the four species of brown algae was followed for 170 h. The pooled data of two different experiments are reported in

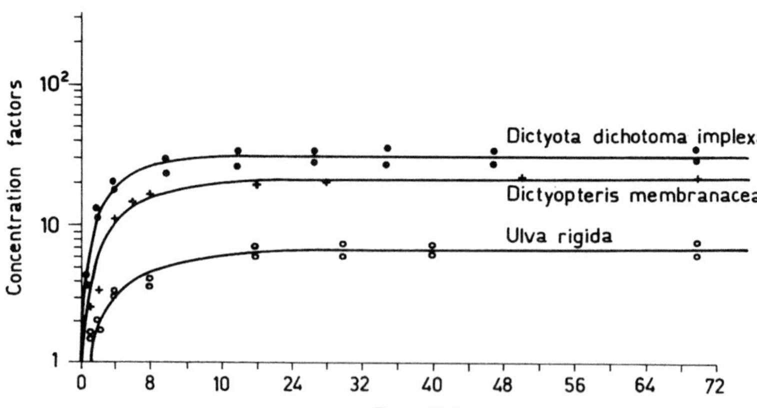

FIG. 4. Accumulation of ^{95m}Tc by brown algae and *Ulva rigida* (green algae).

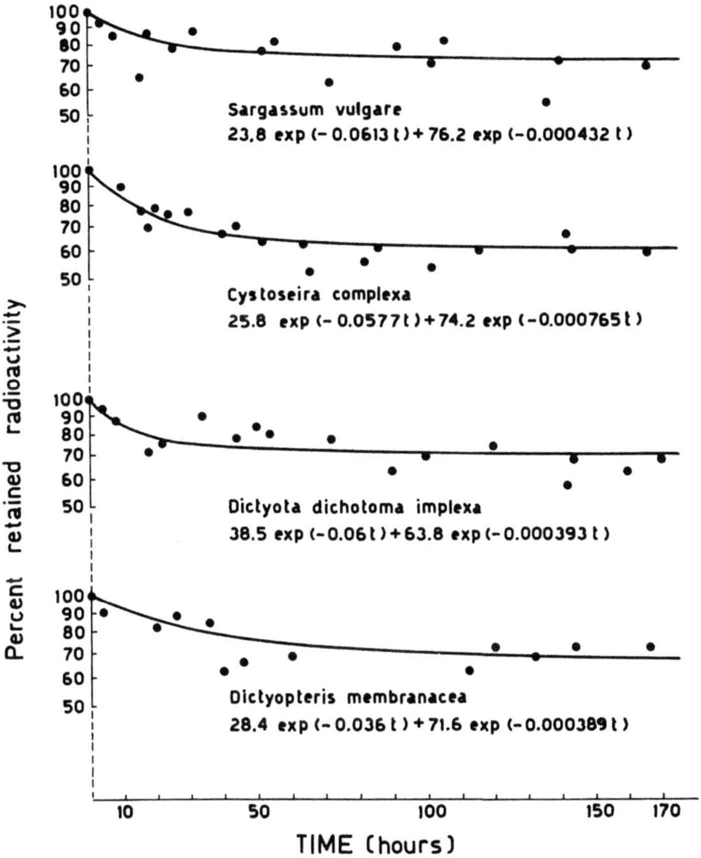

FIG. 5. Elimination of ^{95m}Tc.

Fig. 5. In all cases we have a two-compartment kinetics: one rapid and the other slow. The biological half-lives for technetium turnover were as given in Table II. From the graphs it can be noted that, in general, the quantity lost in the time period considered is not very large, i.e. not more than 30%.

To see if the loss can be enhanced, some samples of *Cystoseira complexa* were resuspended in seawater with decreasing pH (7·95–5·9–3·32–1·59) using HCl, but additional loss was not noted. For the same purpose *Sargassum vulgare* and *Cystoseira complexa* were resuspended in seawater containing Na_2EDTA (10^{-4} and 10^{-5} M). The results are reported in Fig. 6 where one curve only for each experiment is drawn because the difference

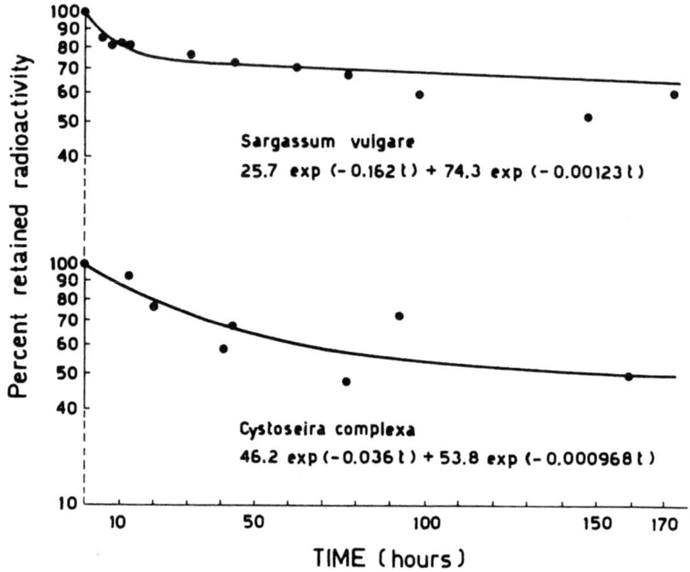

FIG. 6. Elimination of Tc from *Sargassum vulgare* and *Cystoseira complexa* with Na_2 EDTA.

Table I

	CF	$tb_{1/2}(d)$
Cystoseira complexa	532	14
Sargassum vulgare	506	11
Dictyota dichotoma implexa	31	29
Dictyopteris membranacea	21	4
Ulva rigida	7	5

Table II

	Biological half-lives	
	Fast compartment (h)	*Slow compartment (h)*
Cystoseira complexa	12	900
Dictyota dichotoma implexa	11	1 750
Dictyopteris membranacea	19	1 780
Sargassum vulgare	11	1 600

due to the two EDTA concentrations is negligible. It seems that the addition of Na_2EDTA enhances slightly the elimination, probably on account of the existence of two binding energies of Tc algae.

4. DISCUSSION

The uptake results show that among the species of algae tested, brown algae have CF greater than that of a Chlorophyta (*Ulva rigida*). These results agree with those of Jeanmaire *et al.* [16] who found, on the basis of environmental measurements made on algae collected near La Hague out-fall, a greater affinity for technetium for brown algae. Recently Topcuoglu and Fowler [14], in experimental conditions, found a general order of algae affinity for Tc: brown, green, red. In addition Jeanmaire *et al.* [16] report that brown algae accumulate ^{99}Tc to the greatest degree among the plants and animals collected near La Hague.

The in situ results by Jeanmaire *et al.* [16], those in laboratory conditions by Topcuoglu and Fowler [14] and the present work, suggest that brown algae are unmatched in their ability to accumulate Tc. Different species of marine phytoplankton show a rather slow uptake for technetium [9] with a few exceptions; most molluscs, crustaceans and fishes show whole body CF less than ten [2, 8, 15, 10].

The degree of Tc accumulation varies according to different algal tissues. Topcuoglu and Fowler [14] found that in *Sargassum vulgare* young parts such as 'leaves' and air bladders accumulate more than older parts such as the cylindrical main axis. In the course of experiments described in this paper, young stipes of *Cystoseira complexa* concentrate twice as much more Tc than cylindrical main axes. Unfortunately, it was not possible to investigate the concentration of Tc in every tissue during the experiment. Perhaps, such additional information could help to explain the different kinetics found in accumulation (1 compartment) and in loss (2 compartments). In fact, since Tc accumulation seems to be an active process (Topcuoglu and Fowler [14]) the behaviour of one compartment in accumulation or in loss could be masked.

The strong affinity for Tc and other heavy metals (see, for example, Phillips [18]), shown by most brown algae species, seems to be unique among macrophytes. The mechanism of such concentration processes is still to be explained. Metal uptake by these organisms has been attributed partly to polysaccharides in the cell wall including alginates [19]. More recently, Ragan *et al.* [20] have linked the chelation of bivalent metal ions to brown algae polyphenols such as non-dialysable polyphloroglucinals.

In the course of this experience, attempts were made to understand the mechanism of Tc accumulation by brown seaweed. For this purpose, Fucoidan (supplied by the Sigma Chemical Company), which is a pool of high molecular weight polysaccharides extracted from *Fucus sp.*, was put for 48 h with ^{95m}Tc in seawater and then dialysed (dialyser tubing of the A. H. Thomas Co. was used); but no activity was retained by Fucoidan. At present the Tc binding capacity of proteic components of brown, red and green macrophytes is under investigation. Whatever is found to be the process involved in Tc accumulation by brown seaweed, it seems quite clear that some species of *Phaeophyta* are useful bioindicators for Tc pollution.

ACKNOWLEDGEMENTS

This preliminary work has been carried out thanks to the supply of ^{95m}Tc by JRC (Ispra) for which it is well acknowledged. Work carried out under the association contract BIO-B-322-I (S) of the European Communities.

REFERENCES

1. Birks, J. L. (1975). Medical radionuclides in marine environment. *Nature*, **255**, 621–2.
2. Pentreath, R. J. (1981). The biological availability to marine organisms of transuranium and other long lived nuclides. In: *Impacts of Radionuclide Release into the Marine Environment*, Vienna, IAEA, 6–10 Oct. 1980, 242–72.
3. Burkholder, J. C., Cloninger, M. O., Baker, D. A. and Jansen, G. (1975). Incentives for partitioning high-level waste. USERDA Rep. BNWL 1927, Nat. Tech. Inf. Serv., Springfield.
4. Grimwood, P. D. and Wess, G. A. M. (1976). Assessment of the radiological protection aspects of disposal of high level waste on the ocean floor. Report NRPB-R48 Harwell (UK), p. 118.
5. Spies, R. B. (1976). Uptake of technetium from sea water by red abalone *Haliotis rufescens*. *Health Phys.*, **29**, 695–9.
6. Beasley, T. M. and Gonor, J. J. (1980). Biogeochemical studies of technetium in marine and estuarine ecosystems. Progress Report July 1979–June 1980. Doc. DE-AT06-79EV-10251, Oregon State University, USA.
7. Beasley, T. M., Gonor, J. J. and Larz, H. V. (1982). Technetium uptake, organ distribution and loss in the mussel *Mytilus californianus* (Conrad) and the oyster *Crassostrea gigas* (Thumberg). *Mar. Envir. Res.*, **71**, 103–16.
8. Fowler, S. W., Benayoum, G., Parsi, P., Essa, M. W. A. and Schulte, E. H. (1981). Experimental studies on the bioavailability of technetium in

selected marine organisms. In: *Impacts of Radionuclide Releases into the Marine Environment*, Vienna, IAEA, 6–10 Oct. 1980, pp. 319–39.

9. FISHER, N. S. (1982). Bioaccumulation of technetium by marine phytoplankton. *Envir. Sci. Technol.*, **16**, 579–81.
10. SCHULTE, E. H., SECONDINI, A. and SCOPPA, P. (1981). Trasferimento di Tecnezio attraverso le catene alimentari marine. *Atti del XXII Congr. della Associazione Italiana di Protezione contro le Radiazioni*; vol. 1, Brescia, 23–26 June 1981, pp. 563–70.
11. SCHULTE, E. H., SCOPPA, P. and SECONDINI, A. (1982). Accumulo e rilascio del Tecnezio da parte di alcuni organismi marini: *Palaemon elegans*. *Boll. Soc. Ital. Biol. Sperim.*, **58**, 1361–7.
12. SCOPPA, P., SCHULTE, E. H. and SECONDINI, A. (1983). Application of technetium-95m in experimental radioecology: chemical form and behaviour in marine environment. In: *Technetium in Chemistry and Nuclear Medicine* (Ed. E. Deutsch), Verona, Cortina International, pp. 219–26.
13. SCHULTE, E. H., SCOPPA, P. and SECONDINI, A. (1983). Recenti sviluppi delle indagini sul comportamento del Tc negli organismi marini: metodi per lo studio delle cinetiche. *Atti del 3° Convegno Nazionale di Radioecologia*, Bologna, 22–23 June 1983, pp. 259–69.
14. TOPCUOGLU, S. and FOWLER, S. W. (1984). Factors affecting the biokinetics of technetium (^{95}Tc) in marine macroalgae. *Mar. Envir. Res.*, **12**, 25–43.
15. MASSON, M., APROSI, G., LANIECE, A., GUEGUENIAT, P. and BELOT, Y. (1981). Approaches experimentales de l'étude des transfert du technetium à des sediments et à des éspeces marine benthiques. In: *Impacts of Radionuclide Releases into the Marine Environment*, Vienna, IAEA, 6–10 Oct. 1980, pp. 341–59.
16. JEANMAIRE, L., MASSON, M., PATTI, F., GERMAIN, P. and CAPELLINI, L. (1981). Technetium-99 content in some marine organisms collected near La Hague, France. *Mar. Poll. Bull.*, **12**, 29–32.
17. BEASLEY, T. M. (1981). Biogeochemical studies of technetium in marine and estuarine ecosystems. Progress report July 1980–July 1981. Doc. DE-AT06-79EV-10251, Oregon State University, USA.
18. PHILLIPS, D. J. H. (1980). *Quantitative Aquatic Biological Indicators*, London, Applied Science Publishers.
19. HAUG A. and SMIDSROD, O. (1965). The effect of divalent metals on different types of alginates solutions. *Acta Chem. Scand.*, **19**, 341–51.
20. RAGAN, M. A., SMIDSROD, O. and LARSEN, B. (1979). Chelation of divalent metal ions by brown algal polyphenols. *Mar. Chem.*, **7**, 265–71.

19

Technetium in Micro-organisms

B. MANIA,[a] Z. MOUREAU,[a] D. VAN DER BEN,[a] J. VAN BAELEN,[b] C. VERTHÉ,[c] J. M. BOUQUEGNEAU,[c] M. COGNEAU,[d] S. BONOTTO,[b] C. M. VANDECASTEELE,[e] L. PIGNOLET[e] and C. MYTTENAERE[f]

[a]*Institute Royal des Sciences Naturelles de Belgique, rue Vautier 29, B-1040 Brussels, Belgium*

[b]*Département de Radiobiologie, CEN-SCK, Boeretang 200, B-2400 Mol, Belgium*

[c]*Université de Liège, Laboratoire d'Océanologie, Institut de Chimie, Bâtiment 6, B-4000 Liège, Sart Tilman, Belgium*

[d]*Université Catholique de Louvain, Laboratoire de Chimie Nucléaire, 2 chemin du Cyclotron, B-1348 Louvain-la-Neuve, Belgium*

[e]*Université Catholique de Louvain, Laboratoire de Physiologie Végétale, Place Croix du Sud 4, B-1348 Louvain-la-Neuve, Belgium*

[f]*Commission of the European Communities, CEC, DGXII, rue de la Loi 200, B-1049 Brussels, Belgium*

1. INTRODUCTION

The role of micro-organisms in the behaviour of radionuclides in aquatic and terrestrial systems and their possible transfer to man through the food chain has been discussed, on the basis of recent literature, in a previous work [1]. Moreover, multidisciplinary investigations were realised in our laboratories on the uptake of technetium by unicellular organisms, plants and animals [2–9]. Although the extrapolation of results obtained under artificial conditions to situations occurring in the ecosystem may be difficult, experimental work is nevertheless necessary for a better understanding of biological processes, which affect the behaviour of radionuclides.

This paper reports the main results of experimental work on the interaction of technetium with aquatic micro-organisms and with untreated and sterilised sediments. In addition, the utilisation of ^{95m}Tc-labelled bacteria for studying the transfer of technetium through a simple experimental food chain is briefly discussed.

2. MATERIALS AND METHODS

2.1. Micro-organisms, mussels and sediments

Bacteria cultures (*Flavobacterium halmephilum*) were obtained in seawater enriched with $10\,g\,l^{-1}$ bacto-peptone (Difco) at $20 \pm 2°C$ [4, 6]. Protozoa (*Uronema marinum*) were grown by inoculating a single organism into an enriched medium containing a mixed population of marine bacteria or the halophilic species *Flavobacterium halmephilum* only. The protozoa, which feed on bacteria, attain their maximum growth after about 5 days [4].

Microalgae (*Chlamydomonas reinhardtii* and *Dunaliella bioculata*) were cultivated as previously reported [10, 11]. The mussels (*Mytilus edulis*) were collected along the Belgian coast, cleaned and maintained in aerated plastic aquaria. The marine sediments, collected near Coxyde, which contained a mixed population of micro-organisms, were sterilised by repeated autoclaving.

2.2. Technetium

Technetium was used as pertechnetate ($K^{95m}TcO_4$ or $NH_4{}^{99}TcO_4$). ^{95m}Tc was measured by gamma spectrometry and ^{99}Tc by liquid scintillation, after HNO_3 digestion of the samples [2]. The biological effects of technetium on the micro-organisms were studied by adding increasing concentrations of $NH_4{}^{99}TcO_4$ to the culture medium. The levels of NH_4^+ depended on the concentration of pertechnetate and were not adjusted with other ammonium salts to avoid chemical alteration of the medium.

2.3. Electron microscopy

Scanning electron micrographs of micro-organisms were obtained according to the method of Paerl and Shimp [12]. The cells were collected on Nuclepore filters having 0·45 μm pore size.

2.4. Column chromatography

Micro-organism suspensions were extracted in a press and the homogenate was centrifuged for 1 h at 16 000 rpm. The supernatants were directly analysed by column chromatography on Ultrogel AcA54 (LKB), which allows a good separation of compounds having a molecular weight in the range 5000–70 000 [13]. Alternatively, the supernatants were submitted to dialysis (Amicon membrane), for the preparation of two fractions (A and B, containing respectively high and low molecular weight compounds), which were then chromatographed separately. The column was eluted (5 ml fractions; about 30 ml h^{-1}) with 0·05 M ammonium bicarbonate and the fractions were measured for their absorbance at 215 and 280 nm as well as for their technetium activity.

3. RESULTS

3.1. Morphology of the micro-organisms

The morphological features of the microalgae *Chlamydomonas reinhardtii* and *Dunaliella bioculata*, two small biflagellate photosynthetic Volvocales, are well known [14–20]. *Dunaliella bioculata* is a naturally wall-less organism [16–18], whereas *Chlamydomonas reinhardtii* CW15^{+} is a mutant strain, which has lost the capability of synthesising a cell wall [21, 22]. Consequently, no attempt was made to further investigate the morphology of these two algae. On the other hand, the two other micro-organisms

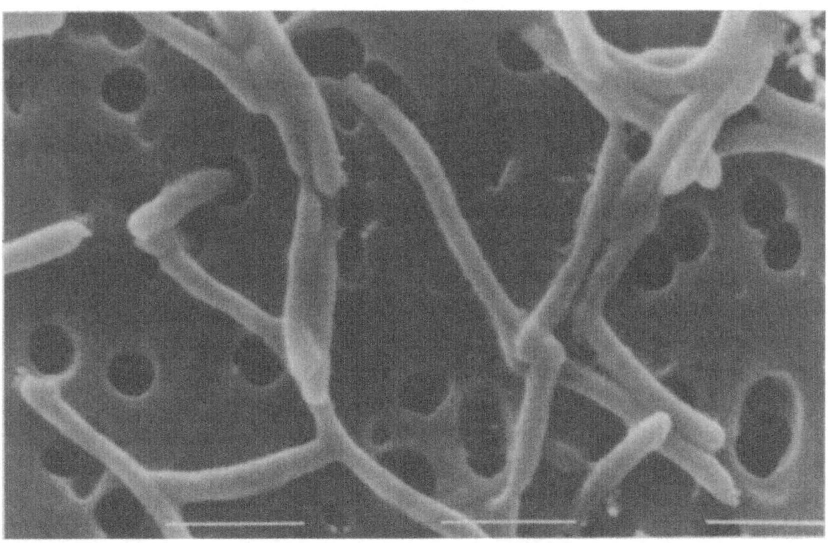

FIG. 1. *Flavobacterium halmephilum*. Scanning electron micrograph of cells grown in the laboratory and collected on a 0·45 μm Nuclepore filter. Scale = 1 μm.

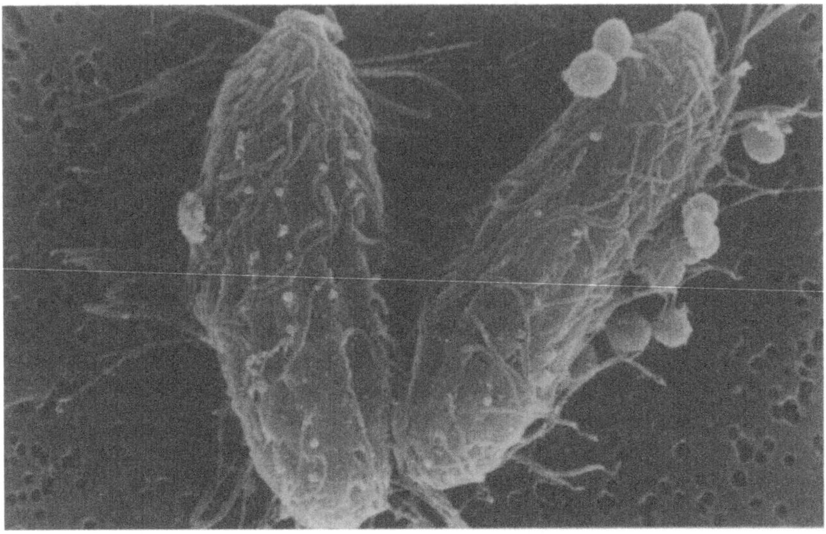

FIG. 2. *Uronema marinum*. Scanning electron micrograph of two cells (length: about 20 μm) fixed as previously reported by Dartevelle-Moureau *et al.* [4] and collected on 0·45 μm Nuclepore filter. Contaminating bacteria (size: about 1·5 μm) are sticking to the numerous cilia.

utilised in this work, *Flavobacterium halmephilum* and *Uronema marinum*, were studied by scanning electron microscopy. *Flavobacterium* cells, cultivated in the laboratory, appear as thin rods (Fig. 1). *Uronema* cells have a large size (20–30 μm), are ovoidal and possess numerous cilia (Fig. 2). Contaminant bacteria may stick to the cilia.

3.2 Uptake of technetium

Several experiments have shown that *Flavobacterium*, *Uronema* and *Chlamydomonas* cells are capable of accumulating technetium. On the other hand, *Dunaliella* cells do not concentrate this element. The concentration factors reported for these four micro-organisms are given in Table I, together with the references. They varied greatly from one experiment to another (for *Flavobacterium* and *Uronema*), probably depending on the growth conditions and age of the culture. Experiments with ^{99}Tc, in which the mass concentration of this element in the medium was higher than that of ^{95m}Tc, produced invariably low concentration factors. The amount of technetium taken up by the micro-organisms was, nevertheless,

Table I
Technetium concentration factors reported in recent works for the four investigated micro-organisms

Micro-organism	*Concentration factor for technetium*[1] ^{99}Tc	*Concentration factor for technetium*[1] ^{95m}Tc	*References*
Bacteria			
Flavobacterium halmephilum		2·3–2·4	Dartevelle-Moureau *et al.* [4]
Flavobacterium halmephilum		31·9	Verthé *et al.* [6]
Flavobacterium halmephilum		88	Moureau *et al.* [7]
Flavobacterium halmephilum[2]		42	Moureau *et al.* [7]
Ciliates			
Uronema marinum	11[3]	1·4–5·5	Dartevelle-Moureau *et al.* [4]
Uronema marinum		65·7	Verthé *et al.* [6]
Uronema marinum		126[4]	Moureau *et al.* [7]
Algae			
Chlamydomonas reinhardtii CW 15⁺		90	Moureau *et al.* [7]
Dunaliella bioculata		0·3	Moureau *et al.* [7]

[1]The concentration factor is defined as the ratio

$$\frac{\text{pCi g}^{-1}\text{ fresh matter}}{\text{pCi ml}^{-1}\text{ medium}}$$

[2]Grown together with *Uronema marinum* and separated by differential centrifugation (10 min, 12 000 × g).
[3]The concentration of ^{99}Tc in the culture medium was 10 μg ml^{-1}.
[4]The concentration of ^{95m}Tc in the medium was 1·15 pg ml^{-1}.

sufficient to allow biochemical analyses and transfer experiments through an experimental food chain.

3.3 Distribution of technetium

Suspensions of ^{95m}Tc-labelled micro-organisms were extracted in a press and the postmitochondrial supernatants were analysed by column chromatography as described in Materials and Methods. This technique allows separation of large molecules (mostly proteins) from low molecular weight compounds and pertechnetates. The comparison of optical density (215 and/or 280 nm) with radioactivity profiles has shown that ^{95m}Tc was present in a few main peaks, which correspond respectively to a first group of high molecular weight compounds (eluted approximately at the void volume of the column), to a second group of lower molecular weight substances, and to a third group, constituted by ^{95m}Tc-pertechnetate and/or ^{95m}Tc bound to a class of small organic molecules. Here are

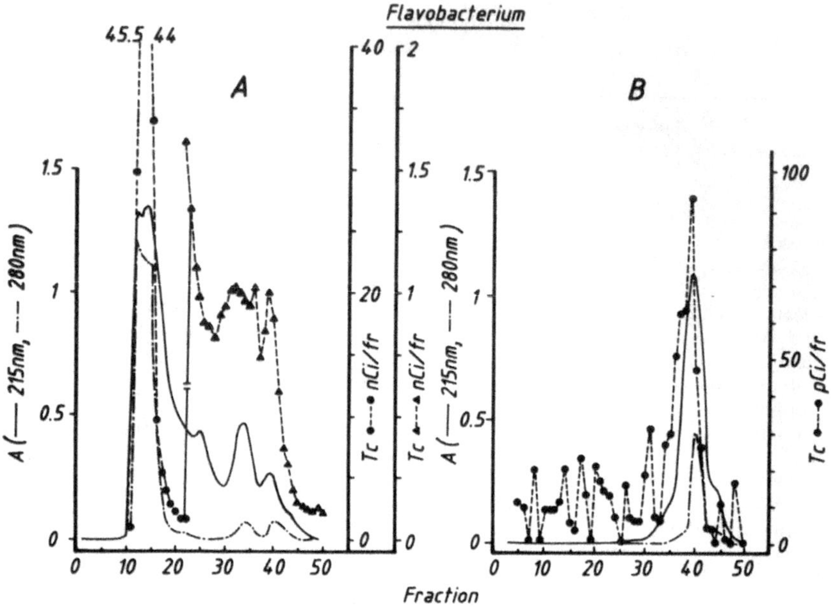

FIG. 3. *Flavobacterium halmephilum*. Elution profiles of materials absorbing at 215 and 280 nm and of ^{95m}Tc activity from an Ultrogel AcA54 (LKB) column for two samples of the postmitochondrial supernatant. (A) Compounds non-dialysable through an Amicon membrane ($\geq$ 10 000 daltons); (B) dialysable compounds ($\leq$ 10 000 daltons).

reported the results obtained with *Flavobacterium* (Fig. 3) and with *Uronema* (Fig. 4). Since *Uronema marinum* does not grow in the absence of living bacteria, mixed cultures were labelled with ^{95m}Tc pertechnetate. The chromatographic analysis of a mixture of protozoa and bacteria is shown in Fig. 5. The comparison of the radioactivity profiles of Figs. 3(A) and 5(A) show that they are rather similar. This finding suggests that in Fig. 5(A), most of the radioactive material is of bacterial origin. The fact that a different radioactivity profile was found for *Uronema* seems to exclude a significant contribution by contaminating bacteria.

The distribution of technetium in the micro-organisms was studied also by enzymatic (lysozyme) or ultrasonic treatment. Some results obtained with these procedures have already been communicated [7]. They are in good agreement with those reported in this paper, showing that technetium enters the cell and becomes incorporated, at least in part, into organic compounds.

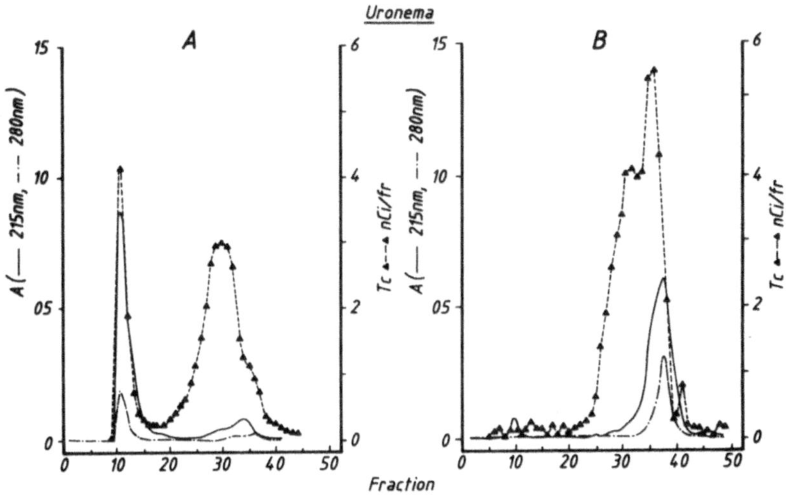

FIG. 4. *Uronema marinum*. Elution profiles of materials absorbing at 215 and 280 nm and of ^{95m}Tc activity from an Ultrogel AcA54 (LKB) column for two samples of the postmitochondrial supernatant. (A) Compounds non-dialysable through an Amicon membrane (≥ 10 000 daltons); (B) dialysable compounds (≤ 10 000 daltons).

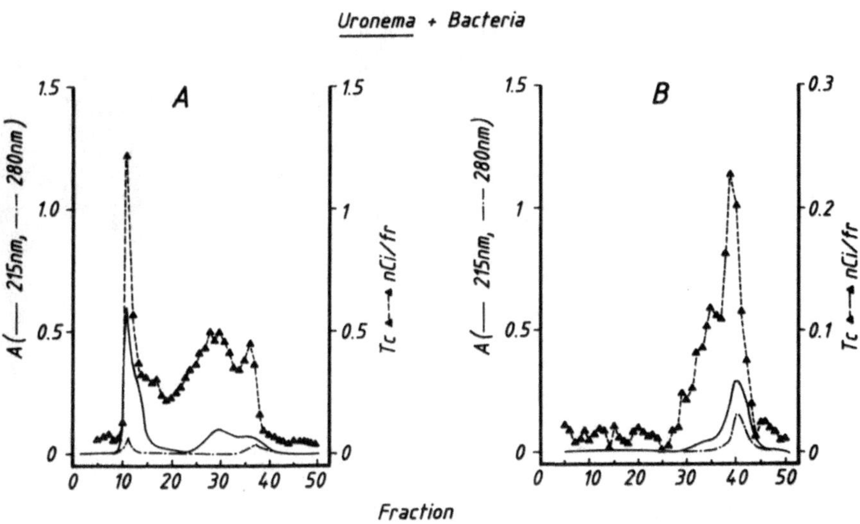

FIG. 5. Mixed population of ciliates (*Uronema marinum*) and bacteria (*Flavobacterium halmephilum*). Elution profiles of materials absorbing at 215 and 280 nm and of ^{95m}Tc activity from a Ultrogel AcA54 (LKB) column for two samples of the postmitochondrial supernatant. (A) Compounds non-dialysable through an Amicon membrane (≤ 10 000 daltons).

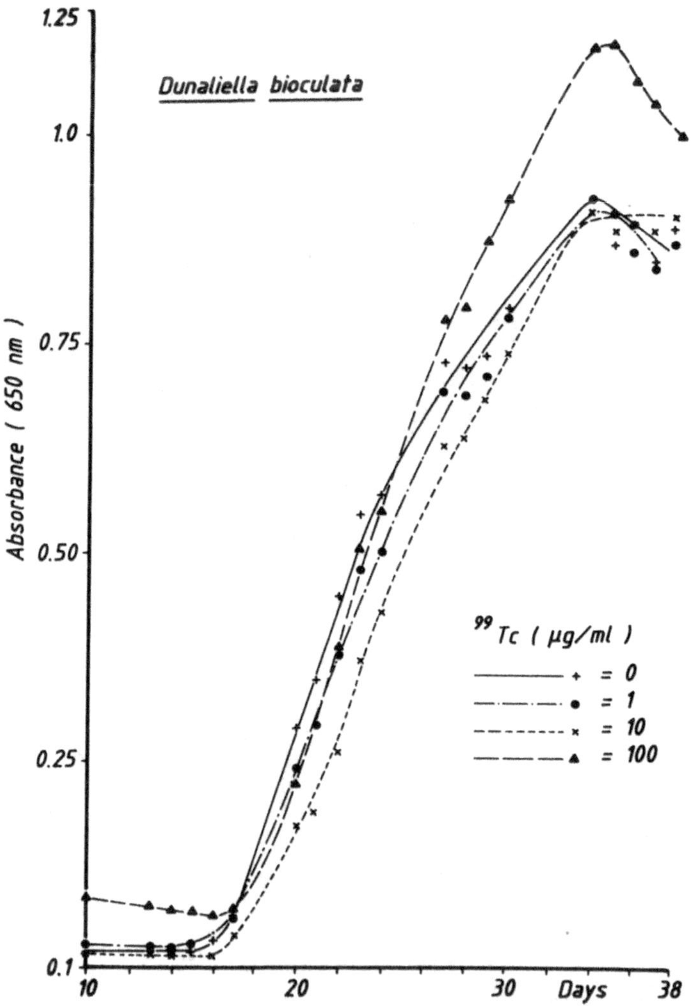

FIG. 6. *Dunaliella bioculata.* Effect of technetium, supplied as ammonium pertechnetate, on the growth of the algal population, measured at 650 nm.

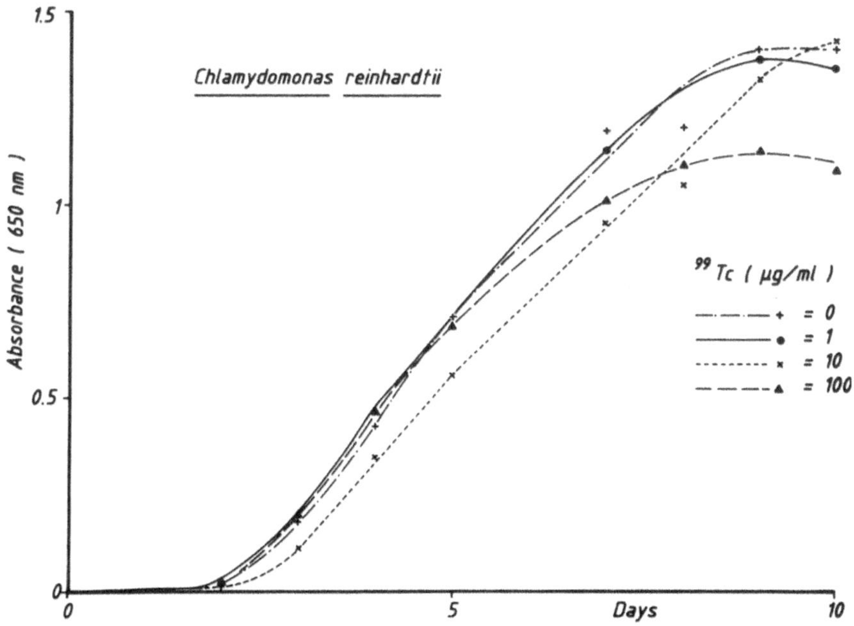

FIG. 7. *Chlamydomonas reinhardtii.* Effect of technetium supplied as ammonium pertechnetate on the growth of the algal population, measured at 650 nm.

3.4. Biological effects of technetium

As already reported [7], the biological effects of technetium were studied by adding to the culture medium $NH_4{}^{99}TcO_4$, at concentrations ranging from 0 to 10 $\mu g\,ml^{-1}$, for *Flavobacterium* and *Uronema*, and from 0 to 100 $\mu g\,ml^{-1}$, for *Chlamydomonas* and *Dunaliella.* At the concentrations utilised, no clear-cut effects on the growth curves were observed (Figs. 6 and 7). A variability was noticed, between different trials, in the response of the micro-organisms to the technetium treatment.

3.5. Transfer of technetium through the food chain

^{95m}Tc-labelled bacteria (*Flavobacterium halmephilum* or other strains) were successfully utilised for studying the transfer of this element through an experimental food chain [6]. When labelled bacteria are supplied to mussels (Fig. 8(A)), they are rapidly ingested, most of the radioactivity (70–80%) disappearing from the medium already after 1–2 h (Fig. 8(B)). After 5·5 h, only 11·3% of the initial radioactivity remained in the medium (Fig. 8(C)). Twenty-four hours later, a small increase of radioactivity was

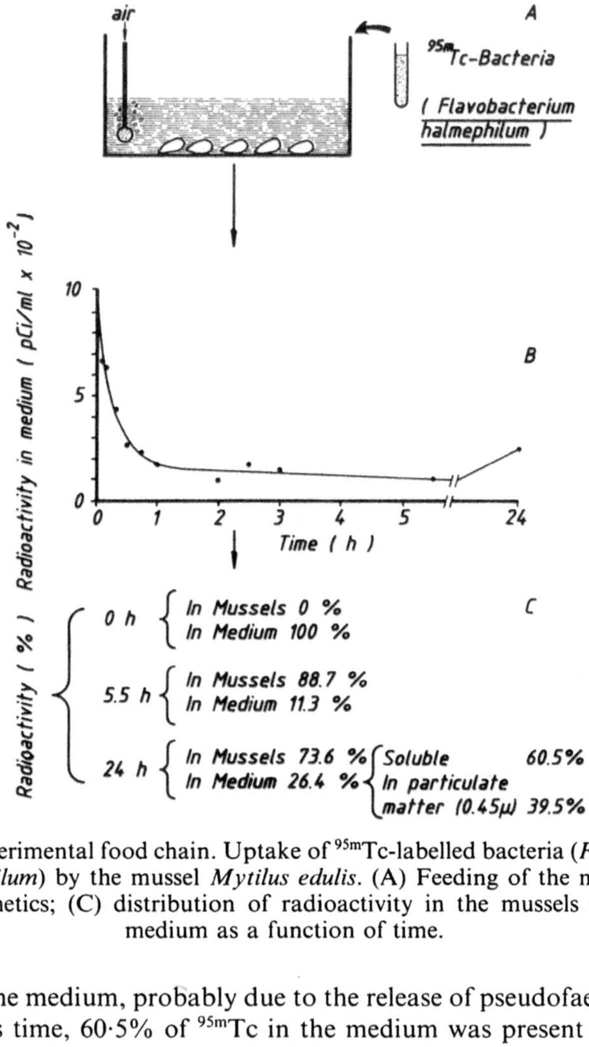

FIG. 8. Experimental food chain. Uptake of ^{95m}Tc-labelled bacteria (*Flavobacterium halmephilum*) by the mussel *Mytilus edulis*. (A) Feeding of the mussels; (B) clearance kinetics; (C) distribution of radioactivity in the mussels and in the medium as a function of time.

noticed in the medium, probably due to the release of pseudofaecal pellets [23]. At this time, 60·5% of ^{95m}Tc in the medium was present in soluble form, whereas 39·5% was contained in particulate matter, retained on 0·45 μm Millipore filters (Fig. 8(C)).

3.6. Technetium in sediments

The role of micro-organisms in the fixation of technetium was studied by labelling untreated and sterilised sediments with ^{99}Tc traced with ^{95m}Tc. The results, reported in Table II, show that sterilisation strongly reduces

Table II
Effect of sterilisation on the fixation of technetium by marine sediments of the Belgian coast

Sediment	*Time (h)*			
	23·5	*73*	*187*	*451*
Untreated*				
nr 08	4·53 ± 0·88	2·97 ± 0·50	2·34 ± 0·08	1·19 ± 0·25
nr 11	6·60 ± 1·00	2·38 ± 0·24	1·56 ± 0·22	1·05 ± 0·03
Sterilised*				
nr 08	97·0 ± 1·49	90·1 ± 4·17	62·6 ± 17·1	21·4 ± 3·81
nr 11	98·7 ± 1·86	96·2 ± 3·37	89·6 ± 3·84	76·7 ± 20·2

*The values for untreated and sterilised sediments represent the percentage of technetium activity remaining in the liquid phase.

the fixation of technetium by the sediments, most of the element (90–99%) remaining in the liquid phase even after 3 days of incubation. Later on, some fixation occurs. Bacteriological analysis of the sterilised sediments failed to detect living micro-organisms. It seems thus that technetium fixation by sterilised sediments, as a function of time, depends on some unknown physico-chemical process, which occurs slowly and which favours the immobilisation of this element.

4. DISCUSSION

Micro-organisms are believed to play a significant role in the behaviour of radionuclides in the aquatic and terrestrial ecosystems [1, 6, 7]. However, information on the interaction of micro-organisms with long-lived radionuclides is still limited. The results presented in this work show that four species of aquatic micro-organisms (*Flavobacterium halmephilum*, *Uronema marinum*, *Chlamydomonas reinhardtii* and *Dunaliella bioculata*) are capable of fixing technetium. In spite of the fact that both *Chlamydomonas reinhardtii* CW15$^+$ and *Dunaliella bioculata* are wall-less, the first micro-alga accumulates technetium (concentration factors (CF) up to 90), whereas the second one does not (CF < 1). This finding suggests the cell wall is probably not a barrier to the penetration of technetium inside algal cells. This conclusion is supported by the observation that the unicellular alga *Acetabularia*, which possesses a thick cell wall, strongly accumulates technetium (concentration factors up to 400 [3, 5]). Uptake of technetium would, thus, depend on other cell constituents and/or metabolic pathways. The accumulation of technetium by *Chlamydomonas*, as well as by *Flavobacterium* and *Uronema*, poses the question of how technetium enters the cell. It has been suggested that an anion such as TcO_4^- would not be passively taken up by algal cells that carry a negative surface charge [24]. Since some anions (Cl^-, SO_4^{2-}) are actively transported into *Acetabularia* [25–28], it is possible that TcO_4^- is also taken up by active transport in this and other algal cells. Previous studies on technetium uptake by phytoplankton species have furnished contradictory results [24, 29–31]. Possibly, the different results reported in the literature for phytoplankton depend on the particular physiological and biochemical features of the algal species investigated. Since numerous species are at present cultivated under laboratory conditions, screening work on their capability of fixing technetium becomes feasible.

The reason why concentration factors in bacteria and protozoa vary considerably from one experiment to another is not clear. Probably, this variability depends on the different physiological conditions of the cultured micro-organisms. The concentration factors reported for technetium in marine brown algae are much higher under natural [32] (field observations) than under artificial conditions (laboratory experiments) [3, 5, 33]. This behaviour is not yet understood. It might be due to integration of technetium uptake over the life span of brown algae (see discussion in ref. 32) or to other factors, as for example the presence of particular chemical compounds, which would form complexes with technetium. The very low concentration of technetium in the sea may also constitute a favourable condition for the attainment of high concentration factors. If a similar phenomenon occurs also for micro-organisms, the concentration factors for technetium would be higher in nature than in laboratory cultures.

The experiments carried out with untreated and sterilised sediments clearly suggest a significant role of the microflora in the fixation of technetium, in agreement with previous work [33, 34]. However, the precise mechanisms by which microbes favour the immobilisation of technetium into the sediments remain to be elucidated. Further work is necessary, firstly to obtain a better knowledge of the microbial population living in the sediments, and secondly to verify which of the suggested mechanisms (modification of the organic matter, creation of reductive conditions by the microbial metabolism inducing incorporation of technetium into insoluble compounds, excretion of particular metabolites modifying the chemical form of technetium, and accumulation by the micro-organisms themselves) are relevant in the fixation process [34].

The results obtained with column chromatography analysis of microbial postmitochondrial supernatants, as well as previous observations on lysozyme mobilisation of technetium [7], suggest that technetium is transported into the cytoplasm where it becomes incorporated into organic compounds having a wide molecular weight range.

This presupposes that, once in the cell, technetium changes its state of oxidation. Work is in progress to show if and to what extent the pertechnetate anion is reduced by the microbial cells and by their homogenates.

The fact that technetium binds to microbial organic molecules is important from the radioprotection point of view. Though grazers consume the whole cells (cell wall and/or cell membrane, organites and cytoplasm), the fate of the ingested radioactivity may depend on its chemical form [35]. Taking into account the results obtained with other radionuclides [36], a longer retention time of technetium in living organisms and man may be expected if this element is present in organic form. Since *Flavobacterium halmephilum* and *Uronema marinum*, as well as microalgae, are eaten by other organisms, part of technetium taken up by these micro-organisms might be transferred through the food chain. However, till now, no biomagnification of technetium in food chains has been found [6].

No clear-cut biological effects of $NH_4{}^{99}TcO_4$ were observed in our experiments, in spite of the relatively high concentrations of technetium used. The light stimulatory effect found in some experiments may be due to NH_4^+, as already suggested [7]. The observed resistance of *Chlamydomonas reinhardtii* and *Dunaliella bioculata* to high concentrations of ^{99}Tc-pertechnetate (up to 100 μg ml^{-1}) is in agreement with the observation, by Gearing *et al.* [29], that the growth of the microalgae *Chlorella sorokiniana* and *Dunaliella tertiolecta* was not inhibited even at 600 μg ^{99}Tc ml^{-1}. On the contrary, the non-sulphur purple bacterium *Rhodospirillum rubrum* was strongly inhibited already at 1 μg ^{99}Tc ml^{-1}. Since the response of aquatic micro-organisms to technetium seems to vary widely from one species to another, a screening on the uptake capability of a large number of species would be of interest. As large variations of the technetium concentration factor were observed in aquatic plants and animals, we cannot exclude the possibility that particular species of micro-organisms would be able to accumulate larger amounts of technetium than those reported in this work.

ACKNOWLEDGEMENTS

We thank M. C. Hemelaers-Gielen for typing the text and M. Mattheuwsen (Technical Services) for making part of the graphs. This work was supported in part by contract CEC No. BIO-B-485-82-B (Publication No. 2115) and by the Belgian Ministry of Labour (BTK Project No. 17.003). We thank Mrs M. Tuaux, Mr A. Bossus and Mr G. Nuyts for excellent technical assistance. Mr J. M. Bouquegneau is a 'Chercheur Qualifié' of the 'Fond National de la Recherche Scientifique'.

REFERENCES

1. Bonotto, S., Kirchmann, R. and Myttenaere, C. (1983). Need for research on the role of microorganisms on the behaviour of radionuclides in aquatic and terrestrial systems and their transfer to man. Workshop on the Role of Microflora and Microfauna on the Long-term Availability of the Radionuclides in Marine Ecosystems, Forte di Santa Teresa, La Spezia, Italy, 27–29 April.
2. Vandecasteele, C. M., Delmotte, A., Roucoux, P. and Van Hove, C. (1982). Technétium et organismes diazotrophes: Toxicité, localisation, facteurs de transfert. In: *Environmental Migration of Long-lived Radionuclides*, IAEA, Vienna, pp. 275–86.
3. Bonotto, S., Gerber, G. B., Cogneau, M. and Kirchmann, R. (1983). Behaviour of technetium in several marine algae. *Ann. Ass. Belge de Radioprotection*, **8**, 281–92.
4. Dartevelle-Moureau, Z., Jakubowski-Mania, B., Oleksi-Gebski, M. J., Bossus, A., Vandecasteele, C. M., Cogneau, M., Bonotto, S. and Ben, D. van der (1983). Effects of technetium on marine microorganisms. Workshop on the Role of Microflora and Microfauna on the Long-term Availability of the Radionuclides in Marine Ecosystems, Forte di Santa Teresa, La Spezia, Italy, 27–29 April 1983.
5. Bonotto, S., Gerber, G. B., Garten, C. T., Vandecasteele, C. M., Myttenaere, C., Van Baelen, J., Cogneau, M. and Ben, D. van der (1984). Uptake and distribution of technetium in several marine algae. In: *Proceedings of the International Symposium on the Behaviour of Long-lived Radionuclides in the Marine Environment*, La Spezia, Italy, 28–30 September 1983, EUR 9214 EN, pp. 381–96.
6. Verthe, C., Maquet, M. N., Moureau, Z., Mania, B., Van Baelen, J., Ben, D. van der, Cogneau, M., Vandecasteele, C. M., Bouquegneau, J. M., Myttenaere, C. and Bonotto, S. (1984). Transfer of technetium in an experimental food chain. In: *Proceedings of the International Symposium on the Behaviour of Long-lived Radionuclides in the Marine Environment*, La Spezia, Italy, 28–30 September 1983, EUR 9214 EN, pp. 397–415.
7. Moureau, Z., Mania, B., Tuaux, M., Van Baelen, J., Verthe, C., Maquet, M. N., Bouquegneau, J. M., Cogneau, M., Ben, D. van der, Vandecasteele,

C. M., Myttenaere, C. and Bonotto, S. (1984). Uptake, distribution and biological effects of technetium in aquatic microorganisms. In: *Proceedings of the Workshop on Role of Microorganisms on the Behaviour of Radionuclides in Aquatic and Terrestrial Systems and their Transfer to Man*, Brussels, Belgium, 25–27 April 1984, IUR-CEC, pp. 66–74.

8. Vandecasteele, C. M., Delmotte, A. Henrot, J., Van Hove, C. and Cogneau, M. (1984). Interaction between technetium and nitrogen fixing organisms. In: *Proceedings of the Workshop on Role of Microorganisms on the Behaviour of Radionuclides in Aquatic and Terrestrial Systems and their Transfer to Man*, Brussels, Belgium, 25–27 April 1984, IUR-CEC, pp. 158–72.
9. Vandecasteele, C. M., Mousny, J. M. and Lietart, E. A. (1984). Conséquences radiologiques de diverses pratiques culturales en rivières irriguées contaminées en technétium-99. In: Proceedings of the Seminar on the Environmental Transfer to Man of Radionuclides Released from Nuclear Installations, Brussels, Belgium, 17–21 October 1983, CEC, pp. 711–25.
10. Sueoka, N. (1960). Mitotic replication of deoxyribonucleic acid in *Chlamydomonas reinhardtii*. *Proc. Nat. Ac. Sci.*, **46**, 83–91.
11. Bonotto, S., Gerber, G. B., Arapis, G. and Kirchmann, R. (1982). Modelization of tritium transfer into the organic compartments of algae. *Ann. Ass. Belge de Radioprotection*, **7**, 283–92.
12. Paerl, H. W. and Shimp, S. L. (1973). Preparation of filtered plankton and detritus for study with scanning electron microscopy. *Limnol. Oceanogr.*, **18**, 802–5.
13. Noel-Lambot, F., Bouquegneau, J. M., Frankenne, F. and Disteche, A. (1978). Le rôle des métallothionéines dans le stockage des métaux lourds chez les animaux marins. *Rev. Int. Océanogr. Méd.*, **49**, 13–20.
14. Schor, S., Siekevitz, P. and Palade, G. E. (1970). Cyclic changes in thylakoid membranes of synchronized *Chlamydomonas reinhardtii*. *Proc. Nat. Acad. Sci.*, **66**, 174–80.
15. Gruber, H. E. (1978). X-ray and proton-induced ultrastructural changes in dividing *Chlamydomonas reinhardtii*. *Rad. Res.*, **73**, 137–59.
16. Marano, F. (1976). Etude ultrastructurale de la division chez *Dunaliella*. *J. Microscopie Biol. Cell.*, **25**, 279–82.
17. Marano, F. (1979). Biosynthèse des DNA nucléaire et chloroplastique chez une volvocale sans paroi. *Biol. Cellulaire*, **36**, 65–70.
18. Marano, F., Amancio, S. and Durrand, A. M. (1978). Synchronous growth and synthesis of macromolecules in a naturally wall-less volvocale, *Dunaliella bioculata*. *Protoplasma*, **95**, 135–44.
19. Grizeau, D., Jeanne, N. and Puiseux-Dao, S. (1983). Effects of various factors and inhibitors on the regulation of cell volume in *Dunaliella bioculata*. *Plant Sci. Lett.*, **28**, 213–22.
20. Strack, S., Bonotto, S. and Kirchmann, R. (1980). Radioactive contamination of the marine environment: Uptake and distribution of ^3H in *Dunaliella bioculata*. *Helgoländer Meeresunters.*, **33**, 153–63.
21. Davies, D. R. and Plaskitt, A. (1971). Genetical and structural analyses of cell-wall formation in *Chlamydomonas reinhardtii*. *Genet. Res.*, **17**, 33–43.
22. Davies, D. R. (1972). Cell wall organization in *Chlamydomonas reinhardtii*. The role of extra-nuclear systems. *Mol. Gen. Genet.*, **115**, 334–48.

23. FOSTER-SMITH, R. L. (1975). The effect of concentration of suspension on the filtration rates and pseudofaecal production for *Mytilus edulis* L., *Cerastoderma edule* (L.) and *Venerupis pullastra*. *J. exp. mar. Biol. Ecol.*, **17**, 1–22.
24. FISHER, N. S. (1982). Bioaccumulation of technetium by marine phytoplankton. *Envir. Sci. Technol.*, **16**, 579–81.
25. GRADMAN, D. (1970). Einfluss von Licht, Temperatur und Aussenmedium auf das elektrische Verhalten von *Acetabularia crenulata*. *Planta*, **93**, 323–53.
26. SADDLER, H. D. W. (1970). The ionic relations of *Acetabularia mediterranea*. *J. exp. Bot.*, **21**, 345–59.
27. SADDLER, H. D. W. (1970). Fluxes of sodium and potassium in *Acetabularia mediterranea*. *J. exp. Bot.*, **21**, 605–16.
28. SADDLER, H. D. W. (1970). The membrane potential of *Acetabularia mediterranea*. *J. Gen. Physiol.*, **55**, 802–21.
29. GEARING, P., VAN BAELEN, C. and PARKER, P. L. (1975). Biochemical effects of technetium-99-pertechnetate on microorganisms. *Plant Physiol.*, **55**, 240–6.
30. GROMOV, V. V. and SPITSYN, V. I. (1975). Assimilability of Pu, Ru and Tc by phytoplankton. USERDA transl., ORNL-TR-2907. Nat. Techn. Inf. Serv., Springfield, Va. Translated from *Dokl. Akad. Nauk SSSR*, **215**, 214–17.
31. GROMOV, V. V. (1976). Uptake of plutonium and other nuclear wastes by plankton. *Mar. Sci. Commun.*, **2**, 227–47.
32. HOLM, E., RIOSECO, J. and CHRISTENSEN, G. C. (1984). ^{99}Tc in *Fucus* from Norwegian waters. In: *Proceedings of the International Symposium on the Behaviour of Long-lived Radionuclides in the Marine Environment*, La Spezia, Italy, 28–30 September 1983, EUR 9214 EN, pp. 357–67.
33. MASSON, M., APROSI, G., LANIECE, A., GUEGUENIAT, P. and BELOT, Y. (1981). Approches expérimentales de l'étude des transferts du technétium à des sédiments et à des espèces marines benthiques. In: *Impacts of Radionuclide Releases into the Marine Environment*, Vienna, International Atomic Energy Agency, pp. 341–59.
34. PIGNOLET, L., MYTTENAERE, C., VANDECASTEELE, C. M., MOUREAU, Z. and COGNEAU, M. (1984). Behaviour of technetium in the marine environment: uptake and distribution of Tc into sediments and sedimentary microorganisms. In: *Proceedings of the International Symposium on the Behaviour of Long-lived Radionuclides in the Marine Environment*, La Spezia, Italy, 28–30 September 1983, EUR 9214 EN, pp. 369–79.
35. BONOTTO, S., BOSSUS, A., NUYTS, G., KIRCHMANN, R., MATHOT, P., COLARD, J. and CINELLI, F. (1983). Experimental uptake study of ^{60}Co, ^{137}Cs, ^{125}Sb and ^{65}Zn in four marine algae. In: *Wastes in the Ocean, Vol. 3, Radioactive Wastes and the Ocean*, New York, Wiley-Interscience, pp. 287–300.
36. HIGUCHI, M., KOMATSU, K. and MUKADE, A. (1980). Assessment of tritium effect on brine shrimp, *Artemia salina*, reared in a model ecosystem contaminated with tritiated water. In: *Radiation Effects on Aquatic Organisms* (Ed. N. Egami), Japan Science Society Press/University Park Press, Baltimore, pp. 13–25.

20

Technetium in Marine Animals

C. VERTHÉ,[a] Z. MOUREAU,[b] B. MANIA,[b] J. VAN BAELEN,[c]
D. VAN DER BEN,[b] M. COGNEAU,[d] C. M. VANDECASTEELE,[e]
J. M. BOUQUEGNEAU,[a] C. MYTTENAERE[f] and S. BONOTTO[c]

[a] *Université de Liège, Laboratoire d'Océanologie, Institut de Chimie, Bâtiment 6, B-4000 Liège, Sart Tilman, Belgium*

[b] *Institut Royal des Sciences Naturelles de Belgique, rue Vautier 29, B-1040 Brussels, Belgium*

[c] *Département de Radiobiologie, CEN-SCK, Boeretang 200, B-2400 Mol, Belgium*

[d] *Université Catholique de Louvain, Laboratoire de Chimie Nucléaire, 2 chemin du Cyclotron, B-1348 Louvain-la-Neuve, Belgium*

[e] *Université Catholique de Louvain, Laboratoire de Physiologie Végétale, Place Croix du Sud 4, B-1348 Louvain-la-Neuve, Belgium*

[f] *Commission of the European Communities, CEC, DGXII, rue de la Loi 200, B-1049 Brussels, Belgium*

1. INTRODUCTION

In previous work we have studied the uptake and distribution of technetium in several marine organisms, including microbial species, unicellular and pluricellular algae and molluscs [1–5]. The experimental work on technetium uptake by molluscs (*Littorina littorea*, *Mytilus edulis*, *Patella caerula* and *Patella vulgata*) was continued. In addition, investigations were realised on the marine fish *Serranus cabrilla*, contaminated with ^{99}Tc-pertechnetate (direct contamination) or fed with labelled molluscs (indirect contamination).

The most significant results of the experimental work are summarised in this paper. More detailed information will be reported later on (paper in preparation).

2. MATERIALS AND METHODS

2.1. Molluscs and fishes

The animals were maintained in aerated aquaria and fed as previously reported [1].

2.2. Labelling and counting

The animals were labelled by adding to seawater $NH_4{}^{99}TcO_4$ and/or $K^{95m}TcO_4$. The fish *Serranus cabrilla* was treated with ammonium pertechnetate or fed with labelled molluscs. Technetium activity was measured by liquid scintillation counting (^{99}Tc) or by gamma spectrometry (^{95m}Tc) in a Packard spectrometer.

2.3. Column chromatography

Chromatographic analysis was done on Ultrogel AcA54 (LKB) according to the procedure already described [1].

3. RESULTS

3.1. Uptake of technetium by molluscs

The three species of molluscs *Littorina littorea*, *Patella caerulea* and *Patella vulgata* were found to accumulate Tc to a very high extent. On the other hand, *Mytilus edulis* has a Tc concentration in its soft tissues 6–13 times lower than the three other species. For each species, a decreasing concentration factor with increasing Tc concentration in the medium was observed.

3.2. Uptake of technetium by fish

In the marine fish *Serranus cabrilla*, a concentration factor of only 0·2 was measured after direct contamination. However, in indirect contamination experiments, in which the fishes were fed with labelled molluscs, 20% of the ingested Tc was assimilated.

3.3. Distribution of technetium in molluscs

On the basis of contamination and decontamination experiments and of chromatographic analyses, different patterns of Tc accumulation in molluscs can be distinguished.

In *Mytilus edulis*, a very low uptake of Tc occurs: the influx is very quickly counterbalanced by loss, and no particular binding sites,

Table I
Technetium activity in soft tissues and shell of *Mytilus edulis* supplied during 12 days with $K^{95m}TcO_4$

Mussel	*Radioactivity*			
	Soft tissues		*Shell*	
	(nCi g⁻¹)	*(%)*	*(nCi g⁻¹)*	*(%)*
1	4·39	43·6	5·67	56·4
2	8·38	54·3	7·07	45·7
3	6·29	40·2	9·37	59·8
4	5·62	28·1	14·38	71·9
5	7·15	40·0	10·73	60·0
6	5·94	27·4	15·78	72·6
Mean	38·9 ± 10·11		61·07 ± 10·11	

permitting concentration of the element, appear to exist. An important trapping of Tc in the shells is, however, observed (Table I). In addition, the highest Tc accumulation values are found in the hepatopancreas, as already observed in previous work [1].

In *Littorina littorea*, important Tc accumulation occurs, which is counterbalanced by slow elimination.

Finally, in *Patella caerulea*, Tc is very quickly immobilised, mainly at a metallothionein-like protein level, and no significant elimination is

Table II
Effect of boiling on the loss of technetium in *Mytilus edulis* supplied during 12 days with $K^{95m}TcO_4$

Mussel	*^{95m}Tc activity*						
	Total animal			*Soft tissues*		*Shell*	
	Before boiling	*After boiling*		*After boiling*		*After boiling*	
	(nCi g⁻¹)	*(nCi g⁻¹)*	*(%)*	*(nCi g⁻¹)*	*(%)*	*(nCi g⁻¹)*	*(%)*
1	8·35	7·15	85·7	5·56	33·3	8·75	52·4
2	8·38	6·89	82·2	6·74	40·2	7·04	42·0
3	15·38	10·91	70·9	5·49	19·5	15·82	51·4
4	14·36	12·39	86·2	4·87	16·9	19·90	69·3
Mean		81·25 ± 7·12		27·47 ± 11·12		53·77 ± 11·36	

detectable. This behaviour results in a linear uptake of technetium as a function of time.

3.4. Loss of technetium in boiled mussels

Since sea mussels (*Mytilus edulis*) are usually eaten by man, it was of interest to know to what extent technetium is lost by the edible parts after boiling. ^{95m}Tc-labelled mussels were counted in a gamma spectrometer and then transferred for 10 min to boiling water. The animals were rapidly washed, blotted on filter paper and counted. They were then dissected into soft tissues and shell, which were counted separately.

Table II shows that boiling removes only about one-fifth (18·75%) of the total radioactivity contained in the mussels and that the soft tissues remain contaminated.

4. DISCUSSION

Studies on the uptake, distribution and loss of technetium by marine animals are useful not only for assessing the impact of long-lived radionuclide releases on life in the marine environment but also for evaluating the risk of human contamination through the aquatic food chains. It is interesting to note that the four species of molluscs investigated are all capable of taking up technetium, which becomes incorporated into the different organs of the animals in a more or less stable form.

Fortunately, *Mytilus edulis*, which is largely eaten by man, shows a low concentration factor. However, some parts of the animal (hepatopancreas and shell) may fix significant amounts of this radionuclide [1]. The common practice of boiling the mussels removes only about one-fifth (18·75%) of the radioactivity. This finding suggests that most technetium binding sites are not destroyed at 100°C. It would be of interest to compare the chemical forms of technetium in untreated and in boiled mussels, by using biochemical procedures and chromatography [1].

Chromatographic analysis has enabled detection of the presence of technetium in organic molecules, a fact which suggests that, once in the animal, technetium changes its oxidation state and becomes incorporated into cell constituents. The nature of shell binding sites remains unknown. In soft tissues the presence of a metallothionein-like protein, capable of fixing technetium, is suggested for *Patella caerulea*. No experimental evidence has yet been obtained for the presence of a similar protein in *Mytilus edulis*, though this species is known to synthesise metallothionein molecules after contamination with other metals [6–9].

Our findings on technetium uptake and distribution in *Mytilus edulis* are in good agreement with the results obtained by Beasley *et al.* [10] on another species, *Mytilus californianus*.

Finally, the observation on technetium uptake in a marine fish, *Serranus cabrilla*, is of particular interest from the radiation protection point of view, since it suggests that accumulation via ingested food could be important.

ACKNOWLEDGEMENTS

This work was supported in part by Contract CEC No. BIO-B-485-82-B (Publication No. 2114) and by BTK Project No. 17.003. We thank Mrs M. Tuaux, Mr A. Bossus and Mr G. Nuyts for excellent technical assistance. Mr J. M. Bouquegneau is a 'Chercheur Qualifié' of the 'Fond National de la Recherche Scientifique'.

REFERENCES

1. Verthé, C., Maquet, M. N., Moureau, Z., Mania, B., Van Baelen, J., Ben, D. van der, Cogneau, M., Vandecasteele, C. M., Bouquegneau, J. M., Myttenaere, C. and Bonotto, S. (1984). Transfer of technetium in an experimental marine food chain. In: *Proceedings of the International Symposium on the Behaviour of Long-lived Radionuclides in the Marine Environment*, La Spezia, Italy, 28–30 September 1983, EUR 9214 EN, pp. 397–415.
2. Moureau, Z., Mania, B., Tuaux, M., Van Baelen, J., Verthé, C., Maquet, M. N., Bouquegneau, J. M., Cogneau, M., Ben, D. van der, Vandecasteele, C. M., Myttenaere, C., Bonotto, S. and Kirchmann, R. (1984). Uptake, distribution and biological effects of technetium in aquatic microorganisms. In: *Proceedings of the Workshop on the Role of Microorganisms on the Behaviour of Radionuclides in Aquatic and Terrestrial Systems and their Transfer to Man*, Brussels, Belgium, 25–27 April 1984, CEC-IUR, pp. 66–74.
3. Mania, B., Moureau, Z., Ben, D. van der, Van Baelen, J., Verthé, C., Bouquegneau, J. M., Cogneau, M., Bonotto, S., Vandecasteele, C. M., Pignolet, L. and Myttenaere, C. (1985). Technetium in micro-organisms (this volume, pp. 229–44).
4. Bonotto, S., Van Baelen, J., Arapis, G., Garten, C. T., Vandecasteele, C. M., Myttenaere, C. and Cogneau, M. (1984). Incorporation of technetium in several marine algae and its transfer to a higher trophic level. *Arch. Int. Physiol. Biochim.*, **92**, V14–V15.
5. Bonotto, S., Gerber, G. B., Garten, C. T. Jr, Vandecasteele, C. M., Myttenaere, C., Van Baelen, J., Cogneau, M. and Ben, D. van der (1984). Uptake and distribution of technetium in several marine algae. In: *Proceedings*

of the International Symposium on the Behaviour of Long-lived Radionuclides in the Marine Environment, La Spezia, Italy, 28–30 September 1983. EUR 9214 EN, pp. 381–96.

6. Noel-Lambot, F. (1976). Distribution of cadmium, zinc and copper in the mussel *Mytilus edulis*. Existence of cadmium-binding proteins similar to metallothioneins. *Experientia*, **32**, 324–5.
7. Noel-Lambot, F., Bouquegneau, J. M., Frankenne, F. and Disteche, A. (1978). Le rôle des métallothionéines dans le stockage des métaux lourds chez les animaux marins. *Rev. Int. Océanogr. Méd.*, **49**, 13–20.
8. Köhler, K. and Rusgård, H. U. (1982). Formation of metallothioneins in relation to accumulation of cadmium in the common mussel *Mytilus edulis*. *Mar. Biol.*, **66**, 53–8.
9. Poulsen, E., Rusgård, H. U. and Møhlenberg, F. (1982). Accumulation of cadmium and bioenergetics in the mussel *Mytilus edulis*. *Mar. Biol.*, **68**, 25–9.
10. Beasley, J. M., Gonor, J. J. and Lorz, H. V. (1982). Technetium: Uptake, organ distribution and loss in the mussel *Mytilus californianus* (Conrad) and the oyster *Crassostrea gigas* (Thunberg). *Mar. Envir. Res.*, **7**, 103–16.

21

Le Technétium et l'Ormeau (*Haliotis tuberculata*): Données Expérimentales et '*in situ*'

M. MASSON,[a] G. APROSI[b] and P. GERMAIN[a]

[a] *Commissariat à l'Energie Atomique, IPSN-DERS-SERE, Laboratoire de Radioécologie Marine, BP 270, 50107 Cherbourg, France*

[b] *Electricité de France, Direction des Etudes et Recherches, Service Applications, Electricité, Environnement, 6 Quai Watier, 78400 Chatou, France*

1. INTRODUCTION

Le premier travail expérimental sur les transferts du technétium (Tc) en milieu marin est celui de Spies [1]. Cet auteur trouve que l'ormeau rouge américain *Haliotis rufescens* accumule le ^{95m}Tc à partir de l'eau de mer et obtient des facteurs de concentration (FC) de 135 à 205. En 1979 notre laboratoire commence un programme expérimental de quatre années sur les transferts du technétium aux différents compartiments du milieu marin avec notamment des études sur les mollusques, dont l'ormeau européen *H. tuberculata* [2–11].

A l'échelon mondial la production de l'ormeau est estimée à 22 000 t (13 espèces pêchées) [12]. Ce mollusque est très consommé au Japon et en Californie, son aquaculture est maîtrisée et pleine d'avenir. La production française d'ormeaux est estimée à seulement 5 à 6 t par an et se limite à une pêche à pied puisque la pêche par plongée est interdite [13]. Les régions les plus concernées sont essentiellement la Bretagne, le Cotentin et les îles anglo-normandes.

Les quantités de ^{99}Tc présentes dans le milieu augmentent du fait du développement de l'énergie nucléaire, les effluents des usines de retraitement sont actuellement la source principale en milieu marin [13]. Il est donc intéressant d'un point de vue radioécologique de vérifier si l'ormeau européen *H. tuberculata* est une espèce bioindicatrice du Tc et une voie de transfert de ce radioélément à l'homme. Parallèlement à nos études

Beasley *et al.* entreprennent aux USA des expérimentations avec l'ormeau américain [14]. Nous comparerons dans cette note ces résultats expérimentaux à des données obtenues par l'analyse du ^{99}Tc dans des ormeaux prélevés au voisinage de La Hague.

2. MATERIELS ET METHODES

Des ormeaux juvéniles d'écloserie de 0·5 à 2 g nous sont fournis par le Centre National d'Exploitation des Océans. Des ormeaux adultes de 40 à 120 g sont pêchés en plongée à La Hague. Ils sont acclimatés 4 à 5 semaines aux conditions du laboratoire, à 14°C, et nourris '*ad libitum*' avec des fragments d'algues rouges *Rhodymenia palmata* ou brunes *Laminaria* sp. Pour l'expérimentation les ormeaux sont placés individuellement dans des boîtes en plastique perforées immergées dans des aquariums de 10 à 60 litres, avec un bullage puissant.

Pour étudier l'influence de la température deux lots de juvéniles sont acclimatés à 7° et 21 °C.

Le ^{95m}Tc nous est fourni par 'New England Nuclear', sous forme d'une solution de pertechnétate ou par 'Oak Ridge Laboratory' sous forme solide [8]. Les activités en ^{95m}Tc dans les aquariums sont comprises entre 0·1 et 1 μCi litre^{-1} (3·7 à 37 × 10^3 Bq litre^{-1}) et maintenues constantes en renouvelant l'eau tous les deux jours. La chute de radioactivité due à la période physique de 61 jours du ^{95m}Tc a necessité le comptage régulier d'un standard. Les comptages sont effectués par spectrométrie gamma sur sonde plate NaI (Tl) pour les gros ormeaux et sur sonde puits pour les juvéniles. L'eau est comptée dans des moulages en résine de forme voisine de celle de l'ormeau adulte.

Après l'étude de la fixation les ormeaux sont replacés en eau de mer courante pour suivre la cinétique d'élimination. En cours d'expérimentation quelques individus sont disséqués. Une extraction des principaux constituants biochimiques est faite sur des homogénats de tissus selon la méthode de Schmidt et Thannauser [15], modifiée par Herve [16]. Pour localiser le Tc dans les tissus quelques ormeaux ont été fortement contaminés (30 000 cpm g^{-1}) avec du $^{99}TcO_4^-$ pour être fixés au liquide de Bouin. Des coupes de tissus inclus à la paraffine sont recouvertes d'une émulsion K2 ou K5 Ilford et après un temps de contact de 1 à 3 mois révélées et colorées par le trichrome de Prenant-Gabe pour être observées au microscope [7].

Pour étudier le rôle de la nourriture dans les transferts du ^{95m}Tc des ormeaux juvéniles sont élevés en eau de mer contaminée pendant 50 jours, puis nourris pendant 12 jours avec des fragments d'algues marquées par du ^{95m}Tc et replacés à jeun pendant 30 jours en eau de mer courante. Des analyses du ^{99}Tc selon la méthode de Patti *et al.* [17] ont été réalisées sur des ormeaux adultes prélevés à La Hague.

3. RESULTATS ET DISCUSSION

3.1. Fixations à partir de l'eau

3.1.1. Influence des paramètres taille et température

Que les ormeaux soient juvéniles (0·5 à 2 g) ou adultes (40 à 120 g) nous observons sur la Figure 1 que le facteur de concentration moyen du

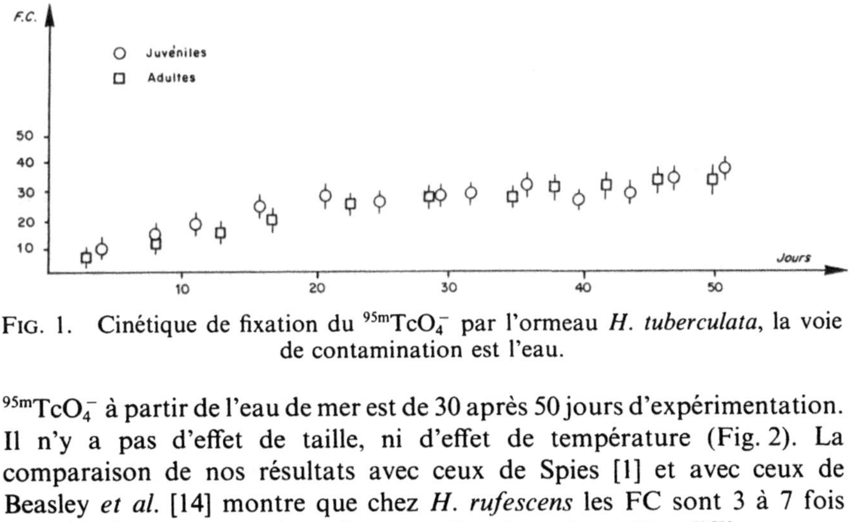

FIG. 1. Cinétique de fixation du $^{95m}TcO_4^-$ par l'ormeau *H. tuberculata*, la voie de contamination est l'eau.

$^{95m}TcO_4^-$ à partir de l'eau de mer est de 30 après 50 jours d'expérimentation. Il n'y a pas d'effet de taille, ni d'effet de température (Fig. 2). La comparaison de nos résultats avec ceux de Spies [1] et avec ceux de Beasley *et al.* [14] montre que chez *H. rufescens* les FC sont 3 à 7 fois plus élevés que ceux observés avec *H. tuberculata*. Ces différences ne s'expliquent pas par des tailles différentes puisque Spies trouve des FC de 135 à 205 avec des mollusques de 500 g à 1 kg qui ne sont pas corrélés avec le poids, et que Beasley *et al.* observent des FC de 100, proches des résultats de Spies, avec des juvéniles de *H. rufescens*.

3.1.2. Répartition par organe

C'est la masse viscérale qui accumule le plus le ^{95m}Tc avec des FC variant de 80 à 140. Les viscères représentent 28% du poids du mollusque jeune et concentrent 80% de la radioactivité totale (Tab. I). La coquille avec

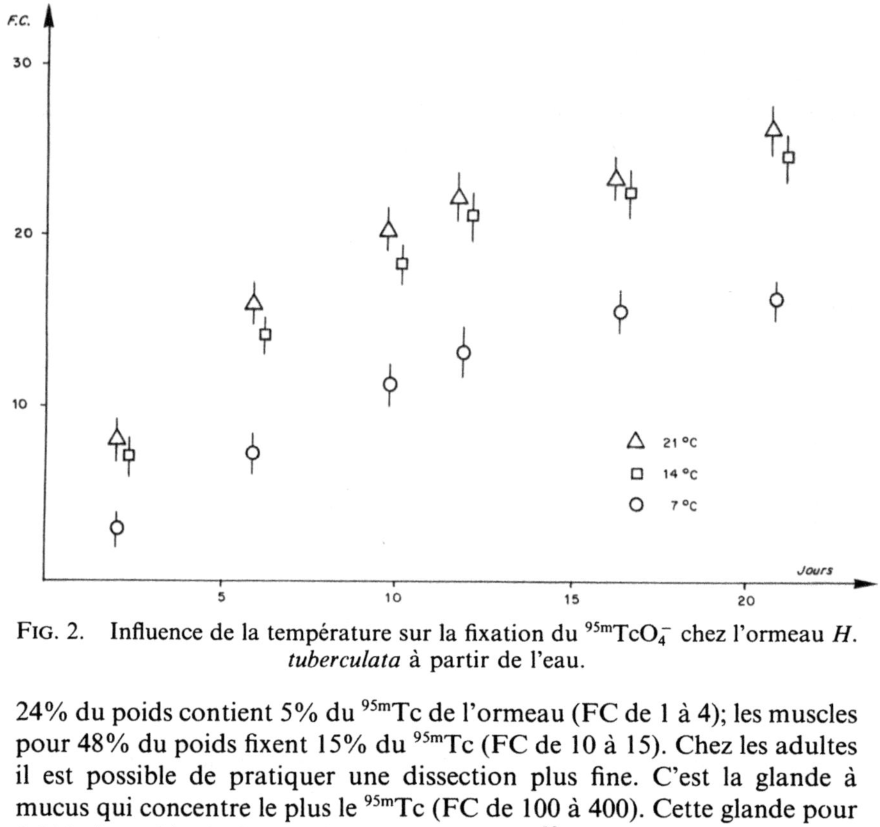

FIG. 2. Influence de la température sur la fixation du $^{95m}TcO_4^-$ chez l'ormeau *H. tuberculata* à partir de l'eau.

24% du poids contient 5% du ^{95m}Tc de l'ormeau (FC de 1 à 4); les muscles pour 48% du poids fixent 15% du ^{95m}Tc (FC de 10 à 15). Chez les adultes il est possible de pratiquer une dissection plus fine. C'est la glande à mucus qui concentre le plus le ^{95m}Tc (FC de 100 à 400). Cette glande pour 1·2% du poids de l'animal contient 5% du ^{95m}Tc accumulé. Le tractus digestif de l'adulte est surtout composé de la glande digestive dans laquelle s'intrique la gonade. Pour 18% du poids de l'animal il concentre 38·5%

Tableau I

Pourcentage moyen de la répartition du pertechnétate, sulfure et pyrophosphate de ^{95m}Tc dans des ormeaux juvéniles contaminés par l'eau

	% du poids de l'ormeau	*% de la radioactivité totale de l'ormeau*		
		Pertechnétate	*Sulfure*	*Pyrophosphate*
Coquille	24	5	10	10
Muscle	48	15	20	20
Viscères	28	80	70	70

Tableau II
Pourcentage moyen de la répartition du $^{95m}TcO_4^-$ dans les organes de l'ormeau adulte contaminé par l'eau

	% du poids de l'ormeau	*% de la radioactivité totale de l'ormeau*	
Coquille	30	49	
Muscle	50	5	
Viscères			
Tractus digestif + gonade	18	38·5	46
Branchies	0·8	2·5	
Glande à mucus	1·2	5	

du ^{95m}Tc (FC de la glande digestive de 100 à 150, de la gonade de 30 à 70). Les branchies concentrent 160 fois le ^{95m}Tc. Les muscles qui représentent 50% du poids de l'ormeau contiennent 5% du ^{95m}Tc (FC de 3 à 10).

On note que la répartition entre organes est différente entre le jeune et l'adulte (Tab. I et II). Cette différence dans les pourcentages de répartition du ^{95m}Tc est due à la coquille qui chez les adultes est plus ou moins perforée par des organismes tubicoles (vers, éponges, algues) alors qu'elle est parfaitement lisse chez le jeune. Avec 30% du poids del'ormeau adulte la coquille contient en moyenne 50% du ^{95m}Tc fixé par le mollusque. Cette observation est en accord avec celle de Spies sur l'ormeau américain. Par autoradiographie de coupes sagittales de coquille, cet auteur a observé que le ^{95m}Tc était associé aux perforations d'éponges et de vers [1]. Nous avons utilisé l'histoautoradiographie pour localiser le ^{99}Tc dans les tissus mous d'*H. tuberculata* juvéniles. Le ^{99}Tc est piégé à la surface ou dans les villosités des épithéliums, dans le mucus, dans les cellules de la glande à mucus, dans les amoebocytes, dans les vacuoles lysosomiales de la glande digestive.

3.1.3. Influence de la forme physico-chimique du ^{95m}Tc sur la fixation
Les contaminations avec des sulfures et des pyrophosphates de ^{95m}Tc montrent que ces formes physico-chimiques sont accumulées plus activement que le pertechnétate par les juvéniles (FC = 100) (Fig. 3). Pour les deux formes 70% du Tc se trouvent dans les viscères, 10% sur la coquille, 20% dans les muscles (Tab. I).

Les extractions séquentielles du ^{95m}Tc dans des homogénats des tissus mous des ormeaux juvéniles (Tab. III) mettent en évidence que le pertechnétate est associé pour 82% à des produits facilement extractibles

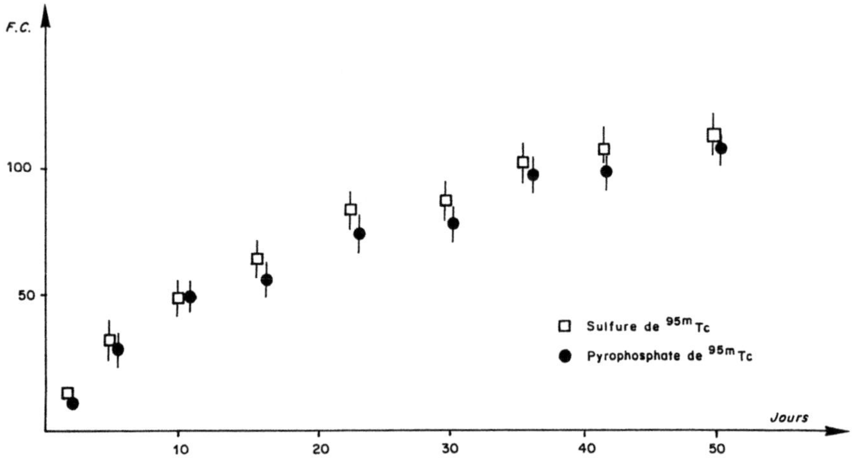

FIG. 3. Cinétiques de fixation du sulfure et du pyrophosphate de ^{95m}Tc par l'ormeau juvénile *H. tuberculata* à partir de l'eau.

et 12% restent liés aux culots protéiques. On retrouve cette tendance avec les organes pris individuellement (Tab. IV). Le pertechnétate ne s'associe pas aux lipides ni aux ADN. Dans le cas des sulfures et pyrophosphates les répartitions sont différentes avec respectivement 19 et 50% dans la fraction acido-soluble, 66 et 26% dans la fraction protéique (Tab. III).

3.2. Elimination après fixation à partir de l'eau

Les cinétiques de perte du pertechnétate sont biphasiques et identiques chez les juvéniles et les adultes (Fig. 4), avec une phase rapide correspondant à

Tableau III

Pourcentages moyens de ^{95m}Tc extraits par fractionnement bio-imique des parties molles de jeunes ormeaux ($n = 4$) après contamination par pertechnétate, sulfure et pyrophosphate de ^{95m}Tc.

Fractions \ *Formes physico-chimiques*	TcO_4^-	*Sulfure*	*Pyrophosphate*
Acido solubles	82·9 ± 2·1	19·5 ± 1	50·3 ± 1·5
Lipides	0·6 ± 0·2	4·3 ± 1·2	4·4 ± 1·8
ARN	3·2 ± 0·8	6·3 ± 0·3	15·2 ± 2·3
ADN	0·8 ± 0·4	3·5 ± 0·8	3·5 ± 0·8
Protéines	12·2 ± 1·2	66·2 ± 2·2	26·3 ± 1·2

Tableau IV

Pourcentages moyens de ^{95m}Tc extraits par fractionnement biochimique des organes de l'ormeau adulte après contamination par le pertechnétate ($n = 4$)

Organes / *Fractions*	*Branchies*	*Glande à mucus*	*Glande génitale*	*Glande digestive*	*Muscle*	*Tractus digestif*	*Haemo-lymphe*
Acido solubles	55·2 ± 0·8	85·3 ± 2·3	36·7 ± 16·3	62·4 ± 2	52·2 ± 2·1	58·8 ± 1	85 ± 1·4
Lipides	4·5 ± 0·4	2·1 ± 1·3	1·3 ± 0·7	4·5 ± 3·2	1 ± 0·6	3·9 ± 0·9	1·5 ± 0·4
ARN	14·7 ± 2·2	2·5 ± 1·7	11·2 ± 0·2	9·4 ± 1·3	10·5 ± 1·2	11·9 ± 1·2	2·5 ± 0·9
ADN	5·8 ± 3·9	1.5 ± 1·3	6·9 ± 3·6	1.4 ± 1	1·2 ± 1	3·6 ± 1·3	3·6 ± 3
Protéines	18·8 ± 1·2	6·6 ± 1·5	42·1 ± 13·4	23·2 ± 0·9	33·5 ± 1·2	20·6 ± 1·9	8·5 ± 2·1

une période biologique de 2·5 jours qui intéresse 20% du ^{95m}Tc fixé, suivie par une phase lente de 150 jours qui intéresse pour les 80% restants. Après 35 jours d'élimination chez l'adulte seules la glande digestive et la coquille contiennent du ^{95m}Tc. Chez le jeune le ^{95m}Tc n'est décelable que dans les viscères. Ces résultats concordent avec ceux de Beasley *et al.* chez *H. rufescens* quant à l'aspect biphasique de l'élimination. Cependant ces auteurs observent une période biologique d'élimination rapide de 7 jours pour 70% du ^{95m}Tc fixé suivie par une période lente de 58 jours pour le reste.

Dans les cas du pyrophosphate l'élimination est biphasique et plus

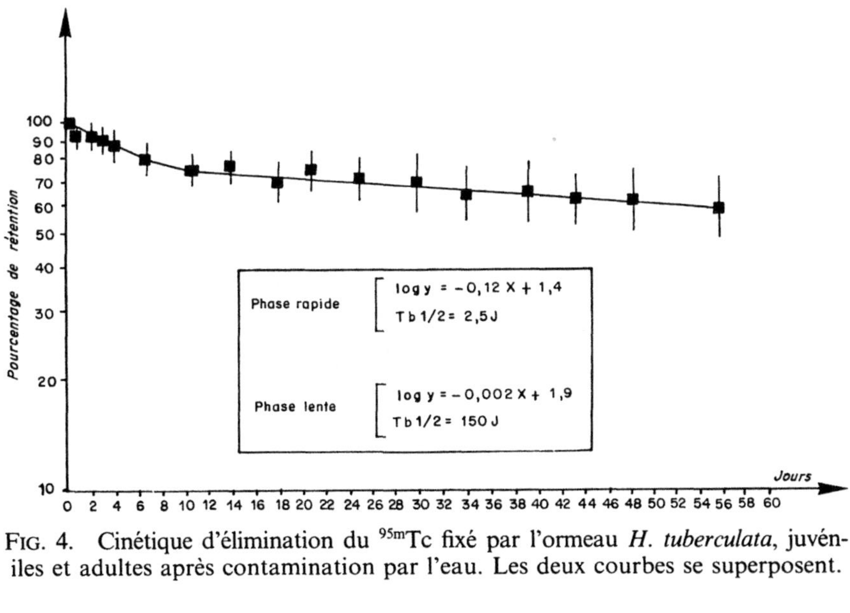

FIG. 4. Cinétique d'élimination du ^{95m}Tc fixé par l'ormeau *H. tuberculata*, juvéniles et adultes après contamination par l'eau. Les deux courbes se superposent.

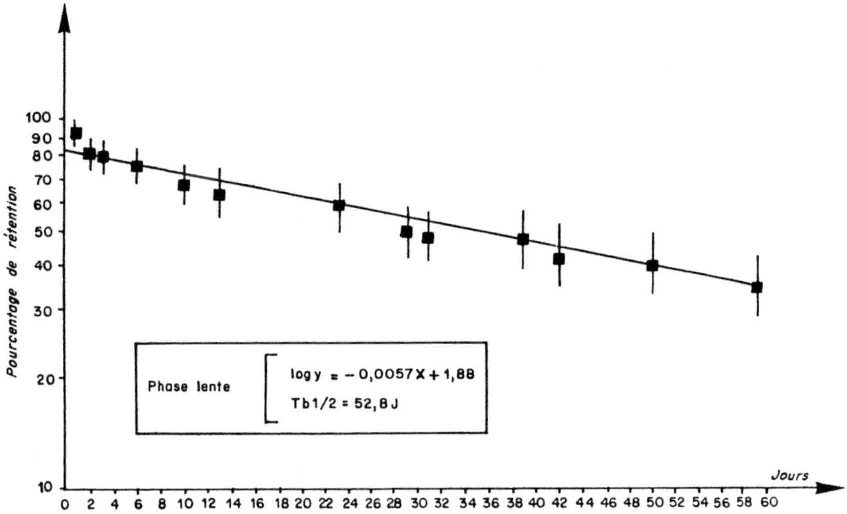

FIG. 5. Cinétique d'élimination du ^{95m}Tc fixé, sous forme de pyrophosphate, à partir de l'eau par l'ormeau juvénile.

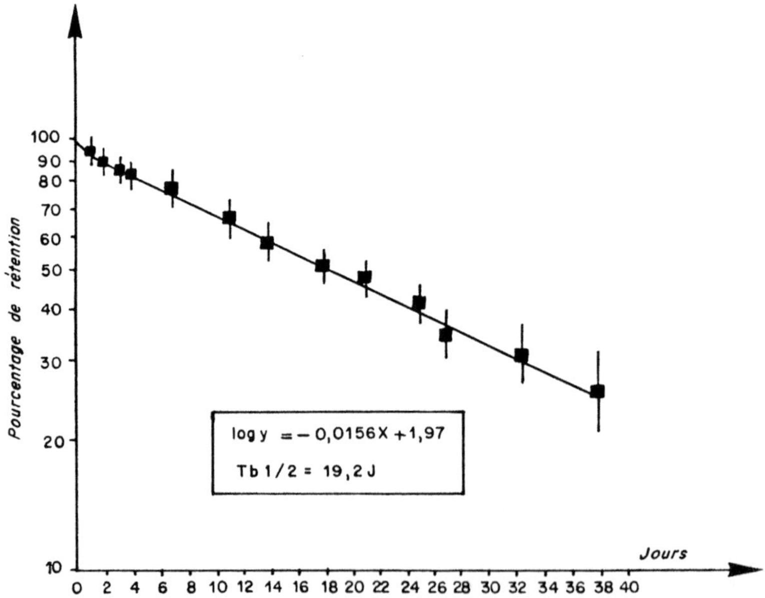

FIG. 6. Cinétique d'élimination du ^{95m}Tc fixé, sous forme de sulfure, à partir de l'eau par l'ormeau juvénile.

rapide que pour le pertechnétate, la période lente étant de 53 jours (Fig. 5). Le sulfure est épuré très rapidement avec une seule phase de 20 jours de période biologique (Fig. 6).

3.3. Fixation et élimination à partir de l'eau et de la nourriture

La cinétique concernant ce protocole est reportée sur la Figure 7. Des *H. tuberculata* juvéniles déjà en équilibre avec le pertechnétate de l'eau de mer (100 ± 20 cpm ml^{-1}) ont une radioactivité moyenne de 3000 ± 450 cpm g^{-1} frais, sont nourris pendant 12 jours avec des fragments de laminaires marqués par du ^{95m}Tc. La radioactivité moyenne des algues évolue de 10 000 cpm g^{-1} à 75 000 cpm g^{-1}, en cours d'expérience. Celle des ormeaux s'élève à 8000 ± 1000 cpm g^{-1} en fin d'essais. On observe une grande variabilité des résultats attributable à des différences notables des quantités de nourriture consommée individuellement. Les ormeaux sont ensuite placés en eau de mer courante, à jeun. L'élimination est biphasique, 20% du Tc sont éliminés en quelques jours, les 80% restants disparaissent avec une période biologique de 170 jours. Après 30 jours de perte seule la masse viscérale contient du Tc.

Si l'on compare ces résultats à ceux de Beasley *et al.* [14] on note une grande différence dans les valeurs des compartiments et des périodes biologiques calculées entre les deux espèces d'ormeaux. Ces auteurs ont développé un protocole différent du notre. Ils ont fourni à 5 jeunes ormeaux un repas d'algue contaminée par ^{95m}Tc, puis nourri les ormeaux pendant 65 jours avec des algues non contaminées. Ils déterminent une phase rapide d'élimination de 10 jours de période biologique pour 65% du Tc ingéré et une phase lente de 60 jours pour le reste. Ils en concluent que les taux de renouvellement du ^{95m}Tc fixé par l'eau et la nourriture sont les mêmes chez l'ormeau *H. rufescens*, ce qui est confirmé par nos observations expérimentales avec *H. tuberculata*.

3.4. ^{99}Tc dans les ormeaux prélevés '*in situ*'

Les résultats d'analyses radiochimiques du ^{99}Tc dans 12 ormeaux de 100 à 230 g prélevés à La Hague au voisinage des rejets de l'usine de retraitement sont reportés dans le Tableau V. Ce sont les viscères qui contiennent le plus de ^{99}Tc, particulièrement l'ensemble glande digestive et gonade (290 Bq kg^{-1} frais) avec plus de 85% du ^{99}Tc accumulé par les ormeaux. La coquille contient peu de Tc 4·3% du total fixé par les ormeaux contre 49% mesurés expérimentalement. Les muscles, parties consommées par l'homme, contiennent 9·9% du Tc soit 11·7 Bq kg^{-1} frais. Si l'on rapporte la quantité de ^{99}Tc par kg frais d'ormeau ou

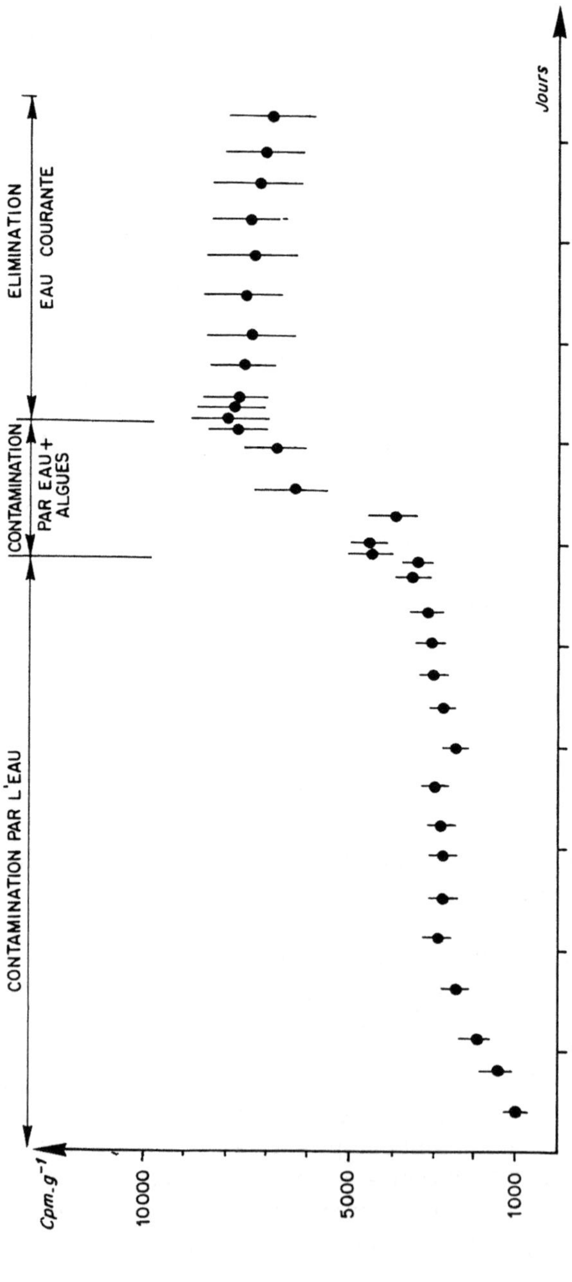

FIG. 7. Cinétique de fixation et d'élimination du $^{95m}TcO_4^-$ à partir de l'eau et de la nourriture chez les ormeaux juvéniles.

Tableau V
Teneurs en ^{99}Tc dans l'ormeau et dans ses organes à Goury (France)
[RR = (^{99}Tc kg^{-1} frais d'organe)/(^{99}Tc $litre^{-1}$ eau de mer)]

	Bq kg⁻¹ frais	*% de la radioactivité totale de l'ormeau*	*RR*
Coquille	8·1	4·3	615
Muscle	11·7	9·9	880
Viscères			
Tractus digestif + gonade	289·4	85·45	21 760
Branchies	10·1	0·15	760
Glande à mucus	6·7	0·2	500
Ormeau entier	57·5	—	4 320

d'organe à la quantité de ^{99}Tc trouvée en moyenne dans l'eau de mer (13·3 mBq $litre^{-1}$) (RR), on observe que ces mollusques contiennent globalement 4350 fois plus de Tc que l'eau, et le tractus digestif 21 760 fois plus. Ces résultats sont 2 ordres de grandeur plus élevés que ceux obtenus '*in vitro*'. Ces observations pour les ormeaux sont confirmées par celles obtenues avec le mollusque *Patella* sp. qui '*in vitro*' concentre moins de 20 fois le Tc et '*in situ*' 3500 fois [24]. Ceci laisserait supposer que la voie de transfert par la nourriture est prépondérante chez les mollusques brouteurs. La nourriture préférée des ormeaux, l'algue *Rhodymenia palmata*, contient 5·25 Bq kg^{-1} frais de ^{99}Tc (FC 395). On peut aussi argumenter que les conditions expérimentales sont loin de refléter la complexité du milieu naturel. Cette remarque est appuyée par le fait que les algues brunes, qui se contaminent directement par l'eau, présentent '*in vitro*' des FC de 1500 à 2500 [5, 25] et de 50 000 '*in situ*' [24].

4. CONCLUSION

Du point de vue de la compréhension des mécanismes biologiques qui régissent les transferts du technétium dans le milieu marin l'ormeau apparaît être un bon 'outil expérimental'. Pour la surveillance de l'environnement cette espèce est aussi un bon bioindicateur du ^{99}Tc.

Cependant si les expériences de laboratoire permettent de cerner l'importance relative des divers paramètres influençant la contamination et l'élimination du technétium, nous avons montré une distorsion entre les données obtenues expérimentalement et celles de l''*in situ*' qui semblerait mettre en évidence une accumulation prépondérante du technétium par la chaîne alimentaire chez l'ormeau. Or on connait très peu de choses sur les chaînes alimentaires naturelles de la majorité des espèces marines, même celles pour lesquelles l'aquaculture est maîtrisée comme l'ormeau. Il convient donc d'être prudent quant à l'utilisation des données expérimentales (FC, $Tb\frac{1}{2}$, taux de transfert par la nourriture, etc.) à des fins de modélisation ou de prévisions se rapportant à l'‘*in situ*’.

REMERCIEMENTS

Nous remercions Monsieur Patti et Mademoiselle Nazard (CEA/IPSN/DPS/SEAPS) pour la réalisation des analyses en ^{99}Tc.

RÉFÉRENCES

1. Spies, R. B. (1975). Uptake of technetium from seawater by red abalone *Haliotis rufescens*. *Hlth Phys.*, **29**, 695–9.
2. Aprosi, G. (1979). Le technétium, informations générales et synthèses bibliographiques. Rapport EDF-DER-HE 31-79.26.
3. Aprosi, G. et Masson, M. (1981). Méthodologie d'étude du transfert du technétium dans quelques organismes marins. Rapport EDF-DER-HE 31-81.15.
4. Aprosi, G. et Masson, M. (1981). Technétium: toxicité et action massique de l'isotope 99, transferts eau—sédiments et sédiments—annélides. Rapport EDF-DER-HE 31-81.16.
5. Aprosi, G. et Masson, M. (1981). Approche expérimentale de l'étude du transfert du technétium à quelques espèces marines benthiques. Rapport EDF-DER-HE 31-81.17.
6. Aprosi, G. et Masson, M. (1983). Les cinétiques d'élimination du technétium chez quelques espèces marines benthiques après accumulation à partir de l'eau et de la nourriture: influence de différents paramètres sur la fixation et l'élimination. Rapport EDF-DER-HE 31-83.06.
7. Masson, M. (1983). Les sites de fixation du technétium dans quelques organismes marins. Rapport EDF-DER-HE 31-83.05.
8. Masson, M., Aprosi, G., Laniece, A., Gueguenia, P. et Belot, Y. (1981). Approches expérimentales de l'étude des transferts du technétium à des sédiments et à des espèces marines benthiques (IAES-SM-248/124). In: *Impacts of Radionuclide Releases into the Marine Environment*, Vienne, 6–10 oct. 1980, Vienne, AIEA, pp. 341–59.
9. Jeanmaire, L., Masson, M., Patti, F., Germain, P. et Capellini, L. (1981). Technetium-99 content in some marine organisms collected near La Hague, France. *Mar. Poll. Bull.*, **12**, 29–32.
10. Masson, M., Patti, F., Capellini, L., Germain, P. et Jeanmaire, L. (1983). Etude de la dispersion du technétium 99 sur les côtes françaises de la Manche

à l'aide de deux indicateurs biologiques: *Fucus* sp. et *Patella* sp. *J. Radioanal. Chem.*, **77**, 247–55.

11. APROSI, G. et MASSON, M. (1984). Bilan des études expérimentales de transferts de technétium à des sédiments et à des espèces marines benthiques et comparaison à des résultats *in situ*. *Radioprotection*, **19**, no. 2, 89–103.
12. OLLEY, J. et THROWER, S. S. (1977). Abalone, an esoteric food. *Adv. Food Res.*, **23**, 143–86.
13. COCHARD, J. C. (1980). Recherches sur les facteurs déterminant la sexualité et la reproduction chez *Haliotis tuberculata* L. Thèse 3e cycle, Université de Bretagne Occidentale, no. 116.
14. BEASLEY, T. M., LORZ, H. V. et CONOR, J. J. (1982). Biokinetic behaviour of technetium in the red abalone *Haliotis rufescens*: a reassessment. *Hlth Phys.*, **43**, 501–7.
15. SCHMIDT, G. et THANNAUSER, S. J. (1945). A method for determination of desoxyribonucleic acid, ribonucleic acid and phosphoproteins in animal tissues. *J. Biol. Chem.*, **16**, 83–9.
16. HERVE, J. Y. (1982). Etude de l'accumulation, de la contamination et de la toxicité du sélénium chez l'huître japonaise *Crassostrea gigas* th. Thèse 3e cycle, Université de Caen.
17. PATTI, F., CAPELLINI, L., JEANMAIRE, L. et MASSON, M. (1982). Détermination du technétium 99 dans les échantillons biologiques marins (IAEA-SM-252/49). In: *Methods of Low Level Counting and Spectrometry*, Berlin, 6–10 avril 1981, Vienna, AIEA.
18. FOWLER, S. W., RENAYOUN, G., PARSI, P. et ESSA, M. W. A. (1981). Experimental studies on the bio-availability of technetium in selected marine organisms (IAEA-SM-248/113). In: *Impacts of Radionuclide Releases into the Marine Environment*, Vienne, 6–10 oct. 1980, Vienna, AIEA, pp. 319–39.
19. SCHULTE, E. H., SECONDINI, A. et SCOPPA, P. (1981). Trasferimento del technezio attraverso le catene alimentari marine. Atti del XXII congresso dell'Associazione Italiano di Protezione contro le Radiazioni, Brescia, 23–26 guiguino, Vol. 1, pp. 563–70.
20. BEASLEY, T. M. et GONOR, J. J. (1980). Biogeochemical studies of technetium in marine and estuarine ecosystems. Progress Report, July 1979–June 1980, Oregon State University, USA, Report DOE/EV/10251.
21. Beasley, T. M. (1981). Biogeochemical studies of technetium in marine and estuarine ecosystems. Progress Report, July 1980–July 1981, Oregon State University, USA, Report DOE/EV/10251-3.
22. BEASLEY, T. M., GONOR, J. J. et LORZ, H. V. (1982). Technetium uptake, organ distribution and loss in the mussel *Mytilus californianus* (Conrad) and the oyster *Crassostrea gigas* (Thunberg). *Mar. Envir. Res.*, **7**, 103–16.
23. FOWLER, S. W., ASTON, S. R., BENAYOUN, G. et PARSI, P. (1983). Bioavailability of technetium from artificially labelled North East Atlantic deep-sea sediments. *Mar. Envir. Res.*, **8**, 87–100.
24. PATTI, F., MASSON, M., VERGNAUD, G. et JEANMAIRE, L. (1984). Activités du technétium 99 mesurées dans les eaux résiduaires, l'eau de mer et deux bioindicateurs (Littoral de la Manche, 1983) (this volume, pp. 37–51).
25. TOPCUOGLU, S. et FOWLER, S. W. (1984). Factors affecting the biokinetics of technetium (^{95m}Tc) in marine macroalgae. *Mar. Envir. Res.*. **12**, 25–43.

22

Plant Root Absorption and Metabolic Fate of Technetium in Plants

D. A. CATALDO, T. R. GARLAND and R. E. WILDUNG
Pacific Northwest Laboratory,
Richland, Washington 99352, USA

1. INTRODUCTION

Technetium-99 (Tc), produced in the fission of U-235 and Pu-239, has a long half-life ($2{\cdot}1 \times 10^5$ year) and has been shown to exhibit a relatively high degree of bioavailability. While Tc can enter the environment from a number of sources [1], the nuclear fuel cycle is the major contributor from the standpoint of point sources. Concentration ratios (CR = μg Tc/g dry wt vegetation per μg Tc/g dry wt soil) for transfer from soils to plants span three orders of magnitude from 1 to 1000 [2–7]. This broad range of CR values presents some difficulty in assessment of dose [8].

Laboratory studies, using soils uniformly amended with pertechnetate (TcO_4^-), result in CR values of 50–1000 [2–5, 7]. To some extent, this range in CR value is affected by plant species [2, 4], soil type [2, 4, 5, 7], and nutrient status of soils [2]. In addition, TcO_4^- amended to soil has been shown to undergo sorption and/or chemical reduction [4–6] reducing its solubility and bioavailability. Recent field studies have been conducted in the vicinity of reprocessing facilities [9] and gaseous diffusion plants [6, 10] to resolve the relative availability of Tc released to the environment over a period of time. In the studies of Garland *et al.* [9], Tc-99 released to an arid environment prior to 1972 was shown to sorb to the upper soil horizon; CR values ranged from 2 to 10 for *Populus trichocrape* (cottonwood), *Asparagus officinalis* (asparagus), *Sisymbrium altissium* (tumble mustard), and a *Salix* species (willow), 150 for *Artemisia tridentata* (sagebrush), and 230 for *Centaurea repens* (Russian knapweed), based on the average Tc concentration in the upper 35 cm of soil. Similarly,

Hoffman *et al.* [10] and Garten *et al.* [6] found that Tc released to eastern United States environs from operating gaseous diffusion plants exhibited a similar soil sorption behaviour, with soil/plant CR values ranging from 2 to 44. While field CR values are somewhat lower than those determined in laboratory studies, the problem of Tc availability to biota will require further investigation to address those processes controlling absorption and chemical fate in plants.

Once absorbed by plants, 50% or more of the absorbed Tc is transported to shoot tissues [7, 12]. The high absorption and root/shoot transfer rates result in toxicity to a range of plant species [3, 7] grown at 0·05 to 1 μg Tc/g dry wt soil. This phytotoxicity is believed to result from chemical rather than radiological damage, since Berlyn *et al.* [11] were unable to demonstrate cytological evidence for radiation-induced damage, nor is the calculated radiation dose from tissue-deposited Tc high enough to induce damage [3, 7].

Recently, research efforts have concentrated on the mechanisms controlling Tc absorption by plants [12–14] and the chemical fate of Tc once deposited in plant tissues [15]. These studies have shown root absorption of Tc in plants to be linear with supplied concentration [12–14] and to exhibit characteristics of multiphasic uptake [13, 14], suggesting that an active regulated absorption mechanism is in place in plant membranes to allow transport of Tc. Relatively little published data concerning the chemical fate of Tc in plant tissues are available, except that of Myttenaere *et al.* [15] indicating a predominance of TcO_4^- in leaves of tomato. However, animal feeding studies [16] have shown whole body retention of Tc in the rat and guinea pig, and transfer to egg in quail to be substantially higher for TcO_4^- compared with ingested plant-incorporated Tc. This would indicate that plant-incorporated Tc exists in a chemical form other than TcO_4^-.

In the present series of studies with TcO_4^-, root absorption characteristics and tissue distributions are discussed and the chemical fate of Tc in plants is described.

2. MATERIALS AND METHODS

2.1. Plant culture

Vegetable and agricultural species, including carrot, cabbage, lettuce, pea, mustard, radish, leek, bunch onion, soybean and alfalfa, were grown from seed. Bulb onion and garlic were grown from bulbs or cloves. Procedures for hydroponic plant culture and maintenance have been described previously [14].

2.2. Ion uptake and competition

Studies of the absorption behaviour of Tc were performed using 10–14 day-old soybean plants. Prior to use, plants were transferred from nutrient solutions to 0·5 mM $CaCl_2$ solutions (pH 6·8) for 12 h to allow for desorption of possible interfering ions from root surfaces. Individual plants were transferred to aerated solutions containing 500 ml 0.5 mM $CaCl_2$ and various concentrations of $NH_4{}^{99}TcO_4$. Following the absorption period, which was 2 h except as noted, plants were transferred to 0·5 mM $CaCl_2$ solutions containing either 10 mM KH_2PO_4, K_2SO_4 or NH_4OH to desorb Tc from root surfaces.

Determination of pH optimum for Tc absorption by intact plants and for transfer from root to shoot was performed in 0·5 mM $CaCl_2$ solutions containing 0·2 μM TcO_4^-. Solution pH was adjusted from 3·0 to 8·5. Experimental conditions for both the metabolic inhibitor and ion competition studies are described in the text. Kinetic constants (K_s, V_{max}) were determined by linear regression of Lineweaver–Burk plots.

2.3. Radioanalysis

Tissues containing ^{99}Tc were oven-dried, weighed, wetted with 10 ml of concentrated NH_4OH, covered and digested for 12–20 h on low heat (60°C). These were brought to dryness, placed in a muffle furnace at 200°C for 1–2 h, then to 450°C for 24 h. Samples were cooled, wetted with 1 ml concentrated NH_4OH, evaporated to dryness, and re-muffled at 450°C for 20 h. Samples were then cooled and suspended in 2–5 ml 0·1 N NH_4OH for analysis. This procedure resulted in no loss of Tc as determined by spike recovery on control plant tissues. Samples containing ^{35}S were acid-digested and ashed as described above. Activity of ^{99}Tc and ^{35}S in samples was determined on an aliquot by liquid scintillation spectrometry, with appropriate quench correction. The activity of ^{75}Se in tissues was determined on entire tissue samples by gamma spectrometry. Uptake rates for Tc were expressed on the basis of root dry weight, which included only lateral roots; fibrous roots were assayed separately, and included in the total for uptake rates.

2.4. Separation procedures

Tissue fractionation procedures, including homogenisation, centrifugation and ultrafiltration using Amicon UM05 and UM20 membranes, have been described elsewhere [17]. The solubles from tissues were separated on a Waters HPLC, using an I-60 GPC column. The running buffer consisted of 20 mM ammonium acetate containing 20 mg sodium azide and 154 mg dithiothreitol per litre. Flow rate was maintained at 1 ml/min. Eluents were monitored for UV absorbance at 254 mμ, and 1 ml fractions collected for radioanalysis.

3. RESULTS AND DISCUSSION

3.1. Root absorption characteristics

The high CR values reported for Tc in both laboratory and field studies would suggest that plants possess either an efficient passive or an active mechanism for its absorption. The latter could employ mechanisms in place for a nutrient species having similar chemical characteristics. The absorption characteristics of TcO_4^-, using hydroponically grown soybean seedlings, were determined on the basis of pH optimum for both absorption and transfer from root to shoot, linearity of absorption, the influence of metabolic inhibitors on absorption, the presence or absence of absorption isotherms, and the effect of chemically similar nutrient and non-nutrient analogues on TcO_4^- absorption.

Soybean seedlings placed on 0·2 μM solutions of TcO_4^- exhibited the highest uptake rates at pH 3·0 (6 μg Tc/g dry wt root/h), with uptake rates decreasing to approximately 1·8 μg Tc/g dry wt root/h at pH 8·5 (Fig. 1). Shoot accumulation exhibited a similar decline from pH 3 to 8·5 (2 to 0·45 μg Tc/g dry wt root/h). The absence of a pH optimum would indicate that the pertechnetate ion is chemically stable under this range of conditions and is the ion involved in at least the first step in membrane transport.

Time course studies (Fig. 2), conducted at solution concentrations of 0·25 μM TcO_4^- (pH 6·8), show uptake to be linear for at least 6 h for the whole plant, root alone and shoot tissues. Extrapolation of the absorption curves for the whole plant and root alone to 0 time indicates the presence of an irreversibly bound component of Tc. This component amounts to approximately 700 ng Tc/g dry weight root, or 18% of the 2-h absorption rate. Attempts to further desorb this fraction using 10 mM NH_4OH failed to remove more than 2% over that removed by 10 mM K_2SO_4. This fraction may represent reduced forms of Tc, not bound TcO_4^-.

The energy requirements for Tc absorption were evaluated by determining the rates of absorption by intact seedlings in the presence and absence

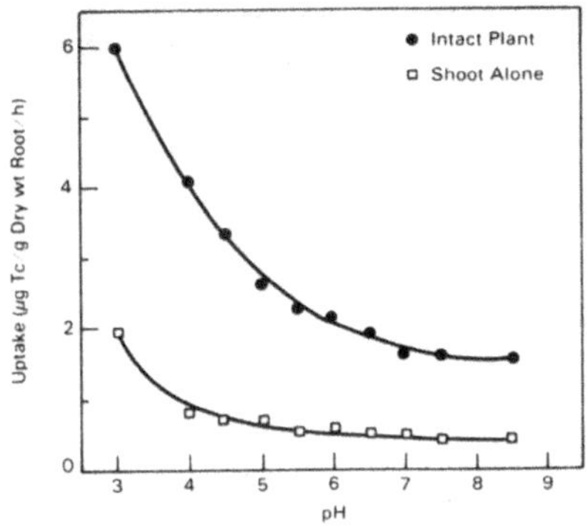

FIG. 1. Effect of pH on absorption of $^{99}TcO_4^-$ from 0·2 μM solutions. Desorption in 10 mM K_2SO_4; values represent average of three replicates.

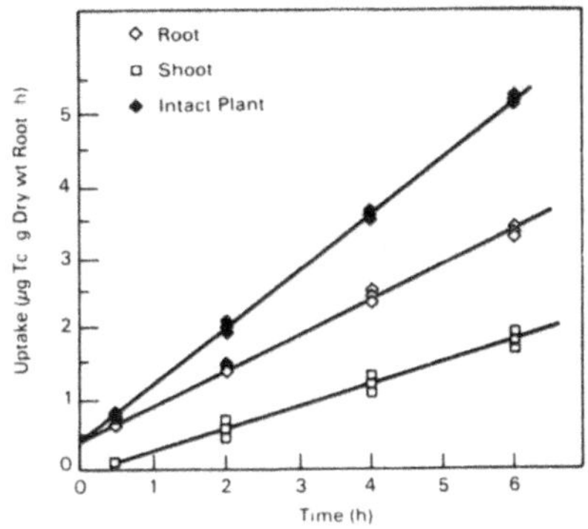

FIG. 2. Time course of $^{99}TcO_4^-$ absorption from 0·25 μM solutions for seedlings and transfer to shoots. Desorption in 10 mM K_2SO_4.

of a series of metabolic inhibitors (Table I). The presence of inhibitors reduced absorption in the intact plant to 7–21% of the control rates. Similarly, the presence of inhibitors reduced the transfer of Tc from roots to shoots to 15–43% of control rates. Based on these data, Tc absorption by plants can be considered an energy-requiring active process, similar in nature to that employed for nutrient ions.

To further resolve the absorption characteristics of Tc, concentration-dependent absorption kinetics were determined using intact seedlings

Table I
Effect of metabolic inhibitors on absorption of $^{99}TcO_4^-$ by 11-day-old soybean plants

Treatment	*Concentration (M)*	*Absorption rate*[1] *Whole plant*	*Root to shoot*
		(% of control[2]*)*	
Control		100 ± 17	100 ± 32
Na azide	10^{-4}	21 ± 2	43 ± 14
Dinitrophenol	10^{-4}	12 ± 4	27 ± 6
Succinate	5×10^{-2}	21 ± 10	38 ± 4
FCCP[3]	10^{-5}	7 ± 1	15 ± 2

[1] Absorption was for 1 h from 0·1 μM $NH_4{}^{99}TcO_4$ in 0·5 mM $CaCl_2$, pH 6·8.
[2] Control absorption rates were 0·574 ± 0·097 and 0·081 ± 0·026 μg Tc/g root dry weight/h.
[3] FCCP, carbonylcyanide-*p*-trifluoromethoxy-phenylhydrazine.

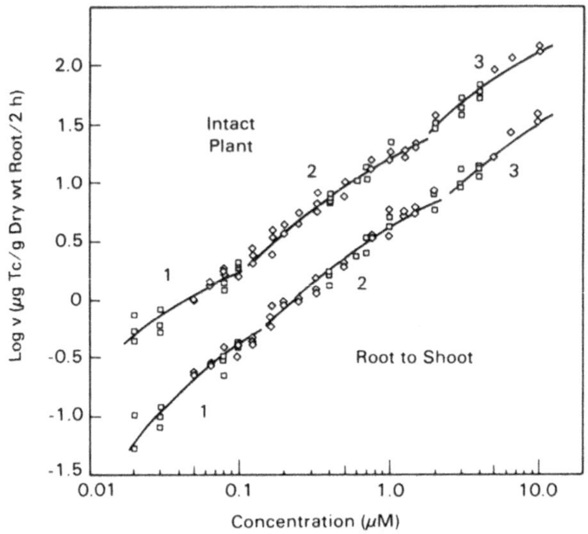

FIG. 3. Concentration dependent absorption of $^{99}TcO_4^-$ by intact plants, and transfer from root to shoot.

(Fig. 3). Absorption patterns for both the whole plant and shoots alone exhibited multiple absorption isotherms, again consistent with the behaviour of nutrient ions. Values of K_s for the three isotherms, based on whole plant accumulation, were calculated at 0·09, 8·9, and 54 µM. Calculated values of V_{max} were 15, 80 and 415 µg Tc/g dry wt root/h, respectively. Similarly, shoot accumulation patterns show multiple isotherms. If kinetic constants are calculated for root-to-shoot transfer, on the basis of the Tc concentrations in root bathing solutions and with the awareness that the Tc concentration in the stellar parenchyma should represent the basis of calculation, isotherms 2 and 3 conform to Michaelis–Menten kinetics. Calculated values of K_s for isotherms 2 and 3 are 7·1 and 77 µM; values of V_{max} are 15 and 135 µg Tc/dry wt root/h, respectively. Isotherm 1 does not exhibit saturation kinetics, possibly due to counting errors at these low Tc levels.

The absorption characteristics shown for Tc, including dependence on metabolic energy, the presence of multiple isotherms and saturation kinetics, suggest that Tc is being absorbed using mechanisms in-place for nutrient ions, ions with similar chemical characteristics. If that is true, Tc could behave as a non-functional nutrient analogue, which would explain its observed phytotoxicity. To address the question of transport analogues,

TcO_4^- absorption rates were measured in the presence and absence of a series of nutrient and non-nutrient species. Of the co-ions tested, borate, nitrate, tungstate and vanadate produced no consistent or significant reduction in Tc absorption. Sulphate, phosphate, selenate, perrhenate and iodate, supplied in excess concentration (10 μM) in the presence of 0·4 μM TcO_4^-, inhibited plant accumulation as shown in Table II. Kinetic

Table II

Influence of nutrient and non-nutrient anions on TcO_4^- absorption by 11-day-old soybean seedlings

Treatment[1]	*Solution species*	*Whole plant accumulation (% control)*	K_i[2] *(μM)*
Pertechnetate control	TcO_4^-	100 ± 9	(9)[3]
Pertechnetate + sulphate	SO_4^{2-}	64 ± 16	7
Pertechnetate + phosphate	$H_2PO_4^-/HPO_4^{2-}$	76 ± 12	14
Pertechnetate + selenate	$HSeO_3^-/SeO_3^{2-}/SeO_4^{2-}$	83 ± 6	61
Pertechnetate + perrhenate	ReO_4^-	66 ± 25	—[4]
Pertechnetate + iodate	IO_3^-	78 ± 13	—[4]

[1] TcO_4^- supplied at 0·4 μM, co-ions at 10 μM.
[2] Calculated from K_s of TcO_4^- and K_{si} in presence of co-ion.
[3] K_s for TcO_4^- alone at solution concentration of 0·25 to 1·33 μM.
[4] Non-linear relationship from double reciprocal plots.

constants associated with Tc absorption in the presence and absence of co-ions were determined from double reciprocal plots and their interactions evaluated graphically [18]. Perrhenate and iodate, although inhibitory to Tc uptake, exhibited variable and non-reproducible kinetics. However, sulphate, phosphate and selenate each exhibited characteristics of a competitive inhibitor (common V_{max}, increased K_s) with respect to TcO_4^- absorption. Therefore, they behave as transport analogues. Calculated values of K_i (Table II), the analogue concentration causing a doubling of the slope in the reciprocal plots, indicate that sulphate is a better analogue for Tc than phosphate, while phosphate is better than selenate in competing for Tc absorption sites.

In general, these absorption studies show the high soil/plant CR values for Tc to result from the presence of an efficient and active root absorption process. Based on the analogue studies, it would appear that Tc absorption employs transport sites normally used for sulphate and phosphate absorption.

3.2. Chemical behaviour and fate of Tc in plants

Assuming that pertechnetate, sulphate, and selenate have sufficient chemical similarities to share membrane transport processes, then these same characteristics may influence the fate of Tc in plant tissues and its observed phytotoxicity. Few data are available on the fate of Tc once accumulated in plants. Aside from its mobility within the plant [5, 12, 13] and its presence as TcO_4^- in xylem fluids [13] and foliar leachates [15], little more is known.

Table III

Comparison of ^{99}Tc uptake rates and leaf concentrations in different plant species[1]

Plant	*Whole plant uptake rate (μg Tc/h/g dry wt root)*	*Plant leaf (48 h) concentration (μg/g dry wt)*
Spanish bunch onion (*Allium cepa*)	0·57	7·7
Garden bunch onion (*Allium cepa*)	0·80	4·1
Yellow bulb onion (*Allium flavum*)	1·1	13
Garlic (*Allium sativum*)	0·16	4·6
Leek (*Allium porrum*)	0·40	5·9
Radish (*Raphanus sativus*)	4·4	62
Mustard (*Brassica juncea*)	3·7	79
Soybean (*Glycine max*)	2·1	130
Alfalfa (*Medicago sativa*)	2·9	94
Pea (*Pisum sativum*)	0·98	46
Lettuce (*Lactuca sativa*)	1·4	12
Cabbage (*Brassica oleracea*)	2·3	28
Carrot (*Daucus sativa*)	0·87	23

[1] Plants grown in hydroponics for 40 days, amended with 100 μM $^{99}TcO_4^-$ for 48 h.

To access the relative extent of Tc absorption and its mobility, plants were grown for 40 days in hydroponics, amended with 100 μM TcO_4^-, and absorption rates and leaf concentrations determined after 48 h. Absorption rates for the 13 plants studied (Table III) ranged from 0·16 to 4·4 μg Tc/g dry wt root/h. Leaf concentrations ranged from 4·1 to 130 μg Tc/g dry weight. The five *Allium* species exhibited both the lowest absorption rates and the lowest leaf concentrations. These data indicate that even when the form and concentration of Tc are controlled, there are substantial differences in absorption rates for Tc among plant species. This would at least partially explain the relatively large range in CR values observed for plants grown on soils. Similarly, there is no apparent correlation between absorption rate and leaf concentration; this suggests that plant species partition Tc differently.

The partitioning and chemical fate of Tc was further studied in six of the previous plant species. This included evaluation of the distribution in plant tissues, distribution within tissues, and the apparent molecular weight (MW) distribution of Tc in the soluble fraction of tissues. These are shown in Table IV. The distribution of accumulated Tc in plant tissues varies with plant species. Roots of onion, leek and garlic contain 28–60% of the Tc after 24 h of uptake, while the leaves of these plants contain much of the remainder. Interestingly, the storage organ (bulb) has a much greater mass than the other tissues, yet contains only 4–7% of the Tc. In radish, mustard and soybean, roots contain 6–26% of the Tc, while leaves contain 74–92%. These differences in distribution may be related to either solubilisation/immobilisation processes associated with transport within the plant or differences in their metabolic ability to incorporate or store absorbed Tc. Physical fractionation of these tissues should indicate whether Tc is being sorbed and immobilised, remains soluble as pertechnetate as indicated by previous data for xylem exudates [13] and leaf leachates and extracts [15], or is being metabolically incorporated.

The tissue distributions in Table IV show some interesting trends. First, with the exception of radish root and soybean leaves, less than 7% of root-associated Tc is found in the insoluble residue, indicating that Tc remains soluble following root absorption and transport to distal tissues. Secondly, the organelle fractions contain less than 4% of the tissue Tc. With the exception of radish root and soybean leaves, greater than 90% of the tissue Tc is extractable and soluble. If the soluble fractions are further fractionated by ultrafiltration, resolution can be achieved for the fractions of Tc associated with components of high MW (greater than 20 000), intermediate MW (20 000–500) and less than 500 MW. It is the latter fraction which should contain any TcO_4^- present, in addition to any low MW complexes of Tc. Based on the observed MW distributions in Table IV, it appears that the majority of the Tc accumulated in tissues is in a form other than TcO_4^-; particularly if one assumes that only a portion of the Tc found as less than 500 MW components is pertechnetate. Again, there appear to be differences in the behaviour of Tc in *Allium* species as compared with the other three plant species. Technetium associated with the solubles of the *Allium* species is primarily found in lower MW while being associated with those of higher MW in radish,

Table IV
Partitioning of Tc in hydroponically grown plants

		Tissue distribution (%)				Distribution in solubles (%)		
		Plant distribution	*Insoluble residue*	*Organelle fraction*	*Solubles*	*>20 000 MW*	*20 000 to 500 MW*	*<500 MW*
Radish	Leaves	79·3	4·3	1·3	94·4	64	23	13
	Bulb	14·2	4·1	2·4	93·5	31	29	40
	Root	6·5	23·6	4·2	72·2	60	9	31
Mustard	Leaves	92·0	6·4	2·2	91·4	56	38	6
	Root	8·0	6·8	1·5	91·3	56	34	10
Soybean	Leaves	74·0	31·9	0·9	67·2	55	14	31
	Root	26·0	4·2	2·1	93·7	38	37	25
Bulb onion	Leaves	64·3	2·7	0·6	96·7	18	48	34
	Bulb	7·4	0·9	1·2	97·8	15	61	24
	Root	28·3	5·3	2·8	92·0	41	36	23
Garlic	Leaves	35·6	3·7	0·6	95·7	24	44	32
	Bulb	4·5	6·7	0·7	92·6	15	53	32
	Root	59·9	4·1	0·9	95·0	23	39	38
Leek	Leaves	48·5	1·5	1·1	97·4	17	54	29
	Bulb	4·1	2·3	1·1	96·6	14	63	23
	Root	47·4	4·2	1·4	94·4	18	47	35

mustard and soybean. These data suggest that Tc accumulated by plants is not subject to chemical reduction and insolubilisation, and that it appears to be either incorporated into soluble plant macromolecules or complexed by plant metabolites.

In an effort to resolve the fate of Tc in plants following uptake, tissue distributions for onion and soybean were compared for sulphate, selenate and selenite which were shown to be transport analogues for Tc. Partitioning of TcO_4^-, SO_4^{2-} and SeO_4^{2-} between roots and leaves was comparable to that shown for Tc in onion and soybean (Table IV); however, approximately 80% of the SeO_3^{2-} absorbed was retained in the roots of both plant species.

Table V

Comparative distribution of TcO_4^- and its analogues in soybean and onion plants[1]

Treatment	*Tissue component*	*Soybean distribution (%)*		*Onion distribution (%)*	
		Leaves	*Root*	*Leaves*	*Root*
$^{99}TcO_4^-$	Solubles	84·7	97·7	92·6	92·1
	Organelle fraction	7·1	1·2	5·7	5·6
	Residue	8·2	1·1	1·7	2·3
$^{35}SO_4^{2-}$	Solubles	66·5	63·8	76·7	80·7
	Organelle fraction	23·8	25·8	21·4	16·5
	Residue	9·7	10·4	1·9	2·8
$^{75}SeO_4^{2-}$	Solubles	88·0	83·0	71·6	80·6
	Organelle fraction	7·1	6·6	21·9	9·7
	Residue	4·9	10·4	6·5	9·7
$^{75}SeO_3^{2-}$	Solubles	58·2	51·6	77·5	53·7
	Organelle fraction	19·8	22·7	16·7	24·9
	Residue	22·0	25·7	5·8	21·4

[1] One-week uptake from 10 μM solutions, pH 6·5.

Table V shows the distribution of these four anions within leaves and roots of onion and soybean. In onion, only 5% of the Tc is associated with the organelle fraction (primarily mitochondria in roots, and mitochondria and chloroplasts in leaves) compared with the 9–25% found with the other three anions. This suggests that Tc may not be metabolised as well as the analogues. In soybean, a similar situation persists. This is consistent with the fact that both sulphur and selenium amino acids exist and have required functions. It is also possible that Tc, based on its

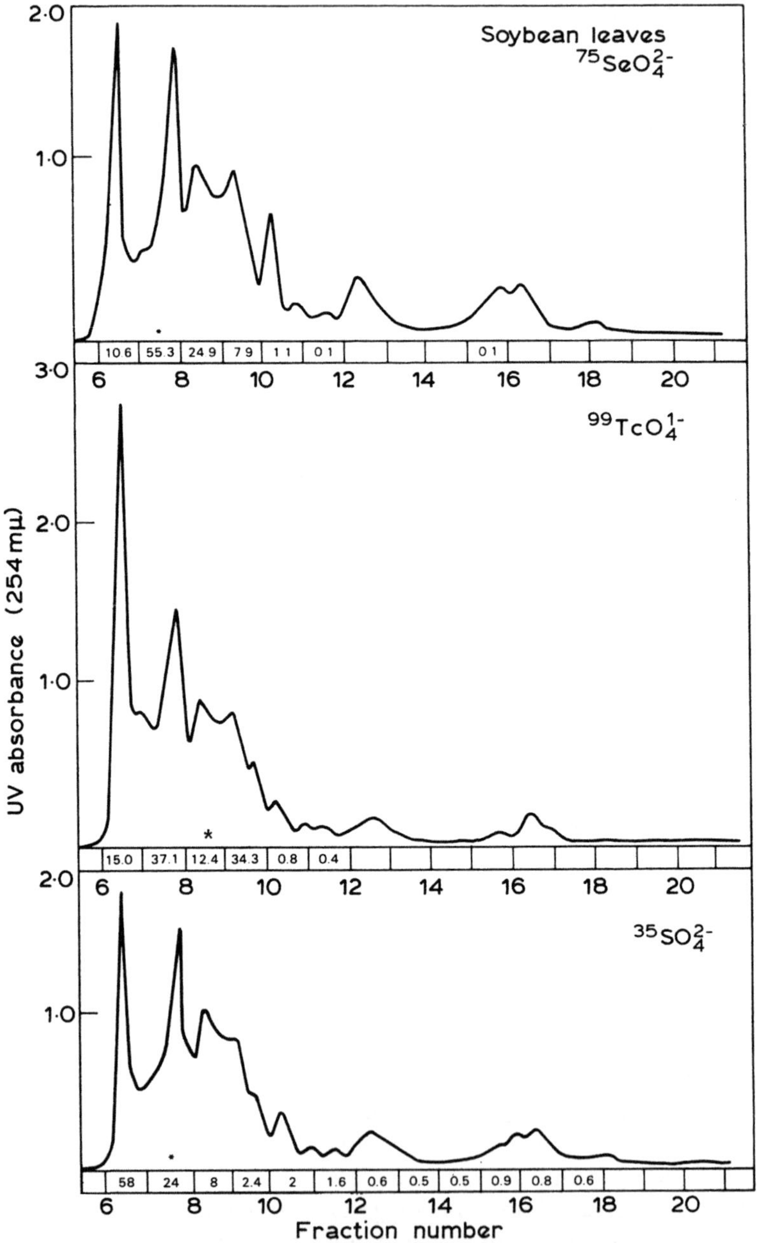

FIG. 4. Molecular weight distributions for protein found in soluble fraction of soybean leaves. Values in lower blocks represent the percentage of ^{35}S, ^{99}Tc or ^{75}Se found in each fraction. (* indicates elution fraction containing inorganic anion.)

observed solubility and presence in higher MW fractions (Table IV), may undergo only partial anabolism and is unable to be synthesised into functional macromolecules.

In an effort to resolve the fate of Tc, a comparison was undertaken of the molecular weight distribution of Tc, selenium and sulphur in the soluble fraction from roots and leaves of soybean plants. Figures 4 and 5 show both the MW distributions and the percentage of radioactivity associated with each fraction for leaves and roots, respectively. It should be noted than the I-60 HPLC column employed is strongly cationic under the separation conditions used; and therefore the inorganic anion species present in the soluble fractions elute in either fraction number 8 (TcO_4^-) or fraction number 7 (SO_4^{2-} and SeO_4^{2-}). The remainder of the soluble or organically complexed components are resolved based on MW (column exclusion is approximately 60 000 MW) and degree of charge interaction with the column. Also, since the plants were supplied with tracer three days prior to fractionation, the partitioning of radioactivity should be indicative of each element's chemical fate. The behaviour of S is as one would expect, with the vast majority being associated with high MW compounds, and smaller amounts associated with intermediate MW proteins and peptides. The 24% of the ^{35}S found in fraction number (FN) 7 represents an undefined mixture of incorporated and inorganic sulphur. The protein distributions observed with the Se-treated plants are generally similar to those seen for S, except for a noticeable increase in the protein peak associated with FN 10. Unlike S, the majority of the complexed Se activity is associated with intermediate MW proteins; a substantial fraction of the Se activity is found in FN 7 and may represent the inorganic anion. In comparison, MW distributions for Tc are quite similar to that seen for S, except for a noticeable increase in higher MW proteins. However, the distribution of ^{99}Tc activity is more like Se, in that Tc is associated with intermediate MW components.

The MW and radioactivity distributions for these three anions in roots are substantially different from those observed for leaves (Fig. 5). Sulphur is distributed over the entire MW range, with only half as much in the high MW fraction compared with leaves. This may simply reflect the slower rate of protein turnover in roots compared with photosynthetic tissues. Selenium distributions are very similar to those found for leaves. The behaviour of Tc is quite similar to that of S in that Tc is distributed between compounds of high and intermediate MW, although the low MW associations are absent. In addition, there is some indication of an accumulation of intermediate MW components.

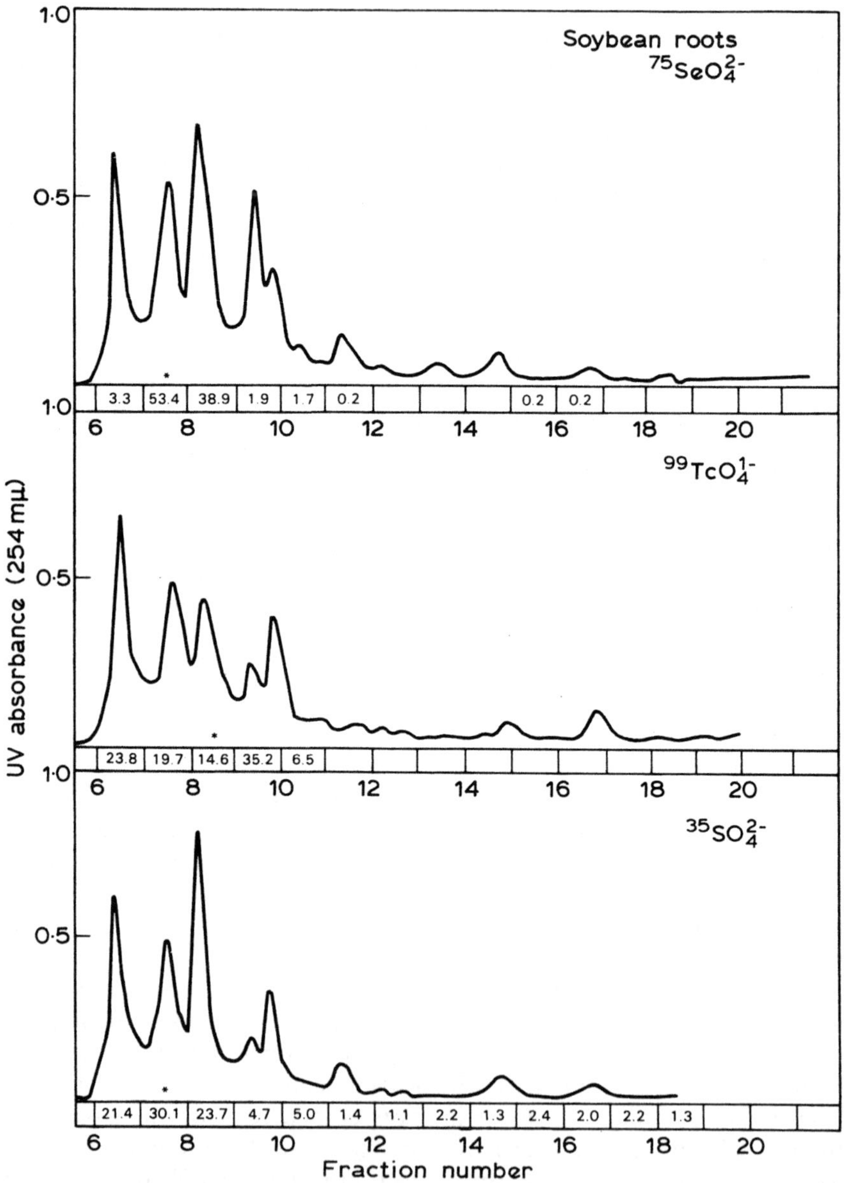

FIG. 5. Molecular weight distribution for proteins found in soluble fractions of soybean roots. Values in lower blocks represent the percentage of ^{35}S, ^{99}Tc or ^{75}Se found in each fraction. (* indicates elution fraction containing inorganic anion.)

4. CONCLUSIONS

While these data define the comparative distribution of Tc, Se and S, they do not provide a great deal of insight into the biological mechanisms controlling Tc fate. However, we can formulate a number of hypotheses based on our current knowledge of S and Se metabolism. First, 'assimilatory sulphur reduction' is known to occur in leaves and to be associated with the photosynthetic apparatus. Secondly, it is known that Se can substitute for S in biological molecules when its concentration is above the normal physiological concentration range compared with S [19]. In this case, sufficient substitution for S can disrupt enzyme function and metabolism. Thirdly, at least in microbial and animal systems, Se-dependent enzymes exist and are required for biological function [20]. What, then, can we deduce concerning the behaviour of Tc? First, on the basis of absorption and ion competition studies, phosphate, selenate and sulphate are analogues of technetium. Secondly, only selenate and sulphate can be incorporated into high or intermediate MW proteins. Thirdly, it has been shown that Tc is toxic at relatively low concentrations, and that young, growing tissues are most susceptible. This suggests an interaction with metabolic processes. Finally, based on the MW and tracer distributions for these three elements, it is likely that Tc is substituting for S, allowing for only partial synthesis of proteins, as indicated by the accumulation of Tc in intermediate MW compounds, resulting in dysfunction. Unfortunately, insufficient data exist to make this assumption, and substantial effort will be required to determine the metabolic processes governing Tc behaviour.

ACKNOWLEDGEMENTS

This work was supported by the US Department of Energy under Contract DE-A606-76RLO 1830. The authors sincerely appreciate the assistance of Marilyn Gority in preparation of this manuscript.

REFERENCES

1. WILDUNG, R. E., MCFADDEN, K. M. and GARLAND, T. R. (1979). Technetium sources and behavior in the environment. *J. Envir. Qual.*, **8**, 156–61.
2. LANDA, E. R., HART THORVIG, L. J. and GAST, R. G. (1977). Uptake and distribution of technetium-99 in higher plants. In: *Biological Implications of Metals in the Environment*, USERDA Symp. Ser., Conf-750959 (Ed. H. Drucker and R. E. Wildung), Springfield, Virginia, National Technical Information Service, p. 390.

3. Wildung, R. E., Garland, T. R. and Cataldo, D. A. (1977). Accumulation of technetium by plants. *Health Physics*, **32**, 314–17.
4. Routson, R. C. and Cataldo, D. A. (1978). Accumulation of ^{99}Tc by tumbleweed and cheatgrass grown on arid soils. *Health Physics*, **34**, 685.
5. Mousny, J. M. and Myttenaere, C. (1981). Absorption of technetium by plants in relation to soil type, contamination level, and time. *Plant and Soil*, **61**, 403–12.
6. Garten, C. T., Hoffman, F. O. and Bondietti, E. A. (1984). Field and greenhouse experiments on the fate of technetium in plants and soils. *Health Physics*, **46**, 647–56.
7. Sheppard, M. I., Vandergraaf, T. T., Thiabault, D. H. and Keith Reid, J. A. (1983). Technetium and uranium: Sorption by and plant uptake from peat and sand. *Health Physics*, **44**, 635–43.
8. Till, J. E., Hoffman, F. O. and Dunning, D. E. Jr (1979). A new look at ^{99}Tc releases to the atmosphere. *Health Physics*, **36**, 21–30.
9. Garland, T. R., Cataldo, D. A., McFadden, K. M., Schreckhise, R. G. and Wildung, R. E. (1983). Comparative behavior of ^{99}Tc, ^{129}I, ^{127}I, and ^{137}Cs in the environment adjacent to a fuels reprocessing facility. *Health Physics*, **44**, 658–62.
10. Hoffman, F. O., Huckabee, J. N., Lucas, D. M., Garten, C. T., Scott, T. G., Walker, R. L., Gauge, P. S. and Holmes, C. V. (1980). Sampling of technetium-99 in vegetation and soils in the vicinity of operating gaseous diffusion facilities. ORNL/TM-7386, Oak Ridge, Tennessee, Oak Ridge National Laboratory.
11. Berlyn, G. P., Dhillon, S. S. and Koslow, E. E. (1980). Technetium: A toxic waste product of the nuclear fuel cycle. Effects on soybean growth and development. *Envir. Managmt*, **4**, 149–56.
12. Mousny, J. M., Roucoux, P. and Myttenaere, C. (1979). Absorption and translocation of technetium in pea plants. *Envir. Exptl Bot.*, **19**, 263–8.
13. Cataldo, D. A., Wildung, R. E. and Garland, T. R. (1978). Technetium accumulation, fate, and behavior in plants. In: *Environmental Chemistry and Cycling Processes* (Ed. D. C. Adriano and T. L. Brisbin Jr), Conf-760429, DOE Symposium Series No. 45. Washington, DC, US Department of Energy.
14. Cataldo, D. A., Wildung, R. E. and Garland, T. R. (1983). Root absorption and transport behavior of technetium in soybean. *Plant Physiol.*, **73**, 849–52.
15. Myttenaere, C., Daoust, C. and Roucoux, P. (1980). Leaching of technetium from foliage by simulated rain. *Envir. Exptl Bot.*, **20**, 415–19.
16. Cataldo, D. A. and Wildung, R. E. (1983). The role of soil and plant metabolic processes in controlling trace element behavior and bioavailability to animals. *Sci. Total Envir.*, **28**, 159–68.
17. Cataldo, D. A., Garland, T. R. and Wildung, R. E. (1981). Cadmium distribution and chemical fate in soybean plants. *Plant Physiol.*, **68**, 835–9.
18. Plowman, K. M. (1972). *Enzyme Kinetics*, New York, McGraw-Hill, pp. 56–73.
19. Stadtman, T. C. (1974). Selenium biochemistry. *Science*, **183**, 915–22.
20. Stadtman, T. C. (1980). Selenium-dependent enzymes. *Ann. Rev. Biochem.*, **49**, 93–119.

23

Processus Biologiques Responsables de l'Accumulation du Tc-99 par les Végétaux*

C. MYTTENAERE,[a] C. M. VANDECASTEELE,[b] P. ROUCOUX,[b] E. A. LIETART,[b] A. ITSCHERT[b] et J. MOUSNY[a]

[a] *Programme Biologie—Radioprotection, Commission des Communautés Européennes, Rue de la Loi 200, B-1049 Bruxelles, Belgique*

[b] *Laboratoire de Physiologie Végétale, Université Catholique de Louvain, Place Croix du Sud 4, B-1348 Louvain-la-Neuve, Belgique*

1. INTRODUCTION

Les résultats obtenus jusqu'à présent en matière de transfert sol/plante semblent indiquer que les plantes jouent un rôle important dans le cycle du Tc. Cependant, dans l'état actuel de nos connaissances reposant principalement sur des essais à court terme, le comportement à long terme du Tc-99 pose encore de nombreux problèmes.

Avant de se prononcer définitivement sur le danger que représente cet élément pour l'homme, certaines incertitudes doivent être levées [2, 7]: un examen des données publiées récemment montre en effet que l'impact radiologique du Tc, calculé sur la base de données obtenues au champ, s'avère de moindre importance que l'impact estimé par l'application des facteurs de transfert sol/plante obtenus en conditions contrôlées et pour des essais à court terme [14, 20, 21]. Ces observations d'une grande importance si l'on se rappelle que les doses aux organes, calculées à l'aide de valeur de transfert comparables à celles obtenues en laboratoire égalent ou dépassent les normes USPA (US Environmental Protection Agency) [15].

En outre, les modèles de transfert qui sont dorénavant requis doivent

* Contribution no. 2124 du Programme Biologie, Radioprotection de la Commission des Communautés Européennes.

répondre aux exigences nouvelles de la CIPR afin de calculer 'aussi exactement qu'il est possible de le faire' les doses reçues par l'homme.

En conditions aérobies et en l'absence d'agents réducteurs et de complexants, le Tc est présent dans le milieu sous forme de TcO_4^- [4]; cette forme peut toutefois évoluer rapidement sous l'influence des conditions biophysicochimiques du milieu [20, 28] avec l'apparition de formes réduites et/ou complexées [20]. Les différences observées entre les comportements à court et à long terme reposent sur un ensemble de mécanismes (biogéochimie du Tc) qu'il est donc important et nécessaire de mettre en évidence. L'étude du comportement à long terme de radionucléides aisément transférés dans la chaîne alimentaire impose donc au radioécologiste de raisonner en termes de cycles biogéochimiques et non plus en terme de simple transfert [3, 10, 23].

Cette étude implique également que soit caractérisée la forme chimique du Tc dans les végétaux afin de permettre une estimation plus précise des transferts aux autres maillons de la chaîne alimentaire et une compréhension des mécanismes responsables de la demi-vie écologique de cet élément [13]. La connaissance de la valeur de ce dernier paramètre est importante lorsque les processus de dilution (croissance, lessivage, translocation...) prédominent sur l'absorption radiculaire et lorsque le temps est suffisamment long entre la contamination et la récolte.

Cette contribution a pour but d'étudier le comportement du Tc en écosystème terrestre, de préciser certaines étapes d'un modèle reposant sur de nombreuses observations, et enfin d'en tirer les conclusions les plus importantes en matière de transfert et de métabolisme.

2. ACCUMULATION DU TECHNÉTIUM PAR LES VÉGÉTAUX

Il est désormais acquis que pour le Tc les facteurs de transfert sol/plante varient en fonction de l'espèce, du mode de contamination, du niveau d'apport et des conditions culturales [7, 21, 29]. Ainsi, les valeurs obtenues en solution nutritive sont en général plus élevées que celles obtenues en sol [7, 20] et de plus, les chiffres relevés à proximité des installations nucléaires (conditions de champ) sont inférieurs d'un ou deux ordres de grandeur à ceux obtenus en laboratoire [15].

L'influence du vieillissement du Tc apporté au sol sur le facteur de transfert sol/plante est loin d'être négligeable: la Fig. 1 illustre des résultats obtenus en nos laboratoire et montre pour différentes conditions

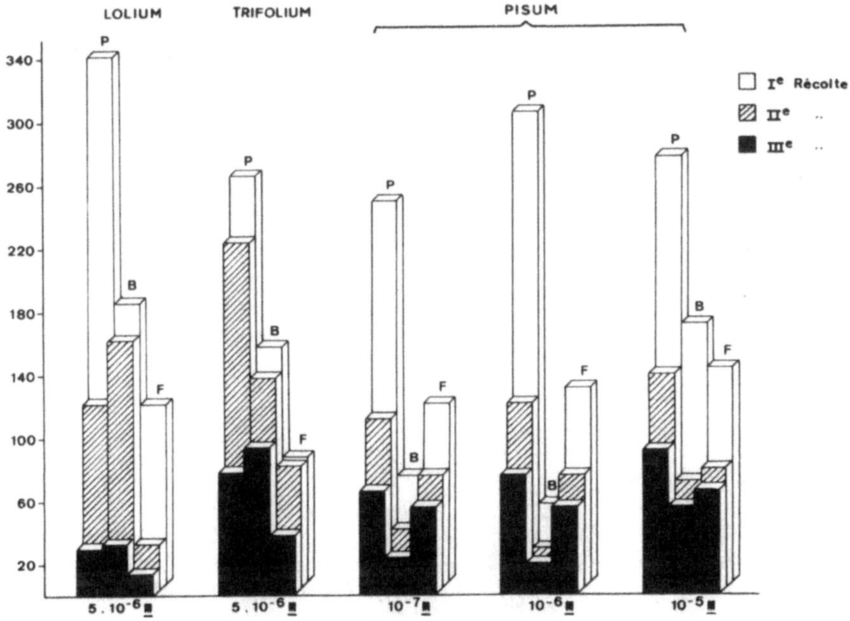

FIG. 1. Influence du vieillissement sur la disponibilité du Tc pour différents sols, espèces végétales et concentrations en Tc-99: facteurs de transfert (cpm g matière sèche/cpm g sol sec).

expérimentales (type de sol (Tab. I) et niveau de contamination) et plantes étudiées une diminution du transfert sol/plante avec le temps.

Il a été, de plus, observé que pour une même espèce végétale, les propriétés biologiques et physiologiques du matériel considéré jouent également un rôle important sur l'accumulation du technétium [32]: les cellules dotées d'un pouvoir réducteur élevé (fixation de l'azote atmosphérique; photosynthèse) sont le siège d'une plus grande accumulation en Tc (Tab. II et III; Fig. 2; autoradio 1). Dans le cas des plantes supérieures (*Phaseolus vulgaris*) les niveaux de transpiration des plantes

Tableau I

	Mat. org. (%)	*Argile (%)*	*CEC (meq/100 g)*
Podzol (Hannovre, D)	7·9	4·2	17
Loess (Braunerde, Allonville, F)	2·9	20·4	18
Fen (Emmen, ND)	23·3	3·0	46

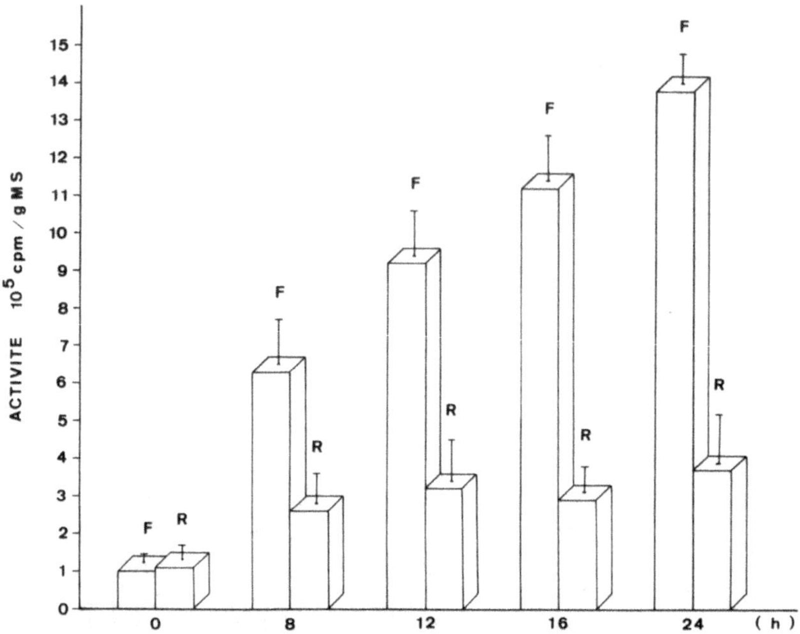

FIG. 2. Influence de la longueur de l'héméropériode sur la teneur en Tc-99 de jeunes plantes de maïs cultivées sur nutrition nutritive (cpm Tc-99/g mat. sèche).

Tableau II

Facteurs de transfert solution-organisme ou organe végétal (cpm poids frais/cpm/ml solution nutritive)

Organisme—organe ou source d'Azote		*$[TcO_4^-]$ en M (10^{-6})*
Soja (20 jours)	Feuilles sup.	242
	Feuilles inf.	154
	Tiges	19
	Gousses	16
	Racines	17
	Nodules	50
	Bactéroides	207
A. cylindrica	$-NO_3^-$	176
	$+NO_3^-$	8
A. chroococcum	$-NO_3^-$	73
	$+NO_3^-$	2
R. japonicum	+ N org.	0·7
S. obliquus		1·5

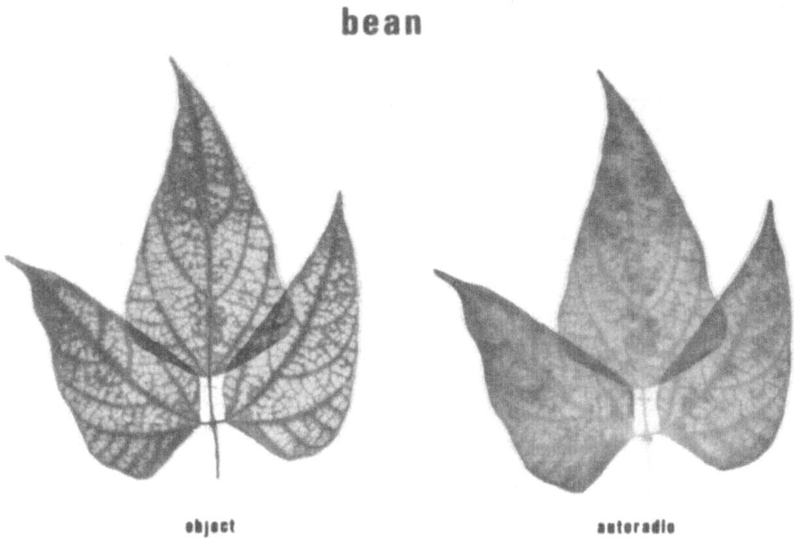

Auto. 1. Autoradiographie d'une feuille de haricot montrant des symptômes de chlorose (plante récoltée sur sol contaminé en Tc-99).

Tableau III

Teneurs (cpm/Tc-95m/g mat. sèche) et distribution relative (%) du Tc chez *Phaseolus vulgaris* (var. nannus L.) cultivé sur solution nutritive contaminée en Tc-99 (+Tc-95m) en conditions d'obscurité ou de lumière (4 jours prétraitement–4 jours de traitement)

	[TcO_4^-] en M			
	10^{-7}		*10^{-6}*	
	Lumière	*Obscurité*	*Lumière*	*Obscurité*
Activité plante entière (cpm/^{95m}Tc/g MS)	174 704 0·19*	64 646 0·17	169 044 0·09	87 931 0·19
Organes	Distribution rel. du Tc (en % de l'act. tot.)			
Racines	21 (0·13)	87 (0·02)	21 (0·09)	83 (0·19)
Feuil. cotylédonnaires	9 (0·22)	1 (0·37)	8 (0·41)	2 (0·99)
Feuil. non cotyl.	66 (0·05)	6 (0·12)	68 (0·04)	9 (1·07)
Tiges	3 (0·27)	5 (0·30)	3 (0·26)	5 (0·78)

* Coefficient de variation.

Tableau IV

Influence de la nutrition en soufre sur les facteurs de transfert du Tc pour des plantes de *Phaseolus vulgaris* (var. nannus L.) cultivées sur solution nutritive (cpm Tc-95m/g mat. sèche/cpm Tc-95m/ml solution nutritive; Tc-99/10^{-7} *M*; 7 jours prétraitement, plantes âgées de 19 jours)

Prétraitement $[SO_4^=]$ *en M*	*Traitement* $[SO_4^=]$ *en M*	*Durée du traitement*							
		2 h				*71 h*			
		Racines	*Tiges*	*Feuilles*	*Pl. ent.*	*Racines*	*Tiges*	*Feuilles*	*Pl. ent.*
8×10^{-4}	8×10^{-4}	29·7	6·3	15·2	16·9	85·9	70·0	528·2	337·7
8×10^{-4}	8×10^{-7}	24·4	4·0	14·5	14·1	78·4	53·8	490·6	318·5
8×10^{-7}	8×10^{-7}	12·4	2·5	3·4	6·3	49·0	46·8	304·4	168·1
8×10^{-7}	8×10^{-4}	15·0	3·2	3·5	7·3	66·1	56·5	400·0	238·2

Facteurs de transfert (cpm ^{95m}Tc/g MS/cpm/ml).

cultivées en lumière continue et à l'obscurité ne peuvent expliquer à eux seuls les différences observées [16].

L'alimentation minérale et, en particulier, l'alimentation en soufre conditionnant la formation de composés organiques soufrés: cystéine méthione, diverses protéines riches en cystéine modifie également la teneur en technétium (Tab. IV) des végétaux: les plantes cultivées sur milieu nutritif liquide riche en soufre accumulent plus de technétium et, en outre, le passage d'une solution pauvre à une solution riche se traduit également par une augmentation significative de la teneur en Tc des plantes. Ces effets qui se marquent très rapidement (2 heures) s'amplifient avec le temps; le phénomène inverse s'observe lorsque les plantes sont transférées d'un milieu riche en soufre à un milieu pauvre.

Des chromatographies sur couches minces ont montré que le TcO_4^- se lie à la cystéine ainsi qu'à la méthionine [9]: *in vitro*, la cystéine peut être marquée directement au Tc grâce à ses groupements—SH qui réduisent TcO_4^- tandis qu'elle est oxydée en cystine. Par contre, le marquage de la méthionine exige une réduction préalable du TcO_4^-, par exemple par l'acide ascorbique, avant la formation du complexe avec l'acide aminé [31]. C'est à ce stade qu'interviendrait l'anabolisme cellulaire et l'influence du niveau de la photosynthèse [12, 16]. A l'appui de ces résultats, il faut rappeler que les images autoradiographiques des feuilles montrent que le Tc s'accumule préférentiellement dans les tissus chlorophylliens (autoradio 2). Par ailleurs, le Tc accumulé par les feuilles se répartit de la façon suivante $\pm 30\%$ dans les parois cellulaires (complexes Tc-polysacharides) et $\pm 70\%$ dans la fraction cytoplasmique soluble [11, 25]. La chromatographie sur gel de cette dernière fraction a montré que le Tc est lié à des molécules de poids moléculaire faible ($4000 < MW < 10\,000$). Ces diverses observations permettent d'entrevoir un lien étroit entre le métabolisme du soufre et la fixation du Tc chez les végétaux. Confirmation de ces résultats nous a été partiellement donnée (Tab. V) par un essai à plus long terme (51 jours) réalisé sur divers sols irrigués à l'aide de solutions riche ou pauvre en $SO_4^=$ et contaminées en TcO_4^- ($10^{-7}M$ Tc-99 + Tc-95m). Les différences observées entre traitements sont fortement amenuisées par rapport à celles observées sur solution nutritive (Tab. IV). La réserve en soufre des sols d'une part, ainsi que l'intégration du $SO_4^=$ apporté par la solution nutritive dans le cycle biogéochimique justifient les résultats obtenus (Fig. 3).

La contradiction apparente relevée entre les résultats obtenus par Cataldo *et al.* [6] et les nôtres peut par conséquent s'expliquer par le fait que l'antagonisme entre anions TcO_4^- et $SO_4^=$ observé à très court terme

Auto. 2. Autoradiographie d'une feuille de *Coleus* cultivé sur milieu nutritif contaminé en Tc-99.

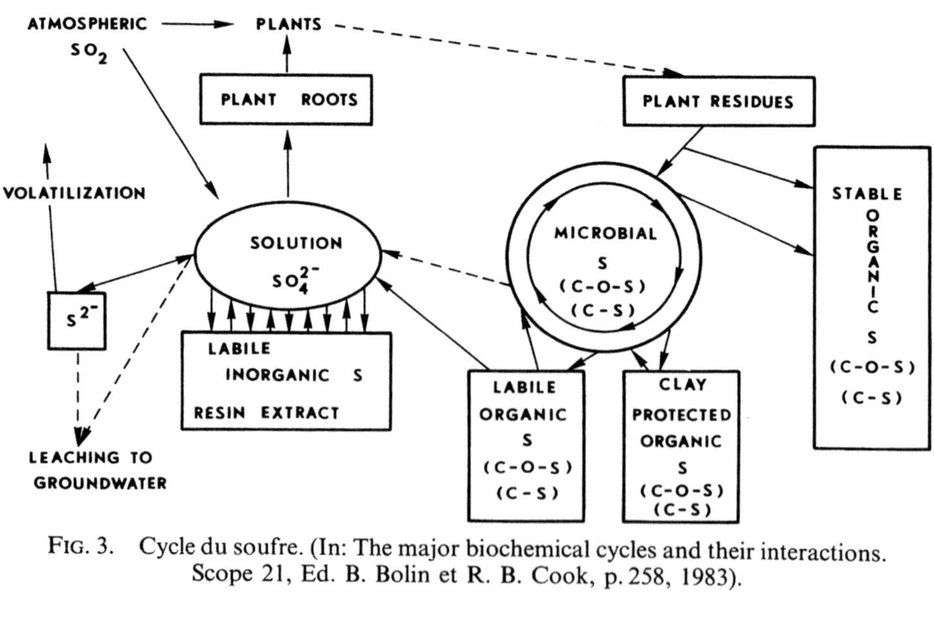

FIG. 3. Cycle du soufre. (In: The major biochemical cycles and their interactions. Scope 21, Ed. B. Bolin et R. B. Cook, p. 258, 1983).

Tableau V

Influence de la nutrition en soufre sur les teneurs relatives (en % de l'activité des feuilles récoltées sur FEN) en technétium d'organes de plantes de *Phaseolus vulgaris* (var. nannus L.) cultivées sur différents sols contaminés en Tc (Tc-99 $10^{-7}M$ + Tc-95m)

Organes / SO_4^{2-} *en M* / *Type de sol*	*Racines*		*Tiges*		*Feuilles*		*Gousses*	
	8×10^{-4}	8×10^{-7}	8×10^{-4}	8×10^{-7}	8×10^{-4}	8×10^{-7}	8×10^{-4}	8×10^{-7}
Podzol	4·96	4·91	10·46	8·74	141·02	128·42	3·12	2·81
Fen	4·71	5·04	8·47	8·87	107·57	100·00	2·10	3·60
Loes	4·10	4·16	9·86	9·81	127·79	116·75	2·78	3·60

Teneurs en Tc dans les différents organes rapportées à la teneur des feuilles du traitement $8 \times 10^{-7}M$ SO_4^{2-} sur fen (en %).

est supplanté dans nos essais par des effets biologiques secondaires qui prennent le pas sur le premier. La production d'acides aminés soufrés peut, à court terme, modifier la vitesse de pénétration et de fixation du TcO_4^- en augmentant le nombre de sites de fixation et éventuellement la teneur en 'transporteur' assurant le passage du TcO_4^- du milieu externe au milieu cellulaire. Le phénomène d'induction de 'transporteur' par un ion présent dans le milieu a été mis en évidence lors d'études précédentes [24]. Ce mécanisme pourrait expliquer les variations de la vitesse d'absorption du Tc observées lors d'études de cinétiques en fonction de la concentration du milieu [8]. A long terme, ces effets sont dominés par la modification de la disponibilité des éléments en présence.

Il est opportun de signaler que d'autres éléments tels que le Se qui, apporté sous forme de sélénates, s'incorpore dans les acides aminés soufrés ou le Cd [25] qui en modifie le teneur [19, 30] peuvent aussi interagir avec le Tc.

Ces résultats révèlent la nécessité d'envisager le comportement à long terme des radionucléides dans des conditions aussi proches que possible de la réalité et montrent tout le danger de l'extrapolation des résultats obtenus en conditions contrôlées. Ils montrent en outre que les modèles de transfert des radionucléides à vie longue doivent prendre en considération les cycles biologiques des éléments en présence [27].

3. DISCUSSION

Il est généralement admis que le Tc est absorbé sous forme de TcO_4^- (Fig. 4). Toutefois, il est important de se rappeler qu'en conditions normales de culture, le Tc apporté sous forme de TcO_4^- change rapidement de forme chimique [20, 34] sous l'influence des conditions édaphiques et, qu'après un certain temps, il est présent pour sa plus grande partie sous forme organique. Cela implique [35] qu'il a dû être réduit chimiquement ou biologiquement (variations des conditions de pH et de Eh, réduction en présence de cystéine, réduction par divers métabolites, utilisation du TcO_4^- comme accepteur d'électrons en l'absence d'oxygène par les microorganismes, incorporation dans les tissus microbiens et libération ultérieure du Tc). Cette réduction entraîne une diminution de la disponibilité du technétium par les végétaux. En sols, on ne peut espérer affecter significativement cette biodisponibilité en jouant sur un enrichissement en soufre des sols étant donné l'importance du pool organique en cet élément assurant un renouvellement continu en soufre de la solution du

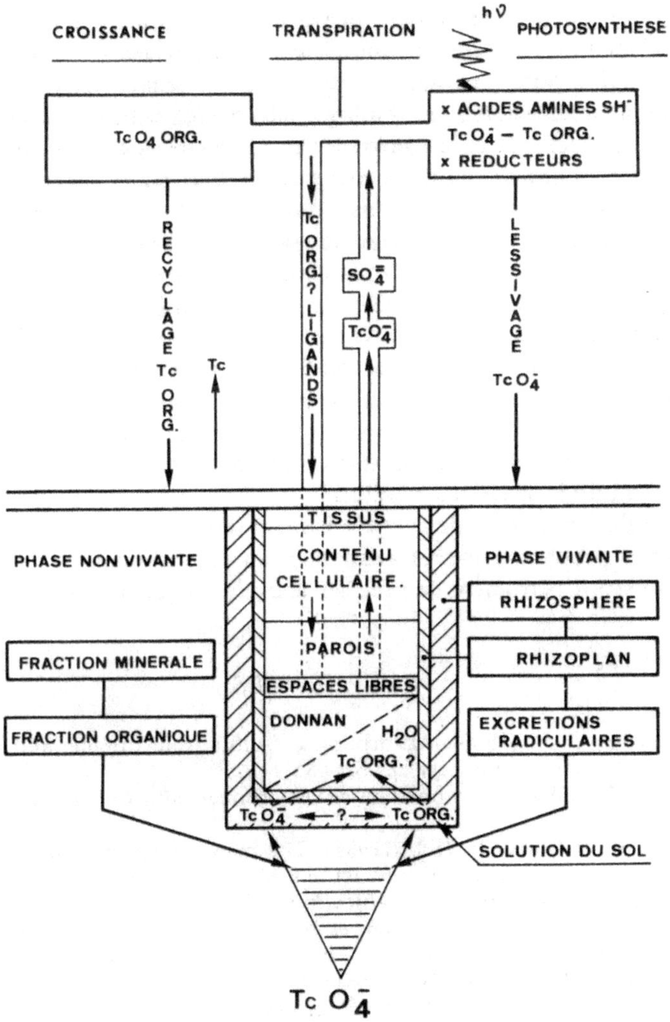

FIG. 4. Transfert sol–plante–atmosphère—cycle du technétium.

sol. Dans les conditions actuelles, le pool organique du sol en technétium en est à sa genèse d'où l'importance de l'étude des phénomènes d'organification et de minéralisations biologique et biochimique de la matière organique technétiée. Il n'est d'ailleurs pas impossible que certains acides aminés-Tc soient directement absorbés par les racines [1] et ces divers mécanismes demeurent à étudier.

Lorsqu'il est absorbé sous forme de TcO_4^-, cet élément transite principalement sous la forme anionique et est peu fixé par les parois des vaisseaux. Il atteint les feuilles et est alors réduit dans les cellules chlorophylliennes (réducteurs biologiques) et fixé définitivement (ligands non labiles). A ce stade, la synthèse de réducteurs (photosynthèse) et de ligands (acides aminés soufrés par ex.) régit la fixation du Tc par les organes chlorophylliens. Les résultats obtenus à l'aide d'inhibiteurs spécifiques confirment cette façon de voir [6, 17]. Dans ces conditions, la compétition entre TcO_4^- et d'autres ions perd une grande partie de sa valeur au profit de l'organification de ces éléments.

Le Tc métabolisé se redistribue peu dans le végétal et son devenir est fonction du type de végétal, des conditions climatiques et des méthodes culturales. Une faible fraction du Tc peut être lessivé en cours de culture [29] tandis que le Tc incorporé au substrat par enfouissement du matériel végétal peut soit être réémis dans l'atmosphère ou réabsorbé après dégradation de la matière organique.

De nombreux aspects de ce cycle exigent d'être clarifiés et quantifiés et les travaux prochains menés aussi bien en conditions contrôlées qu'au champ devront permettre de préciser le cycle biogéochimique du Tc.

RÉFÉRENCES

1. Bardsley, C. E. (1960). Absorption of sulfur from organic and inorganic sources by bush beans. *Agr. J.*, **52**, 485–6.
2. Bittel, R. (1980). Le technétium et l'environnement. *Radioprotection*, **15**(3), 143–6.
3. Bolin, B. et Cook, B. (Ed.) (1983). The major biogeochemical cycles and their interactions. Scope 21, Interactions of biogeochemical cycles in grassland ecosystems, New York, John Wiley, pp. 247–69.
4. Carlsen, L., Skytte Jensen, B. et Nilsson, K. (1984). Basic geochemical research for migration studies. The migration chemistry of technetium. Contract CCE No. 194-81-6 WASDK, Topical Report, 21 pp.
5. Cataldo, D. A., Wildung, R. E. et Garland, R. T. (1976). Technetium accumulation, fate and behaviour in plants. *Proceedings of Symposium on the Environmental Chemistry and Cycling Process*, Augusta (Georgia), pp. 538–49.
6. Cataldo, D. A., Wildung, R. E. et Garland, T. R. (1983). Root absorption and transport behaviour of technetium in soybean. *Plant Phys.*, **74**, 849–52.
7. Coughtrey, P. J., Jackson, D. et Thorne, M. C. (1983). Radionuclide distribution and transport in terrestrial and aquatic ecosystems III, technetium, Balkena Rotterdam, pp. 210–28.
8. De Becker, R. (1983). Absorption et désorption du technetium-99 par le

riz irrigué en solution et en sol. Thèse Licence Spéciale Application de Radioisotopes et Radioprotection, UCL, Fac. Sci., Unité de Physiologie Végétale, Louvain-la-Neuve, Belgium, 92 pp.
9. Delmotte, A. (1981). Comportement écophysiologique d'*Anabaena cylindrica* Lemn (cyanobactérie en présence de technétium-99 (polluant nucléaire)). Thèse de Doctorat en Sciences Botaniques, UCL, Fac. Sci., Unité de Physiologie Végétale, Louvain-la-Neuve, Belgium, 126 pp.
10. Freney, J. R. (1971). Sulfur containing organics. In: *Microbial Ecology* (Ed. Martin Alexander), New York, John Wiley, pp. 209–59.
11. Garten, C. T. (1983). Chemical form of the radioelement Tc in plants and its gastrointestinal absorption into animals from plant foods. Foreign Trip Report ORNL/FTR-1662.
12. Gearing, P., Van Baalen, C. et Parker, P. L. (1975). Biochemical effects of technetium-99 pertechnetate on microorganisms. *Plant Phys.*, **55**, 240–6.
13. Hoffman, F. O., Garten, C. T., Huckabee, J. W. et Lucas, D. M. (1982). Interception and retention of technetium by vegetation and soil. *J. Envir. Qual.*, **11**, 134–41.
14. Hoffman, F. O., Garten, C. T., Lucas, D. M. et Huckabee, J. W. (1982). Environmental behaviour of technetium in soil and vegetation. Implication for radiological assessments. *Envir. Sci. Technol.*, **16**, 214–17.
15. Hoffman, F. O., Huckabee, J. W., Lucas, D. M. *et al.* (1980). Sampling of Tc-99 in vegetation and soils in the vicinity of operating gaseous diffusion facilities. ORNL/TM-7386, 51 pp.
16. Itschert, A. (1984). Etude de l'absorption du technétium-99 par le haricot (*Phaseolus vulgaris*, var. nannus L.). Mémoire Licence en Sciences Botaniques, UCL, Fac. Sci., Unité de Physiologie Végétale, Louvain-la-Neuve, Belgium, 68 pp.
17. Lietart, E. (1981). Etude des mécanismes responsables de l'absorption radiculaire du technétium-99 chez glycine max lemm cv caloria. Mémoire Licence en Sciences Botaniques, UCL, Fac. Sci., Unité de Physiologie Végétale, Louvain-la-Neuve, Belgium, 112 pp.
18. McLaren, A. D. et Peterson, G. H. (1967). Soil biochemistry, microorganisms and biogeochemistry, New York, Marcel Dekker Inc., pp. 391–485.
19. Mengel, K. et Kirkby, E. A. (1978). *Principles of Plant Nutrition*, Bern, Int. Potash Inst., 593 pp.
20. Mousny, J. M. (1981). Contamination de la chaîne alimentaire. Influence du type de sol sur le transfert du technétium-99. Thèse Licence Spéciale Applications Radioisotopes et Radioprotection, UCL, Fac. Sci., Unité de Physiologie Végétale, Louvain-la-Neuve, Belgium, 71 pp.
21. Mousny, J. M. et Myttenaere, C. (1981). Absorption of technetium by plants in relation to soil type contamination level and time. *Plant and Soil*, **61**, 403–12.
22. Myttenaere, C., Daoust, C. et Roucoux, P. (1980). Leaching of technetium from foliage by simulated rain. *Env. Exptl Bot.*, **20**, 415–19.
23. Myttenaere, C. et Scoppa, P. (1984). Il ciclo biogeochimico del technezio. 2° Congresso della Società Italiana di Ecologia, Padova, 25–28 Giugno.
24. Petit, C. M., Ringoet, A. et Myttenaere, C. (1978). Stimulation of cadmium uptake in relation to the cadmium content of plants. *Plant Phys.*, **62**, 554–7.

25. Roucoux, P. (1980). Etude du comportement du technétium (polluant radioactif) chez le soja en milieu contaminé ou non en cadmium (polluant conventionnel). Thèse Doctorat Sciences Agronomiques, UCL, Fac. Sci., Unité de Physiologie Végétale, Louvain-la-Neuve, Belgium, 127 pp.
26. Saas, A., Denardi, J. L., Colle, C. et Quinault, J. M. (1979). Cycle du molybdène et du technétium dans l'environnement. Evolution physico-chimique et mobilité dans les sols et les végétaux. IIe Symp. Int. Radioécologie, CEN Cadarache, 19–22 juin 1979, pp. 443–89.
27. Schwarz, G. et Hoffman, F. O. (1980). An examination of the effect in radiological assessments of high soil plant concentration ratios for harvested vegetation. *Health Physics*, **39**, 983–6.
28. Scwochau, K. (1983). The present status of technetium chemistry. *Radiochimica Acta*, **32**, 139–52.
29. Sheppard, M. I., Vandergraaf, T. T., Thibault, D. H. et Keith Reid, J. A. (1983). Technetium and uranium sorption by and plant uptake from peat and sand. *Health Physics*, **44**, 635–43.
30. Shrift, A. (1961). Biochemical interrelations between selenium and sulfur in plants and microorganisms. *Federation Proceedings*, **20**, 695–702.
31. Tubis, M. et Endow, J. S. (1968). The preparation of Tc-99m technetium-labeled cystime methionine and a synthetic polypeptide and their distribution in mice. *Int. J. appl. Radiat. Isot.*, **19**, 835–40.
32. Vandecasteele, C. M. (1981). Influence de la radiocontamination du sol sur l'activité de la microflore. Comportement du *Rhizobium* et d'*Azotobacter* en présence du technétium-99. Thèse Licence Spéciale en Application des Radioisotopes et Radioprotection, UCL, Fac. Sci., Unité de Physiologie Végétale, Louvain-la-Neuve, Belgium.
33. Vandecasteele, C. M., De Becker, R. et Myttenaere, C. (1982). Comportement du riz (*Oryza sativa* L.) en présence de technetium-99. 2nd Workshop on Technetium Behaviour in the Environment, La Baule, 25–26 février.
34. Vandecasteele, C. M., De Becker, R., Tang Van Hai et Myttenaere, C. (1983). Etude de l'absorption du Tc-99 par le riz irrigué (*Oryza sativa* L.). *Radioprotection*, **18**(1), 19–30.
35. Wildung, R. E. et Garland, T. R. (1980). The relationship of microbial processes to the fate and behaviour of transuranic elements in soils, plants and animals. In: *Transuranic Elements in the Environment* (Ed. W. C. Hanson), pp. 300–35.

24

Accumulation of Tc–Bio-organic Complexes in Spinach Plants in Relation to Growth

J. F. M. M. LEMBRECHTS and G. M. DESMET

NIPHEH, Laboratory of Radiation Research, c/o Association EURATOM-ITAL, Keyenbergseweg 6, 6704 PJ Wageningen, The Netherlands

1. INTRODUCTION

This study aims to contribute to a consistent model on the biochemical behaviour of Tc—administered as NH_4TcO_4—in plants, using spinach (*Spinacea oleraces* L. cv. Verbeterd breedblad) as a test object.

A general idea about the subcellular distribution of Tc [1,2], a description of the diversity of the labelled compounds produced and a rough quantification of each of them are first requirements for a characterisation of Tc–bio-organic complexes [2]. On the other hand, the variability of the observed pattern [3] has to be studied, because of its importance for the selection of appropriate material, and since kinetic studies might be considered a starting point for the explanation of the behaviour of an element.

Results on the influence of the age of the plant material and of the period, time and level of contamination upon the distribution pattern are given.

2. METHODS

Plants were germinated on moistened perlite (seven days in darkness followed by one in the light). They were transferred on to a modified Hoagland–Arnon nutrient solution [4] contaminated with subtoxic levels of NH_4TcO_4. Four different experiments were distinguished with respect to the time, period and level of contamination (Fig. 1).

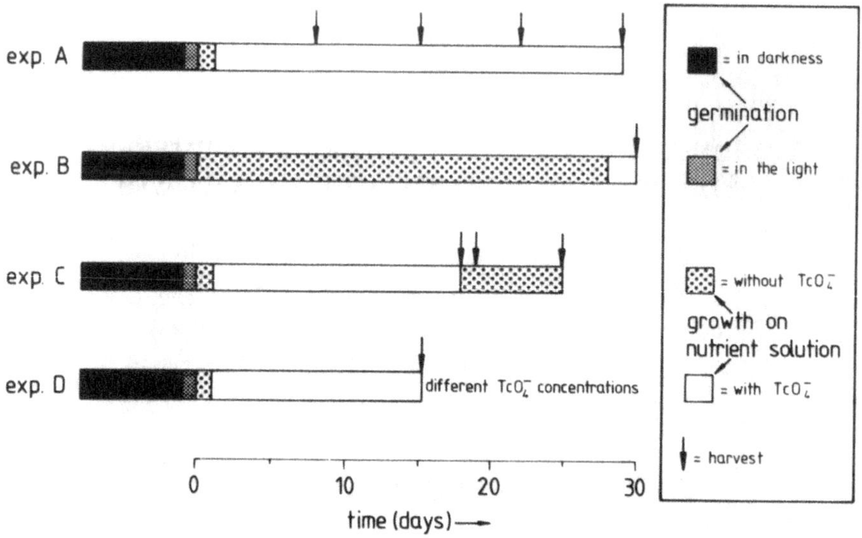

FIG. 1. Cultivation and sampling schedule used to determine the uptake and incorporation of $^{99}TcO_4^-$ by spinach plants as a function of time.

At each harvest, the leaves—without petiole—were weighed and comparable pairs of opposite leaves from different plants were pooled. The leaves were ground in a Virtis-23 homogeniser in an extraction medium (2 ml/g fresh weight) composed of 50 mM TES, 500 mM saccharose, 1 mM 2-mercaptoethanol and 0·5 mM disodium-EDTA, adjusted to pH 7·5 with NaOH. The homogenate was filtered through one layer of nylon cloth (60 μm) and centrifuged at 12 000 *g* for 20 min.

3. RESULTS AND DISCUSSION

The ^{99}Tc-labelled compounds of 3 ml supernatants were separated on a gel filtration chromatographic column (φ 1·6 cm, L 61·5 cm) filled with BioGel P6 (M) (polyacrylamide gel, fractionation range 6000–1000 g/mol), using the extraction buffer as eluent (16 ml/h). The ^{99}Tc content of the fractionated effluent was determined by liquid scintillation counting (Packard Tri-Carb 4530). Four Tc-labelled compounds or groups of compounds were distinguished in a BioGel P6 (M) radiochromatogram (Fig. 2). The relative share of each of them was determined for the

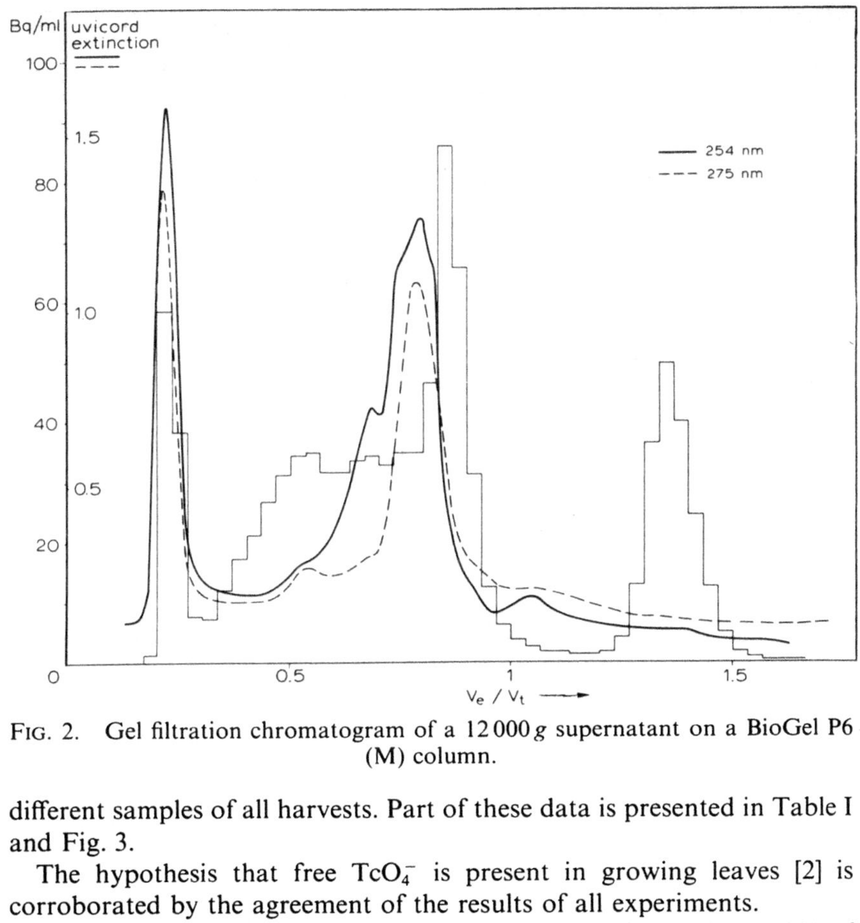

FIG. 2. Gel filtration chromatogram of a 12 000 *g* supernatant on a BioGel P6 (M) column.

different samples of all harvests. Part of these data is presented in Table I and Fig. 3.

The hypothesis that free TcO_4^- is present in growing leaves [2] is corroborated by the agreement of the results of all experiments.

In the roots and petioles of plants of four weeks of age, about 60% of all Tc is present as TcO_4^-, suggesting Tc is transported through the xylem vessels as TcO_4^- [5, 6].

In experiment A it is shown that the relative importance of the TcO_4^- fraction decreases as a function of time, due to the continuous metabolisation of the accumulated TcO_4^-, which is also apparent from the results of experiment C. Furthermore, the influence of the position of the leaves upon the metabolisation of Tc is demonstrated: the share of the low molecular weight Tc–bio-organic complexes (< 1000 g/mol) gradually decreases in subsequent emerging leaves, which also appears from the results of the other experiments.

Table I

Total Tc concentration (Bq/g homogenised material) and relative quantity of free TcO_4^- in the supernatant for the different leaves, and initial Tc content (C_0) of the nutrient solutions (Bq/ml)

Experiment	C_0	*Time (days)*	*Cotyledon*	*1° leaf*	*2° leaf*	*3° leaf*	*3° leaf*
A	55·9	8	1 871 38%	902 53%			
A		15	3 085 36%	2 382 33%	959 35%		
A		22	3 915 18%	2 758 29%	1 514 32%	827 28%	368 33%
A		29	4 101 17%	3 509 19%	2 559 19%	1 288 27%	700 31%
B	71·5	30	145 78%	298 66%	609 63%	834 54%	555 57%
C	51·4	18	2 641 37%	2 060 42%	988 52%		
C		19	2 488 27%	1 934 29%	687 33%		
C		25	2 568 13%	1 671 12%	303 12%		
D_1	0·56	15	21·7 39%	21·0 33%	8·4 42%		
D_2	6·02	15	246 39%	207 37%	84·5 52%		

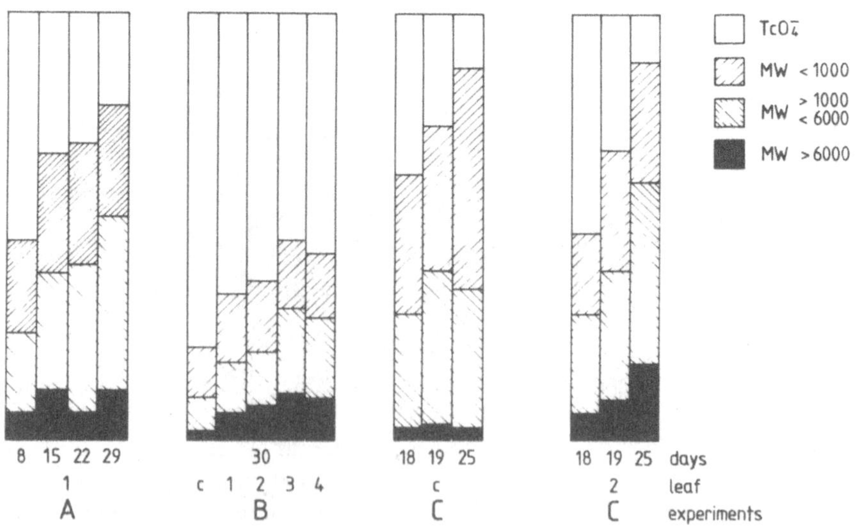

FIG. 3. Relative distribution of four ^{99}Tc-labelled groups of compounds in leaves of spinach plants.

TcO_4^- is the preponderant compound in the leaves harvested in experiment B, because few Tc-labelled metabolites are produced during a two-day contamination period. Furthermore, the influence of the age of the leaves upon the accumulation of Tc is demonstrated: the rates of uptake and metabolisation decrease with the age of the leaves (Table I).

Comparable ratios between free TcO_4^- and bound Tc were found in plants grown on different concentrations of TcO_4^- for 14 days (experiment D), suggesting that the rates of uptake and transformation are directly proportional to the TcO_4^- concentration of the nutrient solution.

In experiment C the TcO_4^- supply was diminished after 18 days of growth, effecting a pronounced decline of the relative TcO_4^- content of the leaves. This can be seen as the result of an evolution towards a new equilibrium between both rate of uptake and transformation rate.

4. CONCLUSIONS

When NH_4TcO_4 is administered to spinach plants:

—technetium is absorbed and transported as TcO_4^-;
—free TcO_4^- is present in growing leaves;
—uptake and incorporation are directly proportional to the concentration in the nutrient solution (between 10^{-8} and 10^{-6} M);
—rates of uptake and metabolisation decrease with increasing age of the leaves;
—different ratios between the various labelled compounds are found in successively emerging leaves.

REFERENCES

1. Vandecasteele, C. M., Delmotte, A., Roucoux, P. and Van Hove, C. (1982). Technétium et organismes diazotrophes: toxicité, localisation, facteurs de transfert. *Proc. Int. Symp. on Environmental Migration of Long-lived Radionuclides*, Knoxville, USA, 27–31 July 1981, Vienna, IAEA, pp. 275–86.
2. Lembrechts, J. F., Desmet, G. M. and Overbeek, H. Subcellular localization of technetium in spinach leaves. *Environmental and Experimental Botany* (submitted for publication).
3. Lembrechts, J. F. and Desmet, G. M. Effects of leaf-aging on the distribution and metabolization of technetium-99 in *Spinacea oleracea* L. *Environmental and Experimental Botany* (submitted for publication).
4. Desmet, G. M. and Dirkse, W. G. (1980). Growth and zinc accumulation in spinach. *Physiol. Plant*, **50**, 314–18.

5. Cataldo, D. A., Wildung, R. E. and Garland, T. R. (1978). Technetium accumulation, fate and behaviour in plants. In: *Environmental Chemistry and Cycling Processes* (Ed. T. C. Adriano and T. L. Brisbin Jr), Dept. of Energy Symp. Series No. 45, CONF-760429. Natl Tech. Info. Serv., Springfield, VA.
6. Roucoux, P. (1980). Etude du comportement de technétium (polluant radioactif) chez le soja en milieu contaminé ou non en cadmium (polluant conventionnel). Thesis, UCL, Louvain-la-Neuve.

25

Speciation of Technetium in Plants Grown on Substrates which Contained Different Chemical Forms of Technetium

L. R. VAN LOON and J. F. M. M. LEMBRECHTS

NIPHEH, Laboratory of Radiation Research, c/o Association EURATOM-ITAL, Keyenbergseweg 6, 6704 PJ Wageningen, The Netherlands

1. INTRODUCTION

When plants are grown on substrates containing different chemical forms of technetium, it is not clear in which chemical form technetium is taken up by the plants. It is important to know this in order to understand the relationship between the amount of technetium in the plant and the amount of available technetium in the growth medium (nutrient solution or soil solution).

Studies on the uptake of technetium by plants grown on soils, contaminated with technetium over a few years, are especially faced with these problems. In these soils, more than 95% of the technetium is present in a chemical form other than TcO_4^- [1]. When technetium is taken up by plants grown on these contaminated soils, it is not clear in which chemical form the technetium is absorbed.

Plant tissue analysis seems to be a promising method for determining whether technetium is taken up as TcO_4^- or in another chemical form. The main idea is that the metabolisation of Tc is dependent on the chemical form of the administered technetium and results in a typical speciation of Tc in the plant [2].

2. MATERIALS AND METHODS

Spinach plants were grown on nutrient solutions [3] containing only TcO_4^- at different concentrations (0·56 Bq/ml, 6·02 Bq/ml, 55·9 Bq/ml) and on a mixture of TcO_4^- and the Tc-cysteine complex (total concentration 58·0 Bq/ml). After 15 days of growth, different leaves were harvested and the cell content was extracted by grinding the leaves in a buffer solution (sucrose 0·5 M, TES 0·05 M, EDTA 0·5 mM, 2-mercaptoethanol 1 mM) at pH 7·5. After filtering the homogenate through a 60-μm nylon cloth, it was centrifuged at 20 000 *g* for 20 min. The supernatant was separated by gel filtration.

3. RESULTS AND DISCUSSION

Spinach plants grown on nutrient solutions containing TcO_4^- at different concentrations show constant relative amounts of Tc-components in their leaves, irrespective of the concentration of TcO_4^- in the nutrient solution [2]. Plants grown on a mixture of TcO_4^- and the Tc-cysteine complex show the same relative amounts of Tc-components (Fig. 1). This is a first indication that the TcO_4^- form is preferred to the Tc-cysteine complex.

There is a linear relationship between the amount of Tc in the different leaves and the TcO_4^- concentration in the nutrient solution (Fig. 2). In the case of the mixture of TcO_4^- and the Tc-cysteine complex, the concentration of TcO_4^- continuously increased as a function of time, due

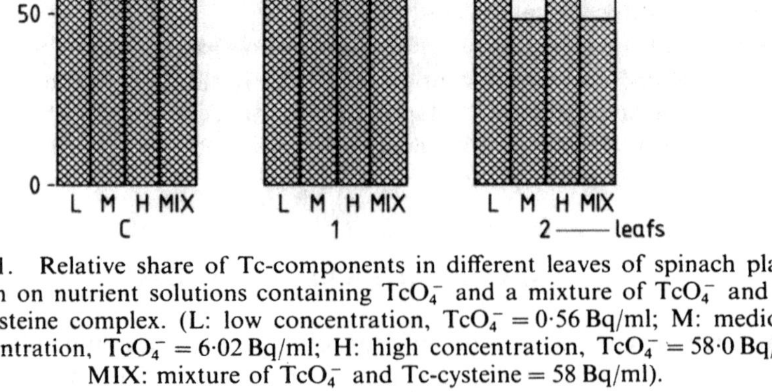

FIG. 1. Relative share of Tc-components in different leaves of spinach plants grown on nutrient solutions containing TcO_4^- and a mixture of TcO_4^- and the Tc-cysteine complex. (L: low concentration, TcO_4^- = 0·56 Bq/ml; M: mediocre concentration, TcO_4^- = 6·02 Bq/ml; H: high concentration, TcO_4^- = 58·0 Bq/ml; MIX: mixture of TcO_4^- and Tc-cysteine = 58 Bq/ml).

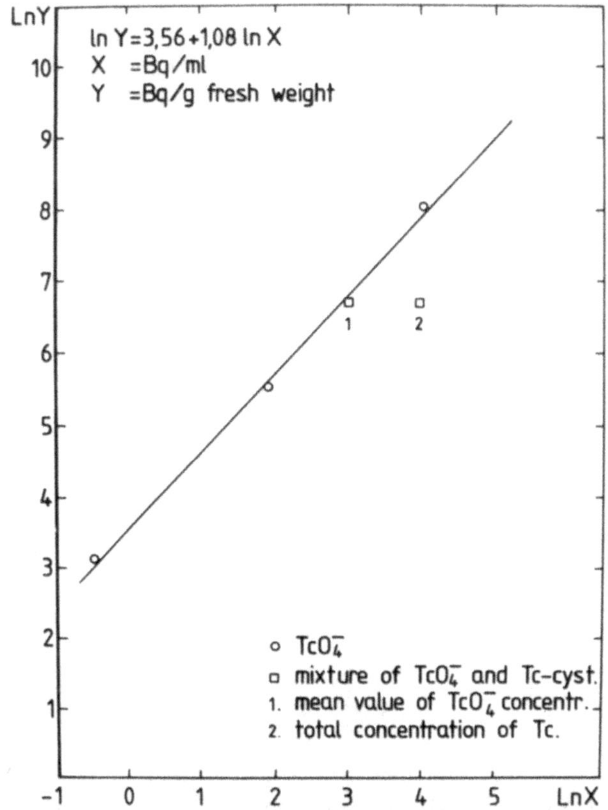

FIG. 2. Uptake of TcO_4^- by the cotyledon of a spinach plant grown on a nutrient solution for 15 days.

to reoxidation of the Tc-cysteine complex. A mean value of TcO_4^- concentration in the nutrient solution was calculated by integrating the concentration function over the growth period of a certain leaf. For each leaf, a different mean concentration was calculated (Fig. 3). When this mean concentration is used, there is good agreement between the calculated and experimentally determined amount of Tc in the leaves. This is not the case when the total amount of Tc in the nutrient solution is used for calculating the amount of Tc in the leaves.

Root analysis shows that much more technetium is absorbed by the plant roots in the case of the mixture of TcO_4^- and Tc-cysteine complex. In the shoot, on the other hand, less amounts were present (Table I). The

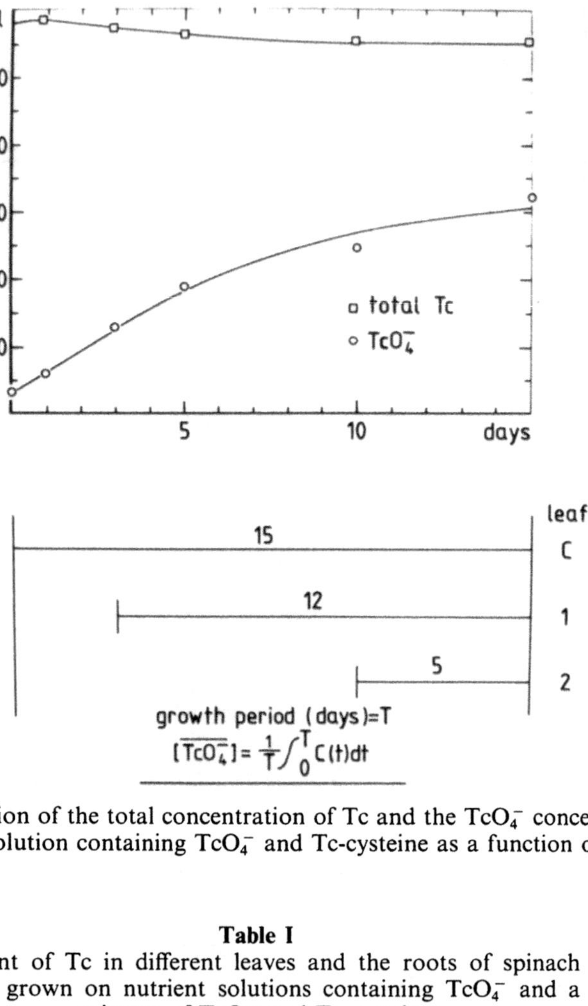

FIG. 3. Evolution of the total concentration of Tc and the TcO_4^- concentration in a nutrient solution containing TcO_4^- and Tc-cysteine as a function of time.

Table I
Amount of Tc in different leaves and the roots of spinach plants grown on nutrient solutions containing TcO_4^- and a mixture of TcO_4^- and Tc-cysteine

	Bq/g dry weight $Tc = TcO_4^-$ $C_T = 58$ *Bq/ml*	*Bq/g dry weight* $Tc = TcO_4^- + Tc$-*cyst.* $C_T = 58$ *Bq/ml*
Cotyledon	58 000	13 988
1st leaf	35 000	14 796
2nd leaf	16 000	7 406
Root	6 300	34 524

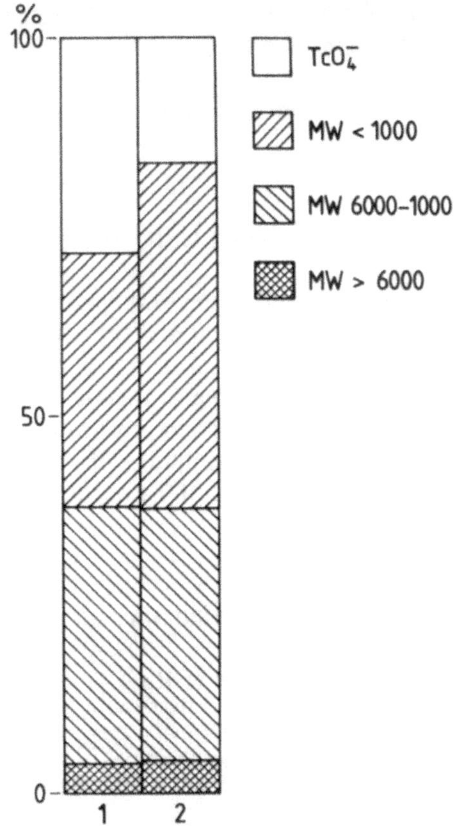

FIG. 4. Relative share of Tc-components in the cotyledon of spinach plants grown on a podzolic soil [1] and a nutrient solution [2].

absorbed Tc-cysteine complex is not translocated to the shoots. If it is translocated to the shoots, it will result in a smaller relative amount of TcO_4^- in the leaves.

These three observations indicate that TcO_4^- is preferred to the Tc-cysteine complex (and perhaps other Tc-components). It is too early to claim that the Tc-complex is not taken up by plants at all. But if it is taken up, the uptake rate is very small and its contribution to the Tc in plants can be neglected.

Spinach plants grown on a podsolic soil containing Tc in a chemical form other than TcO_4^- for over 95% show TcO_4^- in their leaves. The relative amounts of the Tc-components in the leaves are comparable with those plants grown on a nutrient solution containing only TcO_4^- (Fig. 4). The differences can be explained by a difference in physiological age of the plant leaves.

This again is an indication that TcO_4^- is preferred to other Tc-components.

4. CONCLUSIONS

When plants are grown on substrates containing different chemical forms of Tc, plant tissue analysis is a practical tool for determining the chemical form of Tc taken up by the plants.

In future, speciation of Tc in the plant, together with speciation of Tc in the growth medium, should be carried out more frequently. This will result in a better understanding of the uptake of Tc by plants. It will also tell us something about the reactions of Tc occurring in soils and in the root environment or rhizosphere.

REFERENCES

1. Van Loon, L. R. (1984). Unpublished results.
2. Lembrechts, J. F. and Desmet, G. M. Effects of leaf-aging on the distribution and metabolization of Tc-99 in *Spinacea oleracea* L. *Environmental and Experimental Botany* (submitted for publication).
3. Desmet, G. M. and Dirkse, W. G. (1980). Growth and zinc accumulation in spinach. *Physiol. Plant*, **50**, 314–18.

26

Rétention par les Végétaux du Technétium Relaché dans l'Environnement en Association avec d'Autres Radiopolluants

P. ROUCOUX et C. COLLE

Département d'Etudes et de Recherches en Sécurité, CEA-IPSN, Centre d'Etudes Nucléaires de Cadarache, 13115 Saint Paul-lez-Durance, France

1. INTRODUCTION

Les radionucléides les plus importants qui à long terme peuvent contribuer à la dispersion biologique des déchets radioactifs sont: parmi les transuraniens le ^{239}Pu, ^{241}Am et ^{237}Np et pour les produits de fission le ^{99}Tc. Les actinides ainsi que le technétium sont des éléments possédant de multiples états d'oxydation et qui ont tendance à former des complexes spécialement avec des ligands organiques [1]; la prédiction de leur migration dans les sols apparaît comme très complexe. De plus ces éléments se trouvent associés dans le milieu et des interactions entre-eux sont possibles, il peut en résulter des modifications dans leur comportement biogéochimique. Des interactions chimiques existent entre les différents transuraniens, entre transuraniens et certains produits de fission [2] ainsi qu'avec des polluants non nucléaires [3]. Des antagonismes ou des synergies peuvent se créer au niveau des mécanismes d'absorption racinaire ou agir sur le métabolisme du végétal, il en résulte des variations dans les facteurs de transfert sol-plante. Les études relatives au transfert des radionucléides du milieu aux végétaux n'ont établi les taux de transfert que pour chacun des radiopolluants présents isolément dans le milieu et non associés entre-eux. La présence dans le milieu d'un mélange de radionucléides est peut-être une des principales causes de la discordance entre les facteurs de transfert déterminés in-situ sur des dépots réels et ceux mesurés en conditions de laboratoire avec un seul radioélément.

Dans les conditions rencontrées dans l'environnement, le technétium se trouve principalement à l'état d'oxydation +7 (TcO_4^-) et +4 (TcO_2

insoluble ou complexe organique) [4]. Sous forme de pertechnétate, le technétium est très mobile dans les sols et largement disponible aux végétaux [5, 6]. La majorité des complexes du technétium seraient chargés négativement et stables pour un milieu aérobie [7]; dans ces conditions le technétium garde sa mobilité et est toujours biodisponible.

Les facteurs de transfert du pertechnétate (TcO_4^-) sont élevés comparés aux valeurs des produits de fission, des transuraniens et des éléments non nutritifs [8]. Ils sont de la grandeur de ceux observés pour les microéléments comme le Mo, Mn, Cu, Fe, Zn. La forte solubilité du pertechnétate dans les sols, bien que ce soit un point important de son comportement dans le relation sol-plantes n'explique pas totalement la forte affinité de la plante pour cet élément. Les études de cinétique d'absorption ont montré que l'entrée du pertechnétate dans la plante a des caractéristiques d'un transport utilisant un transporteur qui serait spécifique de l'absorption d'éléments nutritifs [8, 9]. Les hauts facteurs de transfert du Tc résultent d'une efficience élevée d'un mécanisme d'absorption qui serait déja en place dans la racine pour des éléments nutritifs essentiels [8]. L'entrée du pertechnétate dans le végétal est donc fonction de son activité métabolique et certains éléments nutritifs entrent en compétition au niveau de l'absorption racinaire (SO_4^{2-}, PO_4^{2-}, ...).

L'étude du comportement des actinides (uranium, neptunium, plutonium) dans l'environnement est compliquée par la variété de leurs états d'oxydation; sauf pour le thorium et l'américium qui existent dans l'environnement principalement aux états +4 et +3 respectivement [10]. Les transuraniens forment des complexes très stables avec les acides humiques et fulviques de la matière organique des sols [7]. De plus les composés inorganiques présents dans la solution du sol comme les carbonates, les sulphates et les hydroxydes peuvent être impliqués dans la formation de complexes et influencer la disponibilité des transuraniens aux végétaux [10]. Il semble évident que la sorption sur les sols plus que la discrimination par la plante limite l'absorption du Pu et que le complexe organo-métallique rende possible la disponibilité du Pu aux racines [11]. Les plantes possèderaient la possibilité d'accumuler du Pu sous forme soluble et de le transporter aux parties aériennes sous forme d'un ou plusieurs complexes organiques. La disponibilité pour les végétaux des différents transuraniens est dans l'ordre Pu < Am < Np ce qui correspond à l'ordre inverse de sorption de ces éléments par les sols [11, 12].

Aussi bien en ce qui concerne le technétium que les transuraniens, leur comportement dans les sols est donc soumis aux réactions de formation de complexes, réactions d'échange et phénomènes de sorption. L'ensemble de ces phénomènes, indépendemment des conditions de milieu, serait influencé par la présence simultanée de technétium et de transuraniens.

Nos travaux ont portés sur l'étude de l'absorption du technétium associé en couples à différents actinides (U, Am, Pu). Essais réalisés sur milieu strictement contrôlé (solution nutritive) ou en conditions voisines des conditions naturelles (différents types de sols et plusieurs âges de dépot). Les résultats de cette étude confirment l'existence d'interaction entre le technétium et les actinides [13] et des hypothèses sont avancées pour essayer d'en comprendre les mécanismes. La connaissance de telles interactions mérite l'attention des programmes visant à étudier la dissémination à long terme des déchets nucléaires.

2. MATERIEL ET METHODES

Les plantes de haricot, mises en germination sur vermiculite, sont repiquées après 10 jours sur solution nutritive (solution de Hoagland diluée 4 fois) à pH voisin de 6·5. Les radioisotopes sont introduits dans la solution quand les plantes sont agées de 20 jours. Les concentrations des radioisotopes sont dans les rapports pondéraux de ceux présents dans les rejets (U/Tc: 2000, Pu/Tc: 0·1, Am/Tc: 0·01) soit: Tc-99: 0·2 mg/litre (0·125 MBq/litre) de TcO_4NH_3; Am-241: 1 μg/litre de $Am(NO_3)_3$; Pu-239: 10 μg/litre de $PuO_2(NO_3)_2$; U-naturel: 0·27 g/litre de $UO_2(NO_3)_26H_2O$. A la récolte les racines et la partie aérienne des plantes sont séparées. Les échantillons (4 plantes) sont minéralisés par voie humide ($HNO_3 + H_2O_2$) à basse température (< 80 °C), ils sont portés à sec, blanchis et repris dans HNO_3 0·02 N. La radioactivité est déterminée en scintillation liquide (β) pour Tc, Am, Pu et en scintillation solide (γ) pour U. Les interférences de comptage β et γ et la valeur du quenching sont déterminées par la méthode de rapport de cannaux.

Les transferts à partir des sols sont établis avec des radis cultivés sur trois sols différents: un sol riche en matières organiques, un sol calcaire et un sol acide. Les caractéristiques de ces sols ont été données par Saas *et al.* [14]. Les bacs de culture contiennent 15 kg de sol et 25 plantes, l'arrosage est assuré de manière à obtenir une percolation minimale. Les concentrations des radioisotopes sont: Tc-99: TcO_4NH_3 21·6 μg/kg (13·3 kBq/kg); U-naturel: $UO_2(NO_3)_26H_2O$ 15·9 mg/kg; Am-241: $Am(NO_3)_3$ 0·21 μg/kg (25·2 kBq/kg); Pu-234: $PuO_2(NO_3)_2$ 2·26 μg/kg (4·44 kBq/kg). Les échantillons végétaux sont traités de la même manière que ceux obtenus sur solution nutritive.

3. RESULTATS

3.1. Transfert à partir de solution nutritive

Pour des cultures de courtes périodes sur solution nutritive, l'absorption du Tc par la plante est largement influencée par la présence de transuraniens (Fig. 1). L'U présent dans la solution induit une diminution d'absorption du Tc d'un facteur 4 dans la racine et d'un facteur 2·2 dans les feuilles. L'Am est pratiquement sans effet mais le Pu provoque une augmentation de la quantité de Tc d'un facteur 2 dans toute la plante. Ces résultats ne peuvent pas être comparés entre eux d'un point de vue de la stoichiométrie de la réaction car les rapports de concentrations entre le Tc et les transuraniens sont différents dans chaque cas. Ces rapports (U/Tc = 1000, Pu/Tc = 0·1, Am/Tc = 0·01) sont voisins de ceux présents dans les rejets. Il faut remarquer que le transfert de l'U dans tout le végétal et du Pu dans la racine est augmenté par la présence de Tc (Fig. 2).

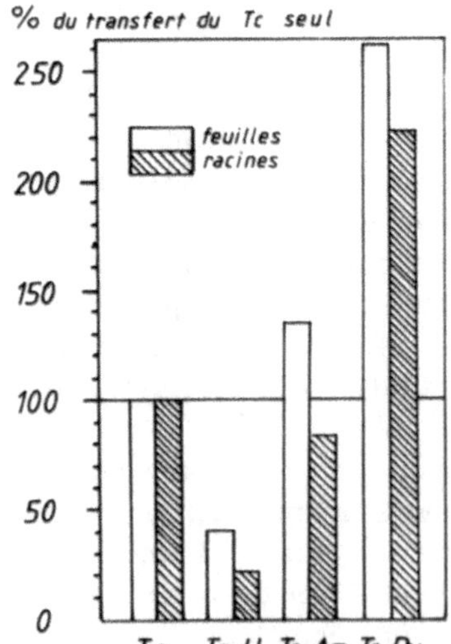

FIG. 1. Concentration en Tc dans la plante. Transfert à partir de solution nutritive. Effet de la présence de U, Am, Pu.

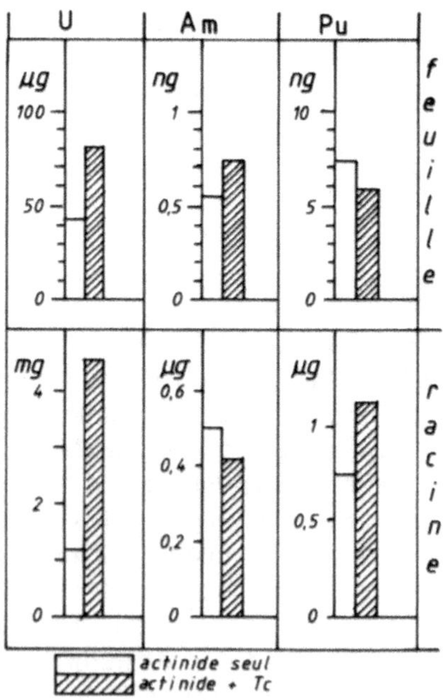

FIG. 2. Concentration en actinides transfert à partir de solution. Effet de la présence de Tc: TcO_4^-: 0·2 mg/litre; UO_2^{2+}: 0·27 g/litre; Am^{3+}: 1 µg/litre; PuO_2^{2+}: 10 µg/litre.

3.2. Transfert à partir des sols

Pour des cultures sur sols, le transfert aux végétaux du Tc est donné en fonction du type de sol et de l'âge de la contamination du sol (Fig. 3). Lorsque le Tc est le seul radionucléide présent, son transfert est le plus faible sur sol organique. Pour un dépot agé de 2 ans, c'est sur sol acide que le transfert est le plus élevé, il est légèrement plus faible sur sol calcaire. En fonction du vieillissement du dépot, le transfert du Tc est diminué pour les trois types de sols d'un facteur 2 à 4. Il reste le plus faible sur sol organique. La présence de transuraniens dans le milieu modifie le transfert du Tc en fonction du type de sols et de l'âge du dépot. Sur le radis pour un dépot agé de 2 mois le seul effet significatif est une augmentation du transfert du Tc sur sol calcaire en présence d'U.

Après 2 ans de dépot et à la quatrième culture, le transfert du Tc est augmenté dans tous les cas par la présence des actinides:

U sur sol calcaire:	4 × dans les feuilles;
Pu sur sol calcaire:	4·5 × dans les feuilles et 3 × dans les racines;
Pu sur sol acide:	1·5 × dans les feuilles et 3 × dans les racines;
Am sur sol organique:	6 × dans les feuilles et 3 × dans la racine.

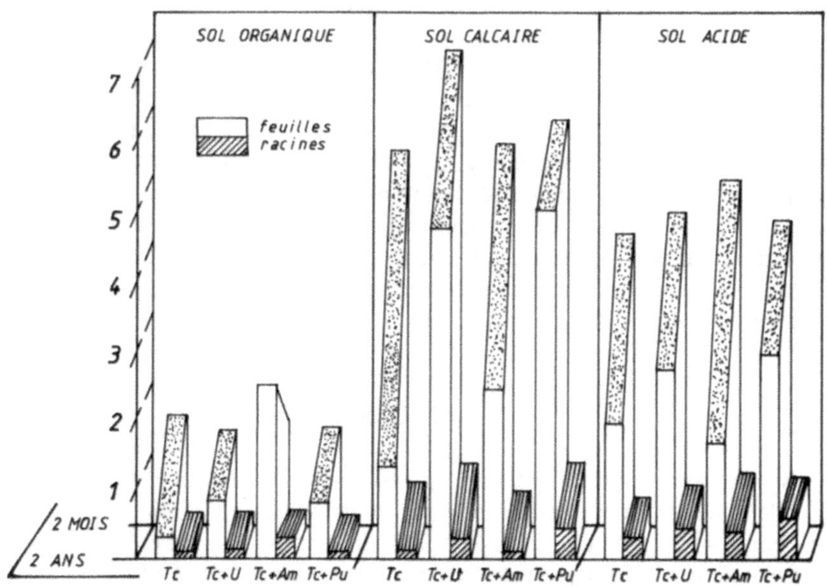

FIG. 3. Facteurs de transfert du Tc à partir de sols. Effet de la présence de U, Am, Pu. Evolution du transfert avec l'âge du dépot.

Tableau I
Translocation du technétium dans les radis
(activité de la partie aérienne/activité totale de la plante, en %)

	Sol organique	*Sol calcaire*	*Sol acide*
Tc seul	74·9	89·9	86·8
Tc + U	90·2	94·6	80·8
Tc + Am	89·3	93·0	87·8
Tc + Pu	85·7	93·4	94·6

D'après le Tableau I, la distribution du Tc dans la plante (rapport de l'activité dans la partie aérienne par rapport à l'activité totale de la plante) varie peu et correspond à la distribution du TcO_4^- dans le végétal (88% du Tc dans la partie aérienne). Cependant sur sol organique le Tc seul est distribué à raison de 74% dans les feuilles et dans le cas des couples Tc–U, Tc–Am, Tc–Pu sur sol calcaire et Tc–Pu sur sol acide (95%) le transfert vers les feuilles est plus élevé. Dans ces quatre cas, la translocation du Tc est favorisée. Les facteurs de transfert des transuraniens n'ont pas pu être déterminés avec précision à cause des faibles teneurs des échantillons et à cause des interférences de comptage avec le Tc. Une estimation a Pu être obtenue et a été confirmée pour l'Am en utilisant une méthode de séparation chimique du Tc et de l'Am suivie d'un comptage α et β séparés (méthode non publiée). Les facteurs de transfert sont compris entre 0·5.E-3 et 2; E-3 pour l'U, entre 2.E-2 et 1.E-1 pour l'Am et entre 2.E-2 et 2.E-1 pour le Pu. Les valeurs pour le U sont les plus faibles et celles du Pu les plus élevées. Ces valeurs sont obtenues dans le cas de la présence de Tc dans le sol et pour des concentrations en actinides très différentes. Dans le végétal les différences de teneurs en U et Pu sont réduites à cause des différences de facteurs de transfert qui peuvent être jusqu'à 200 fois plus élevés pour le Pu que pour l'U.

4. DISCUSSION

Aussi bien sur solution nutritive que sur sols, la présence des actinides dans le milieu a un effet différencié sur le transfert du Tc; effet qui dépend des conditions de milieu et de l'âge du dépot. De manière générale l'influence des actinides est de favoriser le transport du Tc aux végétaux et réciproquement le Tc induit une absorption élevée des actinides. L'explication d'un tel phénomène est liée à des modifications de la disponibilité et de la mobilité des radionucléides dans les sols, ou à une action spécifique au niveau du végétal: soit sur les mécanismes d'entrée des ions à la racine, soit sur les sites d'accumulation des éléments dans tout le végétal. En milieu aérobie et en absence d'agents réducteurs, le Tc se trouve sous forme d'anions TcO_4^- (+7) qui est stable quelque soit le pH et dont la concentration n'est limitée que par la solubilité du pertechnétate [7]. C'est principalement sous cette forme que le Tc se trouve dans l'environnement et il est alors facilement disponible aux végétaux [5, 6]. Sous forme de TcO_4^- le Tc est également accumulé en grande partie dans les feuilles [15, 16].

4.1. Transfert à partir de solution nutritive

Introduit sous forme de TcO_4^- dans la solution nutritive, le Tc est rapidement assimilé et distribué dans la partie aérienne du haricot. En présence de Pu le transfert du Tc est doublé dans tout le végétal par rapport à son absorption quand le Tc est seul dans la solution. De même la présence de Tc provoque une augmentation des teneurs en Pu des racines. Si on compare les augmentations de transfert des deux éléments lorsqu'ils sont présents simultanément, on observe des valeurs équivalents: 300 ng/g pour le Tc et 500 ng/g pour le Pu dans la racine. On peut donc supposer l'existence d'une association Tc–Pu sous forme de complexe qui entre dans le végétal plus facilement que le TcO_4^- seul et qui entraîne le Pu avec le Tc. Ce complexe serait dissocié dans la racine et le Tc serait transporté plus facilement que le Pu dans la partie aérienne. L'existence d'un tel complexe est chimiquement très probable.

Dans le cas d'un apport d'U, une réaction du même type que celle du Pu est possible, le comportement chimique de ces deux actinides étant voisin. Les résultats sont cependant différents, le transfert du Tc diminue de 130 ng/g dans la plante alors que celui de l'U augmente de 3·5 mg/g dans la racine et 20 μg/g dans les feuilles (la concentration de U est 1000 fois plus élevée que celle du Tc). Une expérience complémentaire réalisée avec du Np montre un même effet du Tc sur le transfert du Np. La réduction de l'absorption du Tc peut s'expliquer par une toxicité de l'U qui se manifeste par une inhibition du transport actif du TcO_4^-. C'est un effet comparable à ceux observés par la présence dans la solution nutritive de certains éléments comme le sulphate, le séléniate [8], le Cd [9] et le Th (résultats non publiés), ainsi que des inhibiteurs métaboliques [9]. La toxicité de l'U n'explique cependant pas l'accroissement du transfert de l'U induit par le Tc (qui est en concentration beaucoup plus faible que l'U). La forte accumulation de l'U dans la racine pourrait être due à des phénomènes de sorption importants à la paroi racinaire ou à des modifications de perméabilité des membranes; les deux phénomènes étant induits par la présence de TcO_4^-. Une seconde hypothèse peut cependant être émise. Les études physiologiques de l'absorption du Tc par la plante ont montré que certains anions comme le sulphate, le phosphate, le séléniate et le molybdate sont des inhibiteurs compétitifs de l'accumulation du TcO_4^- [8] et que le TcO_4^- entre dans la plante par un mécanisme existant pour un élément nutritif [8, 9]. D'autre part, les actinides, par exemple UO_2^{2+}, forment des complexes avec les sulphates, les séléniates [7], complexes qui augmentent la disponibilité des transuraniens. Le transfert de l'U par sa complexation au sulphate serait facilité ainsi que celui du TcO_4^- qui n'entrerait plus en compétition avec les SO_4^{2-} pour son entrée dans la racine. L'absorption du Tc et de l'U resteraient indépendantes l'une de l'autre et seraient fonction de leurs concentrations respectives dans la solution. La présence d'Am n'induit aucune modification significative du transfert du Tc ni réciproquement le Tc sur l'absorption de l'Am. Il n'y a vraisemblablement pas d'interaction directe entre l'Am et le Tc; ou s'il existe un complexe Am–Tc, il se comporte dans la plante comme chacun des éléments séparés.

4.2. Transfert à partir des sols

Le transfert du Tc des sols aux radis diminue en fonction de l'âge du dépot. Cette diminution est fonction du type de sol et de la présence des actinides dont l'influence est elle même dépendante des caractéristiques du sol. Dans un sol contaminé en TcO_4^- la présence possible d'agents réducteurs et de complexants peut changer complètement le comportement du Tc [4]. Aucune donnée quantitative n'est cependant disponible sur la complexation du Tc. Ce qui régit le comportement du Tc dans le milieu c'est d'une part sa présence sous forme +7 et d'autre part +4 avec formation de complexes ou liaison à des colloïdes avec des phénomènes de sorption ou de précipitation de ces espèces. La sorption augmentant avec le contenu en matière organique et diminuant avec le pH [17, 18]. La principale interaction serait due aux charges positives de la matière organique, le nombre de charges augmentant quand le pH diminue. C'est ce phénomène qui explique le transfert plus faible du Tc sur sol organique, resultat en accord avec les données actuelles [19–21].

En présence d'Am sur sol organique et pour un dépot agé de deux ans, la disponibilité du Tc est plus élevée que sur les deux autres sols. Une interaction de l'Am avec la matière organique expliquerait ce phénomène. La forme la plus stable l'Am dans les sols est +3, dans cet état d'oxydation, il peut se lier à la matière organique en formant des complexes très stables [11], il pourrait donc libérer en partie le TcO_4^- de certains sites de la matière organique et le rendre plus disponible à la végétation. Les réactions Am-matière organique seraient lentes si l'on observe l'évolution lente du transfert du Tc en fonction du vieillissement du dépot. Comme sur solution nutritive l'existence d'un complexe Tc–Am est possible, il pourrait empêcher la liaison du Tc à la matière organique. Ce complexe devrait être au moins aussi facilement assimilable que le TcO_4^-.

Sur les deux autres sols, comme sur solution nutritive, l'Am est sans effet sur le transfert du Tc, le TcO_4^- étant beaucoup moins fixé au sol (peu de matière organique); ce qui tenterait à prouver que c'est bien la libération du Tc de la matière organique par l'Am qui favorise son absorption par la plante.

Lors de la présence de Pu et de U dans les sols et après vieillissement du dépot on observe un transfert du Tc plus élevé pour l'ensemble des sols et principalement pour le sol calcaire qui est pauvre en matière organique (3·5%). Une interaction du même type de celle de l'Am ne pourrait être que très limitée.

Les charges du milieu autres que les charges de la matière organique influencent également la migration du Tc; par exemple: les oxydes de Fe et d'Al [7]. La solubilité joue alors un rôle important dans la migration plutôt que la sorption. En présence d'hydrate de Fe agissant comme cation de liaison, le Tc est fixé au même titre qu'un molybdate mais reste assimilable par les végétaux [20]. Lorsque les transuraniens sont présent dans le milieu, c'est sous forme cationique, ils peuvent donc jouer le rôle de cations de liaison et pourraient s'associer au TcO_4^-. D'autant plus que le Pu, selon son état d'oxydation a une tendance relative à former des complexes ($Pu+4 > Pu+3 > Pu+6 > Pu+5$) avec les anions $CO_3^{2-} > C_2O_4^{2-} > SO_4^{2-} > F^- > NO_3^- > Cl^-$ [7]. Ces complexes sont alors très faiblement ou non adsorbés sur les colloïdes et les composés organiques. Un complexe $Pu–TcO_4$ serait lui aussi très biodisponible. En aérobiose, le Pu devrait se trouver sous la forme plutonyl ou sous forme de complexe hydroxycarbonate de l'ion tétravalent [7]. La forte adsorption de l'ion tétravalent peut déplacer l'équilibre avec les carbonate de telle manière que les carbonates soient exclus de la solution. La formation de complexes très stable avec les carbonates pourrait augmenter l'absorption et le transport vers les parties aériennes du Pu sur sol calcaire plus que sur sol acide. Les silicates, phosphates vanadates peuvent entrer en compétition avec la formation d'hydroxydes et former des précipités qui selon les conditions sont plus stables que les hydroxydes [7]. La formation de ces complexes, qui dépend des conditions de milieu, peut augmenter la disponibilité des transuraniens. Les réactions décrites si elles s'appliquent au TcO_4^- seraient facilitées sur sol calcaire où la fraction hydrosoluble du TcO_4^- est très élevée et plus importante que sur le sol acide [14, 19]. Les interactions Tc–Pu et Tc–U se produiraient dans cette fraction hydrosoluble et provoqueraient les augmentations importantes du transfert du Tc (4 fois plus dans les feuilles et 2 fois plus dans les racines par rapport au Tc seul).

En présence de Am, U, Pu sur sol calcaire et de Pu sur sol acide, la distribution du Tc vers les parties aériennes (translocation) est augmentée (93% contre 87% pour Tc seul). C'est l'indice que le Tc n'est plus entièrement sous forme de TcO_4^- dans le végétal.

Sur sol organique le Pu et l'U induisent également une augmentation de transfert de Tc (2 fois par rapport au Tc seul). Les deux actinides peuvent donc libérer en partie le Tc de la matière organique et agir comme l'Am mais moins efficacement.

Les valeurs des facteurs de transfert du Tc pour un dépot agé de deux ans et sur les trois sols sont identiques, que le Tc soit accompagné du Pu ou de l'U. Or les concentrations dans les sols de ces deux éléments diffèrent d'un facteur 10 000, leur facteur de transfert étant dans un rapport de 200, leurs concentrations dans le végétal se trouvent dans un rapport de 50. Il faut donc supposer que l'augmentation de disponibilité du Tc ne dépend pas des concentrations respectives de U et Pu. Pour expliquer ce résultat deux hypothèse sont possibles—soit que à la concentration à laquelle se trouve le Pu, tout le Tc libérable devient disponible et qu'une augmentation de concentration des actinides ne libère pas plus de Tc—soit que une caractéristique du sol régit la disponibilité du TcO_4^- comme par exemple la teneur en un anion compétitif de l'absorption du TcO_4^- (SO_4^{2-}?). Et l'on pourrait envisager comme dans le cas de l'interaction Tc–Pu sur solution nutritive la formation d'un complexe Pu–SO_4 qui favoriserait le transfert du Pu en modifiant la compétition SO_4^{2-} et TcO_4^- permettant ainsi une absorption plus importante du Tc. Cette hypothèse doit être vérifiée.

REMERCIEMENTS

Cette étude a pu être poursuivie grâce à un contrat de coopération entre le CEN de Cadarache et le Laboratoire de Physiologie Végétale de l'Université de Louvain. Le financement de cette recherche a été octroyé conjointement par la Commission des Communautés Européennes et le

Commissariat à l'Energie Atomique dans le cadre des contrats BIO-B-318-81-F et BIO-B-315-81-F.

RÉFÉRENCES

1. WILDUNG, R. E. et GARLAND, T. R. (1980). The relationship of microbial processes to the fate and behavior of transuranic elements in soil, plants and animals. In: *Transuranic Elements in the Environment* (Ed. W. C. Hanson), Technical Information Centre, US Department of Energy, pp. 300–35.
2. GMELIN, L. (1975). *Gmelin Handbuch der Anorganischen Chemie*, Band 20, Transurane Teil D1, Berlin, Springer-Verlag.
3. BONDIETTI, E. A., REYNOLD, S. A. et SHANKS, M. H. (1975). Interaction of plutonium with complexing substances in soil and natural waters, *Transuranic Nuclides in the Environment*, Proceedings of a Symposium, San Francisco, 17–21 Nov. 1975, USERDA-IAEA, SM-199/51, Vienna, IAEA.
4. BONDIETTI, E. A. et FRANCIS, C. W. (1980). Geologic migration potentials of Np-237 and Tc-99. *Science*, **203**, 1337.
5. CATALDO, D. A., WILDUNG, R. E. et GARLAND, T. R. (1976). Technetium accumulation, fate, and behavior in plants. In: *Proc. Symp. on the Environmental Chemistry and Cycling Process*, Augusta, Georgia, April 27–30.
6. WILDUNG, R. E., GARLAND, T. R. et CATALDO, D. A. (1977). Accumulation of technetium by plants. *Health Physics*, **32**, 314.
7. JENSEN, B. S. (1980). The geochemistry of radionuclides with long half-lives. Risø National Laboratory, Roskilde, Denmark, Report R-430, novembre.
8. CATALDO, D. A., WILDUNG, R. E. et GARLAND, T. R. (1983). Root absorption and transport behavior of technetium in soybean. *Plant Physiol.*, **73**, 849.
9. ROUCOUX, P. (1980). Etude du comportement du technétium (polluant radioactif) chez le soja en milieux contaminé ou non en cadmium (polluant conventionnel). Thèse Université Catholique de Louvain, p. 127.
10. BONDIETTI, E. A. et TAMURA, T. (1980). Physicochemical associations of plutonium and other actinides in soils. In: *Transuranic Elements in the Environment* (Ed. W. C. Hanson). Technical Information Center, US Dept. of Energy.
11. WILDUNG, R. E. et GARLAND, T. R. (1980). The relationship of microbial processes to the fate and behavior of transuranic elements in soils, plants and animals. In: *Transuranic Elements in the Environment* (Ed. W. C. Hanson). Technical Information Center, US Dept. of Energy.
12. SCHRECKISE, R. G. et CLINE, J. F. (1980). Comparative uptake and distribution of plutonium, americium, curium and neptunium in four plants. In: *Transuranic Elements in the Environment* (Ed. W. C. Hanson). Technical Information Center, US Dept. of Energy.
13. ROUCOUX, P. et SAAS, A. (1983). Modification du transfert sol-plante de certains isotopes en présence d'autres polluants nucléaires et non nucléaires. In: *Séminaire sur la transfert à l'homme des radionucléides libérés dans l'environnement par les installations nucléaires*, IAEA-SR-85/26, Bruxelles, Belgique, 17–21 octobre.

14. Saas, A., Grauby, A., Denardi, J. L. et Quinault, J. M. (1979). Comportement du technétium-99 et du molybdène stable dans un sol brun calcaire. *Radioprotection*, **14**(4), 241.
15. Gast, R. G. (1975). The behavior of technetium-99 in soils and plants, Progress Report April 1974–March 1975, USEAC Contract, COO-2441-1.
16. Mousny, J. M., Roucoux, P. et Myttenaere, C. (1979). Absorption and translocation of technetium in pea plant. *Envir. Exp. Botany*, **19**, 263.
17. Wildung, R. E., McFadden, K. M. et Garland, T. R. (1979). Technetium sources and behavior in the environment. *J. Envir. Qual.*, **8**, 156.
18. Landa, E. R., Thorvig, L. J. et Gast, R. G. (1977). Effect of selective dissolution of electrolytes, aeration and sterilization on technetium-99 sorption by soils. *J. Envir. Qual.*, **6**, 181.
19. Mousny, J. M. et Myttenaere, C. (1982). Transfert du technétium du sol vers la plante en fonction du type de sol, du mode de contamination et de la couverture végétale. In: *Environmental Migration of Long-lived Radionuclides*, Proc. Symposium in Knoxville, USA, 23–31 July 1981, Vienne, IAEA, CCE, NEA.
20. Saas, A., Grauby, A., Colle, C., Denardi, J. L., Nomura, T. et Joyer, Ph. (1982). Evolution physico-chimique du technétium-99 et du sélénium-79 dans les sols, comparaison des transferts sol-plante à long terme. In: *Environmental Migration of Long-lived Radionuclides*, Proc. Symposium in Knoxville, USA, 27–31 July 1981, Vienne, IAEA, CCE, NEA.
21. Balogh, J. C. et Grigal, D. F. (1980). Soil chromatographic movement of technetium-99 through selected Minnesota soils. *Soil Science*, **130**, 278.

27

Chemical Form of Technetium in Corn (*Zea mays*) and the Gastrointestinal Absorption of Plant-incorporated Tc by Laboratory Rats

C. T. GARTEN JR,[a] C. MYTTENAERE,[b] C. M. VANDECASTEELE,[b]
R. KIRCHMANN[c] and R. VAN BRUWAENE*[c]

[a] *Environmental Sciences Division, Oak Ridge National Laboratory, Oak Ridge, Tennessee 37831, USA*

[b] *Laboratoire de Physiologie Végétale, Université Catholique de Louvain, Place Croix du Sud 4, B-1348 Louvain-la-Neuve, Belgium*

[c] *Radiobiology Department, CEN-SCK Laboratories, Boeretang 200, B-2400 Mol, Belgium*

1. INTRODUCTION

Technetium-99, which undergoes radioactive decay by beta emission, has a radioactive half-life of approximately 213 000 years. This radionuclide has a high fission yield associated with the use of all types of reactor fuel, including plutonium, uranium and thorium. Releases of Tc-99 may occur during reprocessing of spent reactor fuel, from the conversion of this reprocessed fuel to UF_6 and from the enrichment of this recycled material. Releases may also occur at sites designated for the disposal of radioactive wastes generated from the preceding operations. Prior research has shown that Tc-99 has an extremely high potential for uptake via plant roots into the terrestrial food chain [1]. The mobility of technetium into plants from contaminated soils can be attributed to the high water solubility and the environmental stability of the pertechnetate anion. Therefore, under circumstances of environmental releases of Tc-99, plants may be an important source of technetium for food chain transfer to animals and humans [2].

In this work, we have used a short-lived (60-d half-life) gamma-emitting

* Deceased 27 August 1984.

isotope, ^{95m}Tc, to make inferences about the behaviour and fate of Tc-99 in the food chain. We have characterised the chemical form of technetium in corn leaves, an important constituent of livestock silage, and conducted feeding studies to determine the availability of plant-incorporated technetium when fed to rats.

2. MATERIALS AND METHODS

2.1. Plant culture

Corn plants (*Zea mays*) were grown under greenhouse conditions for six weeks in 4·5 litres of nutrient solution. After germination, the technetium-95m, in pertechnetate form, was supplied to the plants in fresh nutrient solution. The nutrient solution was changed after the first three weeks of culture with the addition of ^{95m}Tc. The technetium concentration in the nutrient solution was maintained at approximately 60–80 μCi ^{95m}Tc/litre, with a calculated concentration of 0·313 μCi Tc-99/litre (the latter served as a carrier). After six weeks of growth in the contaminated nutrient solution, the corn was harvested and separated into leaves, stalks and roots, and then frozen. Portions of the corn leaves were freeze-dried and used to prepare a contaminated diet for the rodent-feeding studies.

2.2. Chemical extractions

Fresh and freeze-dried samples of corn leaves were successively extracted according to the methods of Bowen *et al.* [3]. Each extraction was replicated once. The plant material was first macerated in boiling 95% ethanol (5 min), followed by centrifugation to separate the residue from the supernatant. The residue was extracted twice more with boiling ethanol (2 min), and the combined supernatants were filtered to obtain a clear ethanol-soluble fraction. The residue was then successively extracted three times with 0·2 M hydrochloric acid (room temperature), three times with 0·5 M perchloric acid (80°C) and once with boiling 2 M NaOH (10 min). Centrifugation (12 000 *g* for 10 min), followed by filtration of the pooled supernatants, was used to obtain a clear extract following each chemical treatment. Equal volumes of acetone were added to the acid extracts to precipitate the 'acetone-insoluble' fractions. The gelatinous precipitates were collected by centrifugation and washed twice with a small volume of 50% acetone. Technetium-95m in extracts, acetone-insoluble fractions and the residue were analysed by gamma spectrometry.

After thawing, frozen corn leaves (chopped into 1-cm long pieces) were also extracted with demineralised water to determine the amount of ^{95m}Tc in the plant tissue that was removable by three successive overnight soakings at 4°C. Extracts were concentrated by evaporation and analysed for ^{95m}Tc.

2.3. Differential centrifugation

Samples of fresh and freeze-dried corn leaves were homogenised at 4°C in 50 ml of 0·02 M tris-HCl buffer (pH 7·2) containing 0·5 M sucrose and 0·01 M magnesium chloride. Each homogenate was filtered through cheesecloth, and the filtrate was centrifuged first at 10 000 *g* for 20 min to collect cell debris, nuclei, plastids and mitochondria, and then at 100 000 g for 70 min to separate the microsomal pellet from the 'post-microsomal supernatant' or cell cytosol. Pellets and aliquots of the supernates were analysed for ^{95m}Tc by gamma spectrometry.

2.4. Chemical form in the post-microsomal fraction

The post-microsomal supernatants (5 ml) from the prior centrifugation of freeze-dried and fresh corn leaves were chromatographed using Sephadex G-50 (MW fractionation range 1500–30 000) and G-25 (MW fractionation range 1000–5000). The column dimensions were 79 × 2·6 cm for the G-50 and 95 × 2·6 cm for the G-25. The columns were equilibrated with the eluent (0·02 M tris-HCl buffer, pH 7·2) and housed in a refrigerated room at 4°C. Quality of the column packing was evaluated, and the void volume was determined twice for each column by chromatographing blue dextran (MW 1 000 000). The flow rate was approximately 0·4 ml/min for all fractionations. Eluent from the columns was monitored continuously for UV absorbance at 280 nm (to detect proteins and polypeptides containing aromatic acids) and collected in 13-ml fractions by an automatic sample collector. A sample of ^{95m}Tc in the form of ammonium pertechnetate was chromatographed on the G-25 column for comparison with chromatographs of the cell cytosol from homogenised corn leaves. The amount of ^{95m}Tc in the various fractions was analysed by gamma spectrometry.

2.5. Animal feeding studies

Adult, female, Wistar rats (R/cnb inbred strain; 200 g body weight) were used in the feeding trials. Rats each received one of the following treatments: (1) a single intravenous injection of a solution containing 2·6 μCi of ^{95m}Tc in pertechnetate form (iv treatment); (2) a single overnight feeding of ^{95m}Tc (pertechnetate) added to a pulverised standard laboratory ration (chow) to give a dose of 3·3 μCi per rat (Tc + chow treatment); (3) a single overnight feeding of ^{95m}Tc contaminated corn leaves that were freeze-dried, milled and mixed in a ratio of 1 g powdered corn leaves to 10 g clean, pulverised chow to give a dose of 8·0 μCi per rat (Tc + corn treatment). The contaminated diets were prepared by adding approximately 30 ml of water to 30 g of the mixed dry food and drying at 40°C to form hardened pellets. The contaminated diets were also extracted according to the methods of Bowen *et al.* (explained above) to characterise the form of ^{95m}Tc in the two types of food. The rat chow contained (RMH-B from Hope Farms) 25% protein, 6·2% fat, 58·4% carbohydrates and 6·1% minerals.

The rats were housed in the laboratory at room temperature in cages with wire screen floors that permitted the collection of urine and faeces. Rats were given access to the contaminated food in the evening. Uneaten food was removed from the cages and weighed the following morning to determine the administered dose. During the first seven days after injection or feeding, excreta were collected and analysed for ^{95m}Tc. There were 12 rats in each treatment group. Three rats from each of the experimental groups were killed on days 1, 3, 7 and 14 after administration of the ^{95m}Tc. The rats were dissected, and 18 different body tissues were sampled, weighed and analysed to determine the concentration of ^{95m}Tc. Blood was obtained by heart puncture. Hair and skin were sampled from a small area on the abdomen. The fat sampled was abdominal fat. The contents of the stomach and intestines were extruded and were not included with the intestinal tissue for analysis. Thyroids, kidneys, liver, reproductive system, lungs, spleen, pancreas, adrenals, heart and brain were removed intact and analysed whole.

3. RESULTS AND DISCUSSION

3.1. Chemical form in plants

Table I shows the percentage of ^{95m}Tc extracted from fresh and freeze-dried corn leaves by successive treatments with boiling ethanol, weak mineral acids and boiling 2 M NaOH. The variation between replicate extractions was small with standard deviations typically being less than 20% of the mean. Boiling 95% ethanol, which extracts lipids, free amino acids, amino sugars and pigments, removed an average of 24% of the ^{95m}Tc from dry corn leaves. A greater amount (40%) of technetium was extracted from fresh corn leaves with the ethanol treatment. This may be

Table I
Fractionation of ^{95m}Tc incorporated in corn leaves (*Zea mays*) by root uptake from a contaminated nutrient solution

Extractant	*Contents of fraction*	*% of total in leaves*[a]	
		Fresh	*Freeze-dried*
95% boiling ethanol	Lipids, amino sugars, free amino acids, lipophilic pigments	40 (1)	24 (4)
0·2 M hydrochloric acid			
Supernatant	Ionic forms, including salts of organic acids, phosphates, carbonates	22 (2)	28 (1)
Precipitate with acetone	Proteins and pectates	1 (0·3)	1 (0·6)
0·5 M perchloric acid	(80°C)		
Supernatant	Remaining ionic forms	6 (0·1)	7 (2)
Precipitate with acetone	Nucleic acids and some proteins	<1	<1
Boiling 2 M sodium hydroxide	Most proteins and polysaccharides	29 (1)	37 (4)
Residue	Cell wall material (cellulose and lignin)	2 (1)	3 (0·5)

[a] Each value is the mean of two replicates and the standard deviation is in parentheses.

due to the water-soluble ^{95m}Tc present in homogenised fresh leaves being diluted into the ethanol fraction. In order to compare fresh and dried leaves, it is better to consider the first two extracts together instead of independently. The two successive treatments with weak mineral acids, which both extract ionic forms, removed an average of 35% of the technetium from the dry corn leaves. Very little of the technetium in the leaves was coprecipitated with proteins, pectates and nucleic acids from the acid fractions by acetone. In both fresh and dry leaves, almost one third or more of the ^{95m}Tc was associated with proteins and polysaccharides extractable with boiling 2 M NaOH. For the most part, these are not free proteins in the cell cytosol. The harsh treatment required for the removal of proteins and polysaccharides indicates that technetium in this fraction was chemically bound in the corn leaves.

The interpretation of the previous extractions is complicated by the possibility that chemical changes in the plant material early in the procedure may lead to later artefacts in the extractions. This problem is minimised in the simplified extraction of fresh chopped corn leaves by three successive overnight treatments with demineralised water at 4°C. Most of the ^{95m}Tc that could be extracted was removed from the chopped corn leaves by a single overnight soaking. The percentages of the ^{95m}Tc removed in the first, second and third extractions with water were 46, 8 and 4, respectively. Approximately 42% of the ^{95m}Tc could not be removed from the leaves by soaking in water. This unextractable portion corresponds approximately to the amount of ^{95m}Tc that is bound in dry corn leaves (i.e. the 40% extractable by boiling 2 M NaOH, plus that present in the final residue (Table I)).

Differential centrifugation studies showed that a high percentage of ^{95m}Tc in freeze-dried corn leaves was present in the post-microsomal supernatant (cell cytosol). The extracted leaves following homogenation in buffer retained 33% of the ^{95m}Tc in the residue while the filtrate contained 67%. In the filtrate alone, 80% of the ^{95m}Tc was found in the clear cell cytosol obtained by centrifugation at 100 000 *g* while only 20% was present in the cell debris and organelles. The distribution was only slightly different in the fresh corn leaves: 29% in the residue, 71% in the filtrate; and in the filtrate, 67% of the ^{95m}Tc was in the cell cytosol while 33% was associated with cell debris and organelles. Therefore, approximately 50% of the ^{95m}Tc in corn leaves was present in ionic form or bound to soluble proteins or smaller molecules in the cell cytosol.

3.2. Chemical form in the cell cytosol

Gel permeation chromatography of the corn leaf post-microsomal fraction on a Sephadex G-50 column resulted in more than 60% of the sample radioactivity being eluted in a single peak at an elution volume to void volume ratio (V_e/V_0) ranging from 2·9 to 3·1. This indicated that most of the ^{95m}Tc in the cell cytosol was associated with molecules <1500 MW and the need for use of a column more suited to this size of molecule. Approximately 40% of the ^{95m}Tc in the post-microsomal fraction chromatographed diffusively on the G-50 column with a range in the V_e/V_0 ratio of 1·0–2·8, which indicated some association of technetium with larger organic molecules in a MW range of 1500–30 000 daltons. Total recovery of the sample radioactivity applied to the G-50 columns was about 98%, and there were no apparent differences between the chromatographs of fresh and freeze-dried corn leaves.

When samples of the post-microsomal supernatant were chromatographed on Sephadex G-25, the total recovery of radioactivity from the column was 82% of that applied. A chromatograph of the post-microsomal supernatant from freeze-dried corn leaves on the G-25 column showed that 50% of the sample radioactivity eluted from the column in a single peak with a K_{av} value of 1·5, while 49% eluted diffusely between K_{av} values of 0 and 1, indicating that some ^{95m}Tc was associated with molecules in the MW range of 1000–5000 (Fig. 1). The partition coefficient, $K_{av} = V_e - V_{0}/V_t - V_0$ (where V_t is $3 \times V_0$), defines the solute behaviour on the gel independently of the bed dimensions and column packing. When ^{95m}Tc as ammonium pertechnetate was chromatographed on the G-25 column, it eluted in a single peak with a K_{av} value of 1·5 and 90% recovery (Fig. 1). Based on the similar K_{av} values, it could be concluded that part of the ^{95m}Tc in the corn leaf cytosol was present as pertechnetate anion. However, in corn leaves the peak at $K_{av} = 1·5$ exhibited appreciable UV absorbance at 280 nm; this, however, was not the case in the chromatograph of pertechnetate alone. The UV absorbance associated

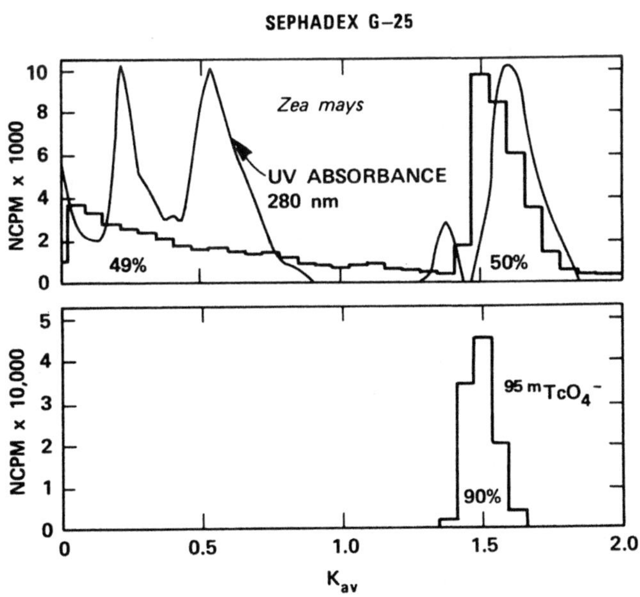

FIG. 1. Chromatograph of ^{95m}Tc in the post-microsomal supernatant from contaminated corn leaves (top) and a chromatograph of anionic ^{95m}Tc (ammonium pertechnetate) alone (bottom). Superimposed on the elution profile for ^{95m}Tc in the corn leaf cytosol is the UV absorbance of the eluted fractions at 280 nm. The percentages show the recovery of the initial ^{95m}Tc applied to the Sephadex G-25 column that was associated with different parts of the elution profile.

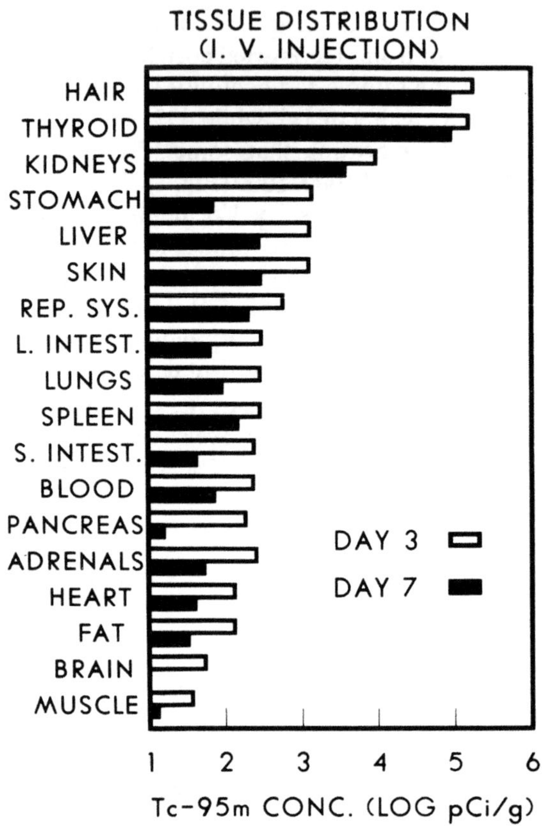

FIG. 2. Average concentrations of ^{95m}Tc, based on three replicate animals, in fresh tissues and organs from adult female albino rats at three and seven days following a single intravenous injection of pertechnetate.

with the chromatography peak at $K_{av} = 1{\cdot}5$ in corn leaves indicated that this peak could be ^{95m}Tc bound to small organic molecules (peptides), in addition to some free pertechnetate anion. Also, some heterocyclic or homocyclic aromatic compounds, including some peptides, are known to interact with Sephadex G-25 gel, resulting in K_{av} values $>1{\cdot}0$. In any case, the ^{95m}Tc in the corn leaf cytosol that eluted at $K_{av} = 1{\cdot}5$ is probably not entirely free pertechnetate anion alone.

3.3. Experiments with laboratory rats

The tissue distribution of ^{95m}Tc in laboratory rats after intravenous injection with pertechnetate was similar to the distribution found in rats that were administered contaminated diets. Tissue concentrations in rats at three and seven days following intravenous injection of pertechnetate are shown in Fig. 2. In every case, the most highly contaminated tissues were rat thyroids, hair, kidneys, liver and skin. Although samples of rat hair were collected by shaving the abdomen, the high ^{95m}Tc concentrations in hair apparently were not due entirely to surface contamination by excreta because washing specimens of the hair twice with water (5 min each) removed a total of only 22% of the radioactivity. The high concentration of ^{95m}Tc in the skin, as well as in the hair, indicated that technetium was being biologically incorporated into hair. Although rats were elevated above their excreta by a wire screen, the possibility of contamination of the abdominal hair by urine or faeces on the screen cannot be ruled out.

The elimination of ^{95m}Tc from the four most contaminated tissues in rats injected with pertechnetate is shown in Table II. There was no significant loss of ^{95m}Tc from rat thyroids or hair over the two-week period of the experiment, as determined by analysis of variance. Concentrations of ^{95m}Tc in kidneys and liver decreased by approximately 50% between one day and three days after injection (Table II) and continued to decline in concentration through 14 days after injection. By 14 days following administration of the ^{95m}Tc to rats via injection or via contaminated food, the concentrations in many tissues (excluding thyroids, hair and kidneys) had fallen below or near our detection limit of approximately 10 pCi/g for 1 g of fresh tissue.

Table II

Elimination of ^{95m}Tc from tissues in rats injected with pertechnetate expressed as mean percentage dose in each tissue at days 1, 3, 7 and 14 after injection. Numbers in the same row sharing the same alphabetic superscript are not significantly different, as determined by analysis of variance and Scheffe's multiple comparison test

Tissue	*Percentage dose at day*				*F-value* ($df = 3,8$)	*Prob.*
	1	*3*	*7*	*14*		
Hair	0·16 (a)	14·6 (a)	7·40 (a)	8·40 (a)	3·80	NS
Thyroid	0·12 (a)	0·15 (a)	0·10 (a)	0·12 (a)	0·85	NS
Kidney	0·92 (a)	0·51 (b)	0·23 (b)	0·14 (b)	21·6	<0·001
Liver	0·66 (a)	0·32 (b)	0·07 (b)	0·02 (b)	20·4	<0·001

NS = no significant difference between the various days.

The chemical form of technetium in freeze-dried corn leaves mixed with the rat chow was different from the chemical form of ^{95m}Tc added to pulverised laboratory rat chow (Table III). The extractability of ^{95m}Tc from both types of contaminated food was different from that observed for ^{95m}Tc incorporated into corn leaves alone (Table I). Mixing ^{95m}Tc,

Table III

Extractability of ^{95m}Tc from two different diets fed to laboratory rats. Diet 'Tc + chow' is ^{95m}Tc (pertechnetate) freshly added to uncontaminated standard laboratory rat chow. Diet 'Tc + corn' is 1 g of dry, powdered corn leaves (containing ^{95m}Tc biologically incorporated in the plant tissue via root uptake) mixed with 10 g of uncontaminated rat chow

Extractant	*Fraction containing*	*% of total Tc-95m in food*	
		Tc + chow	*Tc + corn*
95% boiling ethanol	Lipids, amino sugars, free amino acids, lipophilic pigments	42	12
Weak mineral acids[a]	Ionic forms, including salts of organic acids, phosphates, carbonates	36	35
Boiling 2 M NaOH and residue	Most proteins and polysaccharides, as well as cellulose and lignin	22	53

[a] 0·2 M hydrochloric acid (room temperature) followed by 0·5 M perchloric acid at 80°C.

either as pertechnetate or as technetium incorporated into corn leaves, with the rat chow apparently influenced the extractability of ^{95m}Tc from the two types of food. The main difference between the two diets was the greater percentage of ^{95m}Tc found in bound forms (i.e. proteins and polysaccharides extractable by boiling 2 M NaOH, and cellulose and lignin) in the food made from contaminated corn plus rat chow (Table III). Even though ^{95m}Tc was added to rat chow as pertechnetate anion to make one of the contaminated foods, most of the ^{95m}Tc was not extracted in the weak mineral acid fraction. Based on the extraction data in Table III, technetium simply mixed with the rat chow apparently interacted with the protein and lipophilic components already present in this food. The rat chow contained 25% protein and 16 different vitamins.

The cumulative excretion of ^{95m}Tc in urine and faeces combined during the first seven days after administration of technetium by the various treatments was as follows: iv treatment—91%, Tc + chow—82% and Tc + corn—74%. The distribution of ^{95m}Tc between urinary and faecal excretion varied depending upon the route of administration (Fig. 3). For pertechnetate injected into rats, excretion via the urine was greater than

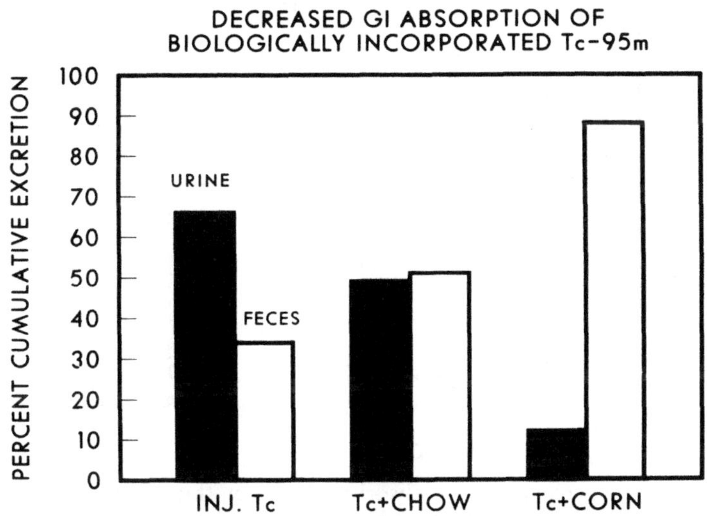

FIG. 3. Average percentage distribution of ^{95m}Tc in the cumulative urine or faeces for seven days following a single dose administered to laboratory rats as intravenously injected pertechnetate (inj. Tc), as pertechnetate mixed with a standard rat food (Tc + chow) or as ^{95m}Tc biologically incorporated into corn leaves and mixed with rat food (Tc + corn).

that via the faeces. Technetium found in the faeces of injected rats may have been due to the wetting of faeces by contaminated urine or the movement of ^{95m}Tc from the body into the intestine by way of the bile. Cumulative urinary excretion was nearly equal to faecal excretion in rats eating ^{95m}Tc mixed with their rat chow, but urinary excretion was greatly decreased in rats fed ^{95m}Tc incorporated into the corn leaves. The decreased urinary excretion of ^{95m}Tc in rats fed Tc + corn as compared to rats fed Tc + chow indicated that technetium was made less available for gastrointestinal absorption by biological incorporation into corn leaves.

The percentage of the administered dose present in rat thyroids, kidneys, liver and carcass (i.e. whole body minus 18 organs in Fig. 2) as a result of the three methods of ^{95m}Tc administration to rats also indicated a

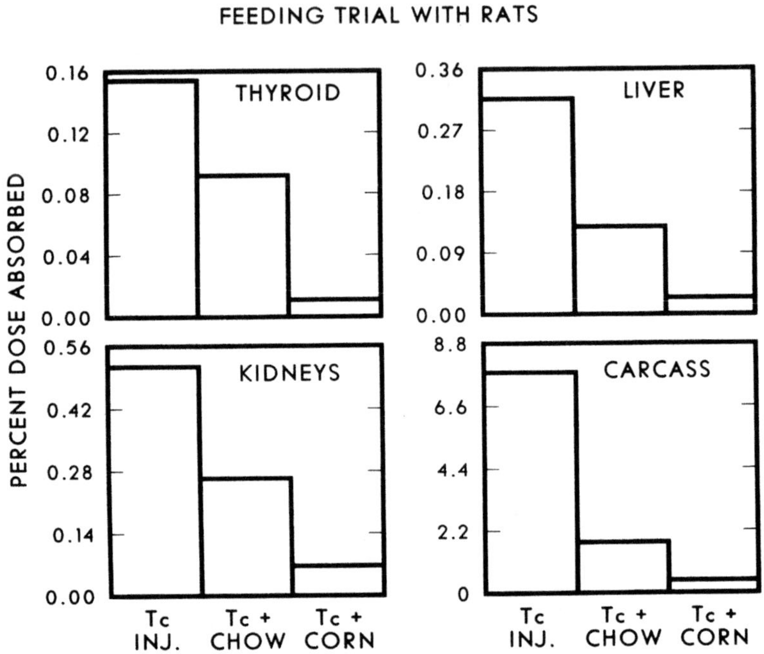

FIG. 4. Average percentage dose absorbed, based on three replicate animals, of ^{95m}Tc in thyroid, kidneys, liver and carcass three days following a single administration of technetium-95m to laboratory rats according to one of the following methods: (1) intravenous injection (inj. Tc), (2) pertechnetate mixed with a standard rat food (Tc + chow) or (3) ^{95m}Tc incorporated into corn leaves mixed with rat food (Tc + corn).

lowered gastrointestinal absorption for technetium biologically incorporated into corn leaves (Fig. 4). There was less incorporation of ^{95m}Tc into thyroids, kidneys, liver and carcasses three days following the administration of the isotope in rats fed the Tc + corn diet as compared to the Tc + chow diet. Analysis of variance showed that the difference between the percentage dose absorbed from the two diets was statistically significant ($P < 0{\cdot}05$) for thyroids and kidneys. The carcass contained a greater amount of ^{95m}Tc than any of the body tissues, probably because of the high concentration of ^{95m}Tc present in the rat hair (most of the hair remained with the rat carcass). The percentage administered dose found in the hair of rats from the various treatment groups followed the same pattern as that in the carcass. However, the amount of ^{95m}Tc in the hair was extremely variable between rats, as reflected by the standard deviations being 75–100% of the mean.

4. CONCLUSION

Based on the results of the chemical extractions, almost 40% of the ^{95m}Tc biologically incorporated into corn leaves via root uptake is present in the plant tissue in forms that are not readily extractable, except by boiling 2 M NaOH. The 'bound' forms of ^{95m}Tc in the leaves appear to be mostly Tc complexes with proteins and polysaccharides of the plant cell walls. Roucoux [4] found that about 38% of the Tc-99 incorporated in soybean leaves was associated with the NaOH extractable fraction and its residue.

Differential centrifugation studies showed that about half of the ^{95m}Tc in the corn leaves was present in ionic form or associated with soluble protein and smaller organic molecules. Chromatography of the cell cytosol demonstrated that ^{95m}Tc was associated with molecules in the MW range of peptides (1500–5000 daltons), as well as with smaller molecules and some larger molecules. About half of the ^{95m}Tc in the cell cytosol co-chromatographed with ^{95m}Tc pertechnetate, but UV absorbance data indicated that the ^{95m}Tc smaller than 1000 MW was not present as pertechnetate anion alone. Roucoux [4] also found that Tc-99 in soybean leaves was associated principally with small molecules of MW less than 10 000 daltons.

Once assimilated into the body, technetium was rapidly excreted from most tissues except thyroid and hair. The present study confirms the result of Sullivan *et al.* [5], which showed that incorporation of ^{95m}Tc into plant tissues (soybeans) resulted in decreased gastrointestinal absorption by rats and guinea pigs relative to the absorption of inorganic ^{95m}Tc. Associations with organic molecules in plants are perhaps partly responsible for the lowered availability of technetium from the GI tract when it is incorporated into vegetation. The diminished availability of ^{95m}Tc incorporated into corn leaves may also be due to the poor digestibility of this particular diet in the rat intestine. Nonetheless, biological incorporation of technetium into plants appears to reduce the potential food chain transfer of this radionuclide through decreased availability for gastrointestinal absorption.

ACKNOWLEDGEMENTS

Research sponsored in part by the Office of Health and Environmental Research, US Department of Energy, and in part by the Radiation Protection Program, Commission of the European Communities, under

Contract No. DE-ACO5-840R21400 with Martin Marietta Energy Systems, Inc. Publication No. 2403, Environmental Sciences Division, Oak Ridge National Laboratory.

REFERENCES

1. Wildung, R. E., McFadden, K. M. and Garland, T. R. (1979). Technetium sources and behavior in the environment. *J. Envir. Qual.*, **8**, 156–61.
2. Till, J. E., Hoffman, F. O. and Dunning, D. E. Jr (1979). A new look at Tc-99 releases to the atmosphere. *Health Physics*, **36**, 21–30.
3. Bowen, H. J. M., Cawse, P. A. and Thick, J. (1962). The distribution of some inorganic elements in plant tissue extracts. *J. Exp. Bot.*, **13**, 257–67.
4. Roucoux, P. (1980). Etude du comportement du technétium (polluant radioactif) chez le soja en milieux contaminé ou non en cadmium (polluant conventionnel). PhD thesis, Faculté des Sciences Agronomiques, Université Catholique de Louvain, Louvain-la-Neuve, Belgium.
5. Sullivan, M. F., Garland, T. R., Cataldo, D. A. and Schreckhise, R. G. (1979). Absorption of plant-incorporated nuclear fuel cycle elements from the gastro-intestinal tract, IAEA-SM-237/58. In: *Biological Implications of Radionuclides Released from Nuclear Industries*, Vol. I, Vienna, International Atomic Energy Agency, pp. 447–57.

28

Absorption and Retention in Sheep of Technetium Administered into the Rumen as Pertechnetate or Bound to Food

R. Van Bruwaene,*[a] G. B. Gerber,[a] R. Kirchmann,[a] C. T. Garten, Jr,[b] J. Vankerkom,[a] S. Bonotto,[a] T. Mathieu[a] and M. Cogneau[c]

[a] *Centre d'Etude de l'Energie Nucléaire, Department of Radiobiology, B-2400 Mol Belgium*

[b] *Environmental Sciences Division, Oak Ridge National Laboratory, Oak Ridge, Tennessee 37830, USA*

[c] *Laboratoire de Chimie Inorganique et Nucléaire, Université Catholique de Louvain, B-1348 Louvain-la-Neuve, Belgium*

1. INTRODUCTION

Technetium is produced with high yields of about 6% at nuclear fission and has been released during nuclear weapons testing. Technetium-99m is used in increasing quantities in nuclear medicine. Due to its long half-life of 210 000 years Tc-99 accumulates in the environment and undergoes a variety of chemical and biological transformations which could profoundly modify its absorption by the intestine and its metabolism in animals. Moreover, the age, sex and iodine content of the diet may markedly influence the fate of technetium. Information on technetium metabolism in large animals, particularly in ruminants, is scanty; we have therefore studied in sheep, in parallel with similar investigations in rats, uptake, distribution and excretion of Tc-99 labelled with Tc-95m given into the rumen as pertechnetate or bound to algae. These investigations complement earlier ones where Tc-95m was given by the intravenous route [1, 2].

*Deceased 27 August 1984.

2. METHODS

Technetium-95m was prepared in the cyclotron of the UCL at Louvain-la-Neuve by irradiating a molybdenum target with 50 MeV protons. It was purified by solvent extraction and column chromatography and converted to pertechnetate. At application it had a specific activity of approximately 10 Ci/mg. Brown algae, *Fucus vesiculosus*, were contaminated with Tc-95m by placing them in a mixture of pertechnetate-95m (480 μCi) and about 1 μg of pertechnetate-99 carrier in sea water for a period of 7 h. The algae were then washed and dried.

Sheep of the Suffolk strain received the pertechnetate Tc-95m/Tc-99 mixture either directly or fixed on the algae via an injection into the rumen. Urine and faeces were collected for periods of up to 93 days, and animals were sacrificed at different times after application as indicated in the figures. A total of 10 sheep were used in these investigations. The activity of Tc-95m in excreta and organ samples was assayed by gamma spectrometry.

Two experiments under somewhat different conditions were carried out. In one case, the sheep were 8 months old and received the technetium in September when the animals had fed from pasture for several months. The other experiment involved 1·5 year old sheep treated in May after spring pasture feeding had just started.

The excretion data and where possible the organ activities were fitted to sums of exponential functions using non-linear regression analysis.

3. RESULTS

Faecal excretion reaches a peak soon after injection and declines thereafter. The bulk of activity is excreted within the initial 4 days after application and less than 1% of the dose is found thereafter (Table I). Activity in faeces remained detectable until the end of the observation period. Regression analysis allowed us to discern several metabolic components (Tables I and II). Urinary excretion is small—less than 1% of the dose—except in one animal where contamination by small amounts of faeces might have occurred. Several metabolic components were also detected in urine.

Among the organs assayed, highest specific activities were found in thyroid followed by kidney and intestine in agreement with the data after intravenous injection [1, 2]. Metabolic replacement in liver and kidney

Table I

Excretion parameters for technetium given to sheep of 8 months age as pertechnetate or fixed to algae at the end of summer. The data are expressed as parameters of the sum of exponential functions $A_i \times \exp(-t/To_i)$ (A_i in amount excreted/day/dose $\times 10^{-3}$, To_i in days)

Excreta	*Form applied*	*Excreted (% dose)*	A_i	To_i
Faeces	Pertechnetate	89·4 (total)	600	1·02
		0·73 (>4 d)	0·040	10·7
		0·027 (>8 d)	0·005	30
		0·003 (>30 d)		
Faeces	*Fucus* bound	68·6 (total)	3 010	0·94
		10·2 (>4 d)	1·43	8·0
		0·45 (>8 d)	0·05	>50
		0·004 (>30 d)		
Urine	Pertechnetate	0·61 (total)	8·87	1·3
		0·046 (>4 d)	0·058	12·5
		0·022 (>8 d)	0·021	25
		0·004 (>30 d)		
Urine	*Fucus* bound	4·25 (total)	40·5	1·26
		0·42 (>4 d)	0·270	18
		0·11 (>8 d)	0·1	25
		0·02 (>30 d)		

Table II

Excretion parameters for technetium given to sheep of 18 months age as pertechnetate or fixed to algae at the end of spring. The data are expressed as parameters of the sum of exponential functions $A_i \times \exp(-t/To_i)$ (A_i in amount excreted/day/dose $\times 10^{-3}$, To_i in days)

Excreta	*Form applied*	*Excreted (% dose)*	A_i	To_i
Faeces	Pertechnetate	74·11 (total)	1 490	0·68
		0·42 (>4 d)	0·63	3·4
		0·028 (>8 d)	0·001	>100
Faeces	*Fucus* bound	117 (total)	980	1·18
		0·34 (>4 d)	0·04	10·3
		0·018 (>8 d)		
Urine	Pertechnetate	0·22 (total)	1·05	1·9
		0·024 (>4 d)	0·008	17
		0·010 (>8 d)		
Urine	*Fucus* bound	0·23 (total)	4·63	1·1
		0·024 (>4 d)	0·056	7·0
		0·011 (>8 d)		

Table III
Turnover of technetium given to sheep as pertechnetate or fixed to algae at the end of summer. The data are expressed as parameters of the sum of exponential functions $A_i \times \exp(-t/To_i)$
(A_i in activity/organ/dose $\times 10^{-3}$, To_i in days)

Tissue	*Form applied*	*First term*		*Second term*	
		A_1	To_1	A_2	To_2
Total Body	Pertechnetate	0·40	1·06	0·011	50
	Fucus bound	0·12	1·60	0·040	20
Liver	Pertechnetate	0·047	0·9	(+component about 50 d)	
	Fucus bound	0·011	7·0		
Kidney	Pertechnetate	0·43	2·9	0·07	32
	Fucus bound	0·09	10·1	(prob. component 40 d)	
Thyroid	Pertechnetate	0·35	1·0	0·18	35
	Fucus bound	? (slight increase with time)			

nad a turnover time of 6–10 days (Table III). Several components may however be present and there are strong suggestions for a component with a turnover time of about 30 days in liver and kidney in agreement with the earlier observations following intravenous injection. The sheep which had received pertechnetate also showed two metabolic components in the thyroid. A clear decrease in thyroid activity with time could not be observed in the animals which had received technetium fixed by the algae, perhaps because differences in absorption were larger in this series. Considerable amounts of technetium were found in the wool with specific activities attaining about the levels in kidney and liver; skin and wool together contain of the order of 10–20% of the body burden of technetium when that in the gastro-intestinal tract has been subtracted. It is unlikely that external contamination is the cause of activity in skin and wool. The samples were taken from the back of the neck; activity in wool remained almost constant or even increased with time, and wool grown after shearing displayed high technetium activity.

4. DISCUSSION

Information on absorption of technetium by mammals is scanty [3–6]. Man apparently absorbs technetium rapidly and with high efficiency since about 35% of an oral or intravenous pertechnetate dose was found in urine within 8 days after application [4]. Absorption in ruminants is smaller and only about 1% of oral pertechnetate were found in the urine of goats after oral administration [6]. These values seemed to be independent of dietary iodine.

The present data confirm that relatively little technetium is absorbed by the ruminant sheep regardless of whether the material was given bound to algae or as pertechnetate solution. Most technetium taken up by way of mouth is excreted within 4 days by the faeces. An exact estimate of the percentage of absorption is difficult since even after an iv injection most technetium is excreted into faeces. Taking into account this percentage one arrives at an absorption of about 1% for three of the animals studied. Technetium taken up is however tenaciously retained, particularly in the thyroid. Wool, and by implication hair in man, show relatively high technetium activities for long periods of time and could thus serve as suitable indicators for technetium contamination. The differences between the treatments are not clear-cut. While the data using technetium absorbed on algae are more variable and, in particular, did not show a clear metabolic replacement of technetium in thyroid, the differences are not significant. Moreover, the experiments were not entirely homogeneous, since some were carried out in spring, others in autumn and the age of the animals differed. Investigations are now under way to study the effect of iodine in the diet and of age in sheep in parallel with similar experiments in rats.

5. SUMMARY

Technetium-99 (1 μg labelled with Tc-95m) was injected into the rumen of sheep as pertechnetate or bound to the brown alga *Fucus vesiculosus*. Faeces and urine were assayed for radioactivity up to 93 days after application, and animals were sacrificed at different times after application. Absorption of technetium from intestine is small, and urinary excretion amounts to only about 1% of the dose. Highest specific activities are found in thyroid followed by liver and kidney. Rather high activities were also found in skin and wool. Hair may thus constitute an easily accessible indicator to assess extent and age of a technetium contamination. No marked differences in absorption of technetium were seen when it was given as pertechnetate or bound to algae; some of these variations may be due to differences in age and dietary iodine of the sheep employed.

ACKNOWLEDGEMENTS

This work was supported by contract BIO-D-378-81-B of the Radiation Protection Programme of the Commission of the European Communities.

REFERENCES

1. Van Bruwaene, R., Gerber, G. B., Kirchmann, R., Vankerkom, J., Colard, J. and Cogneau, M. (1983). Technetium metabolism in sheep. *Ann. Assoc. Belge Radioprot.*, **8**, 244–8.
2. Van Bruwaene, R., Gerber, G. B., Kirchmann, R. and Vankerkom, J. (1983). Technetium (Tc-95m) metabolism in sheep after an intravenous injection, Proc. 7th ICRR. Review and summaries, pp. 603–11.
3. Van Bruwaene, R., Gerber, G. B. and Kirchmann, R. (1980). Technetium transfer through the food chain: an analysis of present knowledge and of perspectives for future research. *Ann. Assoc. Belge Radioprot.*, **5**, 25–33.
4. Beasley, T. M., Palmer, H. E. and Nelp, W. B. (1966). Distribution and excretion of technetium in humans. *Health Physics*, **12**, 1425–35.
5. Hays, M. T. (1973). Pertechnetate transport in man: Absorption after subcutaneous and oral administration; secretion into saliva and gastric juice. *J. Nucl. Med.*, **133**, 331–5.
6. Jones, B. (1983). Metabolism of technetium in goats. *Int. J. Appl. Radiat. Isot.*, **34**, 837–9.

29

Transfer of ^{131}I and ^{95m}Tc from Pasture to Goat Milk

E. A. BONDIETTI and C. T. GARTEN JR
Environmental Sciences Division, Oak Ridge National Laboratory, Oak Ridge, Tennessee 37831, USA

1. INTRODUCTION

The behaviour of technetium-99 (^{99}Tc) in food chains is of interest because of the release of this radionuclide to the environment from the nuclear fuel cycle [1]. Technetium (element 43) was predicted in the periodic table but was not studied until the advent of man-induced fission. Its principal isotope, ^{99}Tc, is produced in high yields (~6%) from U-235 fission. Technetium-99's long half-life ($2{\cdot}1 \times 10^5$ years) and potential mobility [2] make understanding its behaviour important in evaluating the radiological consequences of the operation of the nuclear fuel cycle.

The purpose of this research was to compare the behaviour of ^{131}I ($T_{1/2} = 8{\cdot}0$ d) and ^{95m}Tc ($T_{1/2} = 61$ d), which were sprayed on to pasture that was subsequently grazed by a herd of dairy goats. An important objective was to evaluate the pasture-to-milk transfer under realistic animal management conditions.

2. EXPERIMENTAL METHODS

Twelve dairy goats were independently selected and bred by Windrock International, Morrilton, Arkansas. The does, 2–4 years in age, were delivered in February 1983. Kidding occurred in March, and the first experiments were started in May. The number of kids per doe was eventually reduced to one and the kids were allowed to nurse until August. When the grazing experiments were conducted, the doe–kid pairs were separated during the day and allowed together at night. In September, the kids were weaned and the does hand-milked daily. On 30 June the does weighed between 42·4 and 53·1 kg. Daily milk production ranged from 0·41 to 2·93 litres/goat in August and 0·25 to 1·4 litres/goat in September.

In May, July and September, ^{131}I, as NaI, was applied to 4000 m^2 of mixed herbaceous and grass pasture at rates of 1–5 μCi/m^2. In September, ^{95m}Tc, as $NaTcO_4$, was also applied at the rate of 2·5 μCi/m^2. The radionuclides were sprayed on the pasture, using a commercial pesticide-application boom mounted on a tractor. Spraying was done early in the morning to avoid drift; the application had dried by noon when the does were turned on to pasture. An electric fence was used to contain the animals within the contamination zone.

Each day in May and July the does were fed 2 kg of commercial dairy cow grain mix (16% crude protein). In September this was reduced to 1 kg to encourage more pasture intake as well as to follow dairy goat management guides, which recommended reducing grain intake near the end of lactation. Vegetation and milk samples usually were taken on the same day: random vegetation samples ($N = 11$–13, ~50 g fresh weight each) from designated plots within the contaminated pasture and at least 100 ml of milk from each goat. A simple ratio of the radionuclide concentration (pCi/litre) in milk to the radionuclide concentration (pCi/kg) in pasture was calculated to quantify the relative transfer factors of ^{131}I and ^{95m}Tc from pasture to milk. In related experiments, attempts were made to determine pasture ingestion rates by measuring ^{95m}Tc and ^{7}Be (the latter naturally deposited on vegetation from rainfall) in pastures and faeces. Faecal collections were made for several days, using collectors strapped to several animals.

Radionuclide analysis was accomplished using NaI or germanium detectors. In the September experiment, the transfer of ^{95m}Tc to milk was so low that ^{131}I activity completely swamped the 204-keV ^{95m}Tc photopeak (even using a germanium detector). Consequently, about 30 d were allowed to pass before ^{95m}Tc measurements were made on the samples. Future studies of this type should use a 100:1 ratio (not 1:1) of ^{95m}Tc to ^{131}I in pasture applications.

3. RESULTS

3.1. Iodine-131 and technetium-95m

The concentrations of ^{131}I and ^{95m}Tc on pasture were carefully followed during the course of the three experiments. May showed the highest apparent loss rate, with a 'weathering' half-life for radioiodine of 6·6 d and an effective half-life (weathering + radiodecay) of only 3·6 d. July and September loss rates (weathering half-life) for radioiodine were 10 and 15 d, respectively. These differences reflected precipitation wash-off losses, the faster pasture growth rate in the spring, and a major drought experienced in eastern North America in July, August and September 1983, which slowed pasture growth. No recordable rainfall occurred on this pasture in August and September, and the amount that fell in July was less than 1·25 cm.

The weathering half-life of technetium from the pasture in September was 15 d, the same as ^{131}I and close to the 15·9–18·7 d determined for ^{95m}Tc on fescue at this site in 1979 [3].

3.2. Iodine-131 and technetium-95m transfer from pasture to milk

The observed milk/pasture concentration ratios for ^{131}I transfer in May and July are illustrated in Fig. 1. They were calculated by dividing the milk concentration (pCi/litre) by the pasture concentration (pCi/kg fresh weight). The transfer factor in May reached a steady-state value of 0·55 kg/litre after the first 3 d. Analysis of variance showed that the milk/pasture ratio was not significantly different ($P < 0{\cdot}05$) from day 4 through day 18 after spraying. In July the milk/pasture ratio (mean = 0·17) was not constant. Analysis of variance revealed that milk/pasture ratios during the first week were significantly ($P < 0{\cdot}05$) greater than the ratios observed during the second week. The low overall mean value of

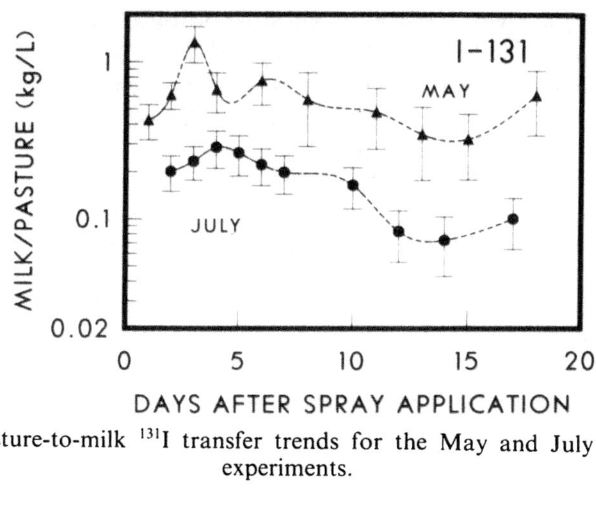

FIG. 1. Pasture-to-milk ^{131}I transfer trends for the May and July 1983 goat experiments.

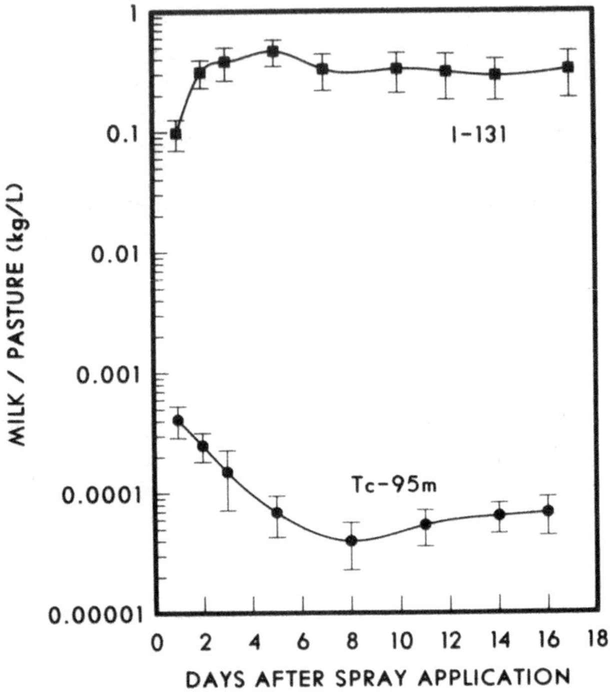

FIG. 2. Comparative transfer of ^{131}I and ^{95m}Tc from pasture to goat's milk (11 animals) in the September 1983 experiment.

0·17 in July, as well as the variability in the ratio, was due, in part, to repeated grazing by does outside the sprayed area due to the kids knocking down the electric fence. In September, 2 d after spray application, the milk/pasture ratio reached a steady-state value of 0·34 (Fig. 2).

The transfer of ^{95m}Tc to milk was much lower than that for ^{131}I under identical field conditions (Fig. 2). While ^{131}I transfer factors reached a steady state in 2 d, technetium did not reach a steady state until 5 d after spraying. In addition, the nature of the trends with time was opposite: ^{131}I built up to equilibrium, whereas ^{95m}Tc in milk declined from the first sampling. The apparent steady-state milk/pasture ratio for ^{95m}Tc was 6×10^{-5} kg/litre.

3.3. Pasture immobilisation of technetium

In late October, an experiment was performed to test for the relative immobilisation of ^{131}I and ^{95m}Tc sprayed on to pasture vegetation.

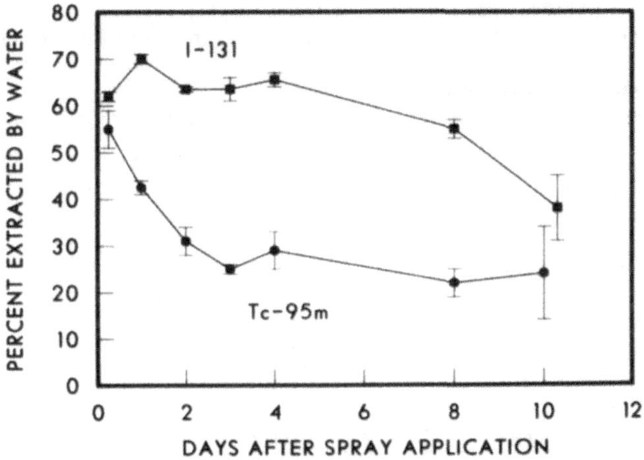

FIG. 3. Comparative water extractability of ^{95m}Tc and ^{131}I applied to fescue as a spray.

Pasture grass (2 m^2 area) was sprayed with an aqueous solution of both radionuclides. Samples of grass (*Festuca* sp.) were collected, cut into 4-cm long sections and washed three times by shaking with tap water for 5 min in a bottle. Analysis of successive washes showed that 95–100% of the technetium or iodine that could be removed was extractable by this treatment. Figure 3 illustrates that the extractability of ^{95m}Tc and ^{131}I from vegetation was initially comparable (~60%), but the extractability of technetium declined to about half the initial value within 3 d after application to pasture. As Fig. 2 indicates, the technetium milk/pasture ratio declined about an order of magnitude during this period. By comparison, the extractability of ^{131}I was unchanged over the same time period. This experiment indicated that a rapid immobilisation on vegetation occurred for ^{95m}Tc, but not for ^{131}I. This immobilisation was probably caused by the translocation and incorporation of technetium into vegetation by foliar absorption.

3.4. Pasture ingestion rates

If an element that is insoluble in the gastrointestinal tract is present on pasture vegetation (but not on stored feed), then the pasture ingestion rate can be evaluated by measuring the element in daily excreted faeces as well as in the pasture (Table I). Two radionuclides served this purpose: naturally-occurring ^{7}Be and ^{95m}Tc. Beryllium-7 was measured in both pasture and goat faeces in late July, while ^{95m}Tc was used after it became

Table I
Estimation of daily fresh pasture ingestion rate using Tc-95m and Be-7

Goat number	*Total faecal output (nCi)*	*Days collected*	*Pasture (pCi/g fw)*	*Daily ingestion[a] (kg)*
^{7}Be (30 June)				
9	2·09	2·9	0·66	1·1
^{95m}Tc (September)				
3	4·310	2·0	828	2·6
14	3·650	2·0	828	2·2
7	3·500	2·0	612	2·9
				$\bar{X} = 2{\cdot}5$

$$^{a}\,\mathrm{kg/d} = \frac{\text{total pCi faecal output per days collected}}{\text{pCi/g pasture} \times 1000\,\text{g/kg}}.$$

apparent that very little was excreted into milk. In both cases, less than 1% of the intake was assimilated. Table I presents the results of the calculation of ingestion (in kilograms per day) based on these measurements. The daily ingestion rate for September was about 2·5 kg fresh weight or 0·88 kg dry weight (dry weight/fresh weight ratio for vegetation in September was about 0·35).

4. DISCUSSION

Technetium metabolism in goats has been examined by Jones [4]. Technetium-95m transfer to blood after oral administration was about 1% of the administered dose per litre of blood plasma after 1 h; this declined to 0·001% per litre after 24 h. A similar pattern occurred for milk, with 1·5% of the administered dose found in a litre of milk after 5 h. A comparison of oral and abomasal injections indicated that passage through the rumen reduced the availability of technetium for absorption into the bloodstream. Most of the ^{95m}Tc was excreted in the faeces. In summary, Jones [4] found that following ingestion, ^{131}I was transferred about 10–100 times more efficiently than ^{95m}Tc.

In the present work, the transfer of ^{131}I to goat's milk was about 5600 times more than that of ^{95m}Tc after 5 d of grazing contaminated pasture. This was considerably less relative to transfer of technetium to milk than observed by Jones [4]. Most of this difference appears to be explained by

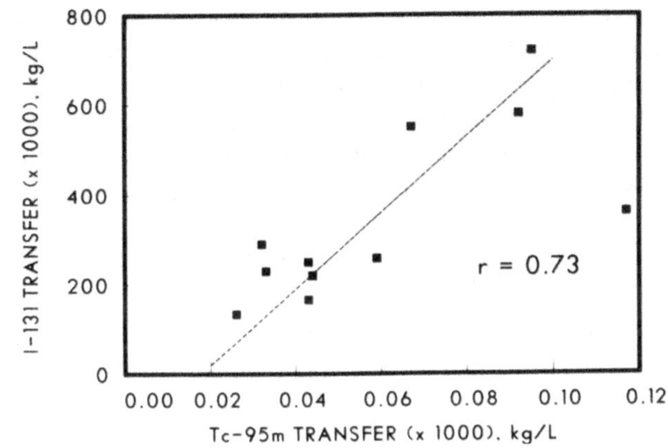

FIG. 4. Comparison of 'equilibrium' ^{95m}Tc and ^{131}I pasture-to-milk transfer factors measured simultaneously for 11 goats.

a progressive immobilisation of technetium on vegetation, which occurred during the first few days of the experiment. This interpretation is also consistent with the findings of Garten *et al.* [5], who examined the availability of plant-incorporated technetium to rats.

The transfer of ^{131}I and ^{95m}Tc to milk was also compared on an individual animal basis. The average ^{131}I pasture-to-milk transfer factor for each animal from day 5 forward ($N = 6$) was plotted against the average technetium transfer factor for the same time period ($N = 5$) (Fig. 4). A good correlation existed between the transfer of iodine and technetium from pasture to milk, except for one goat. An examination of the pasture-to-milk transfer factor and daily milk production showed no significant relationship.

In these experiments, the amount of pasture ingested by most of the animals is unknown. Differences in the pasture-to-milk transfer rates between animals could be due to differences in their intake of pasture relative to grain. Historically, feeding studies have been conducted in stalls where the intakes can be controlled or measured. This is not practical in field experiments unless pasture is harvested by hand and then fed to penned animals. This introduces an artificial condition that may bias the resulting data.

Assuming a daily ingestion of 6 kg fresh weight (all pasture), the goat milk/pasture ratio calculated from data presented in US Regulatory Guide 1.109 is 0·36 for ^{131}I [6]. The mean milk/pasture ratios for ^{131}I in this study ranged from 0·17 kg/litre in the July experiment to 0·55 kg/litre in May. However, our estimates of daily pasture ingestion in September were approximately 40% of the total intake assumed in Regulatory Guide 1.109. Recognising that the ingestion rate varies depending on length of lactation, weather, pasture conditions, etc., we believe that the transfer of ^{131}I from pasture to goat's milk found in this study is in reasonable agreement with data recommended for use in the mathematical model found in Regulatory Guide 1.109.

It is now apparent that the food chain behaviour of technetium is very different from that of radioiodine. The transfer of ^{95m}Tc from pasture to milk is affected not only by the reactions in the rumen that affect absorption [4] but also by reactions with pasture vegetation. Radioiodine does not seem to be so affected. The spray application of ^{95m}Tc to pasture is not necessarily the ideal method to examine the pasture–milk pathway, since the long-term pathway of ^{99}Tc transfer to vegetation is via roots because of its long half-life. The results of this study are most applicable when aerial deposition is the main source of technetium contamination. The relative availability of plant-immobilised technetium derived from root *v.* foliar uptake is unknown, and comparative leaching experiments are suggested. In any case, the low transfer to milk suggests that direct ingestion of technetium-contaminated vegetation by ruminants is not a significant exposure route leading to contamination of milk.

ACKNOWLEDGEMENTS

Research sponsored by the Office of Health and Environmental Research, US Department of Energy, under Contract DE-ACO5-840R21400 with Martin-Marietta Energy Systems, Inc. Publication No. 2413, Environmental Sciences Division.

REFERENCES

1. TILL, J. E., SHOR, R. W. and HOFFMAN, F. O. (in press). Environmental effects of the uranium fuel cycle—a review of data for technetium, NUREG-3738 (ORNL/TM-9150). Springfield, Va, National Technical Information Service.
2. BONDIETTI, E. A. and FRANCIS, C. W. (1979). Geologic migration potentials of technetium-99 and neptunium-237. *Science*, **203**, 1337–40.
3. HOFFMAN, F. O., GARTEN, C. T. JR, HUCKABEE, J. W. and LUCAS, D. M. (1982). Interception and retention of technetium by vegetation and soil. *J. Envir. Qual.*, **11**, 134–41.

4. JONES, BERNT E. V. (1979). Metabolism of technetium-95m pertechnetate in goats and pigs. Department of Clinical Chemistry, College of Veterinary Medicine, Swedish University of Agricultural Sciences, S-750 07, Uppsala, Sweden.
5. GARTEN, C. T., MYTTENAERE, C., VANDECASTEELE, C. M., KIRCHMANN, R. and VAN BRUWAENE, R. Chemical form of technetium in corn (*Zea mays*) and the gastrointestinal absorption of plant-incorporated Tc by laboratory rats (this volume, pp. 319–32).
6. US NUCLEAR REGULATORY COMMISSION (1977). Regulatory Guide 1.109, Rev. 1, Calculations of Annual Doses to Man from Routine Releases of Reactor Effluents for the Purpose of Evaluating Compliance with 10 CFR Part 50, Washington, DC, US Nuclear Regulatory Commission.

30

Distribution of Orally Administered and Chronically Fed ^{95m}Tc in Japanese Quail Tissues and Eggs

J. M. Thomas, L. L. Cadwell, D. A. Cataldo and T. R. Garland
Pacific Northwest Laboratory, Richland, Washington 99352, USA

1. INTRODUCTION

Several studies dealing with the availability of technetium (Tc) in soils, plants, and animals have recently appeared [1–6]. Apparently the work of Wildung *et al.* [7 and 8], which showed a high potential for Tc mobility in terrestrial food chains, was in part an impetus for this interest. Moreover, technetium-99 is long lived ($T_{1/2} = 210\,000$ yr), is produced in quantity in nuclear reactor fuel, and can enter the environment during the recycling of spent fuel [9]. Thus, it is of interest in food chain modelling and dose assessment. Recent research, where rats and mice were injected with a single dose of $Na^{99}TcO_4$, has indicated that the chemical toxicity of technetium may be more important than its radiological hazard [6].

While investigations of Tc transfer in terrestrial food chains are limited (refs. 4 and 5 contain most references), dosimetric and radiopharmacological studies have been conducted [9]. In studies with *Coturnix japonica* quail, oocytes accumulated up to 26% of a single intramuscular injection of $Na^{99m}TcO_4$ (carrier free) in 0·2 ml physiological saline [10]. Moreover, perrhenate, with chemical properties similar to pertechnetate, has been shown to concentrate in the oocytes of the chicken [11].

Food chain models used to estimate human radiation dose employ transfer coefficients and concentration ratios to estimate radionuclide transfer from soil to plant and from plant to the edible parts of large animals, poultry and eggs. Studies with soybean plants have indicated that ^{99}Tc and sulphur behave as chemical analogues with respect to membrane transport and possibly metabolism [12]. Thus, we hypothesised that transfer of Tc from feed to quail eggs and edible tissues (which contain a substantial quantity of sulphur-containing protein) could represent an important food chain pathway.

Since Japanese quail (*Coturnix japonica*) are often considered suitable substitutes for commercial laying hens [13, 14], and because of their reduced feed consumption compared to the domestic chicken, concentration ratios and transfer coefficients for this species were determined. At this writing, transfer coefficients for domestic chicken eggs have been estimated by Ng *et al.* [15] from our initial quail experiment [5] and in a preliminary study using white leghorn hens [16]. Transfer coefficients for edible tissues of chicken have been estimated [15] from our unpublished quail data.

In the present study, male and female Japanese quail were chronically fed alfalfa grown on solutions containing TcO_4^- which was mixed into a commercial turkey starter. Our objective was to estimate concentration ratios and transfer coefficients from Tc incorporated into alfalfa tissue to quail eggs, edible tissues and other organs.

2. METHODS

Male and female quail (ten weeks old), hatched in our laboratory and raised under temperature and lighting regimes similar to those used during the experiment, were used to determine the ratio of Tc in daily food intake to that in egg, edible tissue and other organs. The Tc was incorporated into alfalfa leaves using the procedures detailed in [17] and outlined in [5]. Leaves were subsequently mixed in turkey starter (Ralston Purina). Both feed (0·2 μCi ^{95m}Tc/g) and water were supplied *ad libitum*.

Eggs were collected daily for 15 days during the chronic feeding experiments. All eggs were solidified by boiling, separated into yolk, albumin and shell components, freeze-dried and analysed for ^{95m}Tc. At the end of 15 days, the three chronically fed females were sacrificed by cervical dislocation and the three largest oocytes, ovary and oviduct (with formed eggs removed) were excised. Other tissues and organs were also excised from the females, as well as from three groups of males ($n = 4$, 4 and 3, respectively) which had been chronically fed for 13, 15 and 17 days. All tissues and organs were also weighed and assayed. Activity of ^{99m}Tc was determined on all samples by gamma spectrometry and corrected for background, decay and counting efficiency.

Feed consumption was estimated using the average 15-day production rate per female (about 70%), average egg weights (about 9 g), and the conversion ratio in [13] (3 g feed/g egg) as 19 g/day. Consumption for males was estimated to be 15 g/day (Department of Poultry Science, Oregon State University). Thus, 115 g males and 130 g females consume about 0·13 and 0·15 g feed/day per g of body weight, respectively.

3. RESULTS AND DISCUSSION

Percentage of daily ingested Tc transferred to eggs for quail chronically fed incorporated Tc is shown as a function time in Table I. Calculation of ingested dose was based on the assumption that high egg-producing female quail ingest about 19 g/day (see Methods). These data indicate that biological half-life is about 1–2 days, since each quail reached practical equilibrium (about five half-lives) after 5–7 days of chronic ingestion. Therefore, the transfer coefficient for each egg (amount of Tc in egg divided by the amount of Tc consumed, i.e. day/egg) was averaged for each bird after equilibrium was achieved (7 days). Results (mean ± SD)

Table I
Percentage of total ^{95m}Tc activity in quail eggs as a fraction of total activity in feed (dpm)

Days fed ^{95m}Tc incorporated into alfalfa	*Percent chronic ingested dose/egg*			*Mean*	*Standard deviation*
	Female number				
	1	*2*	*3*		
1	1·4	0·1	—[a]	0·8	—
2	6·3	2·3	0·1	2·9	3·1
3	—	5·4	2·3	3·9	—
4	11·2	—	—	11·2	—
5	12·1	7·6	8·0	9·2	2·5
6	—	7·2	8·5	7·9	—
7	—	—	9·2	9·2	—
8	11·4	8·3	—	9·9	—
9	11·2	7·3	—	9·8	—
10	11·1	7·5	—	9·2	—
11	11·0	8·3	6·1	8·5	2·5
12	10·2	8·2	8·1	8·8	1·2
13	9·9	8·5	6·6	8·3	1·7
14	—	—	8·1	8·1	—

[a] No egg laid.

Table II

Concentration of chronically fed ^{95m}Tc in tissues of Japanese quail (dpm/g wet tissue)

Tissue	*Males (days fed)*[a]						*All males*		*Females*[b]	
	13		*15*		*17*		*13–17*		*15*	
	$\bar{X}$	*CV*[c]	$\bar{X}$	*CV*	$\bar{X}$	*CV*	$\bar{X}$	*CV*	$\bar{X}$	*CV*
Heart	70	53	110	21	150	13	110	40	108	9
Liver	270	28	55	19	434	32	385	33	255	15
Gizzard (empty, lining removed)	1 080	33	1 270	19	1 640	5	1 300	25	1 545	11
Muscle	28	30	45	34	45	11	40	33	30	7
Kidney	304	68	620	37	620	9	505	46	565	19
Proventriculus (empty)	375	48	430	29	495	41	425	37	340	27
Lung	90	48	150	29	160	30	130	40	130	13
Bone	25	88	80	55	70	10	55	65	235	30
Skin and feathers	211	28	340	32	340	7	294	33	420	8
Rest of carcass	831	19	145	35	155	29	126	39	110	5
Testes	164	25	150	30	370	58	215	66	—	—
Uterus									110	16
Oviduct minus uterus									200	15
Ovary minus three largest oocytes									4 110	10
Largest oocyte 1									13 765	7
Smaller oocyte 2									12 565	3
Smaller oocyte 3									12 730	8
Shelled egg									3 900	9
Total minus reproductive organs	116	19	172	25	181	16	154	28	200	8
Total with reproductive organs	117	19	171	26	187	17	156	28	1 000	5

[a] $n = 4$, 4 and 3 male quail, respectively.
[b] $n = 3$ female quail.
[c] Coefficient of variation = (standard deviation/mean) × 100.

were $0{\cdot}11 \pm 0{\cdot}01$, $0{\cdot}08 \pm 0{\cdot}01$, and $0{\cdot}07 \pm 0{\cdot}01$ for each of the three birds, or $0{\cdot}09 \pm 0{\cdot}02$ for all 17 eggs produced beyond 7 days. There was no apparent correlation between body weight, egg weight, egg production and transfer coefficient (day/egg).

After 15 days of chronic feeding, all three females were sacrificed and necropsied. The eleven males were sacrificed for analyses after 13 ($n = 4$), 15 ($n = 4$), and 17 days ($n = 3$). Tissue concentrations (dpm/g) for all male and female groups are shown in Table II, and concentration ratios (dpm/g wet tissue divided by dpm/g air dry food) and transfer coefficients (day/kg and day/tissue) are shown in Table III. The oocytes, shelled eggs and the ovary (principally smaller developing oocytes) contained the major fraction of Tc associated with reproductive tissues (Table III, day/tissue). Percentage of daily intake in these tissues can be acertained by ignoring the exponent for each value (Table III). Shelled eggs contained an average of 8·8% of daily intake, which was very similar to the average of the 17 eggs used to compute daily transfer coefficients for eggs actually produced (Table I). Developing oocytes exhibited diminished percentages of Tc as size of the oocyte decreased (mean weights were 2·5, 1·3 and 0·4 g, respectively; these weights can be obtained from Table IV, based on an average 130-g female), but showed remarkably similar Tc concentrations (dpm/g, Table II). Thus, it appears that the biological processes which regulate Tc concentration in the yolk are constant, that increased Tc content is solely a function of increased oocyte mass, and that Tc is deposited during oocyte maturation.

Egg components, including yolk, white and shells of all eggs produced, were also assayed. Less than 1% of Tc in boiled egg was associated with shell. After about 10 days of chronic feeding, approximately 80% of the Tc was in yolk and 20% in albumin. We also found that albumin concentrations (dpm/g) declined by 20 to 50% after day 3 or 4, while yolk concentrations continued to increase and reached a maximum in about 6 to 7 days. Additional details are in ref. 5.

Estimates of the total percentage of daily-ingested ^{95m}Tc in all tissues (calculated using the original count data) for all eleven males and the three female quail were $4{\cdot}7 \pm 1{\cdot}7$ and $27{\cdot}6 \pm 2{\cdot}4\%$, respectively. About $23 \pm 2{\cdot}0\%$ of the daily female intake was in the developing ova, shelled egg and reproductive organs. Based on the average data in Tables II and IV, the average daily intakes (dpm) and final quail weights (footnote (b) in Table III and footnote (a) in Table IV), the respective indirectly calculated male and female estimates are 4·8 and 27·6% (e.g. $131{\cdot}5$ g/bird $\times\, 0{\cdot}874 \times 1000$ dpm/g $\div\, 417\,100$ dpm/day $= 0{\cdot}276$ or 27·6%

Table III
Mean transfer coefficients and concentration ratios for male and female Japanese quail

Tissue	*Concentration ratio* $\left(\frac{dpm/g\ wet\ tissue}{dpm/g\ air\ dry\ food}\right)$		*Transfer coefficient*[a] *(day/kg)*		*Transfer coefficient*[a] *(day/tissue)* $(\times 10^{-2})$	
	Males[b]	*Females*	*Males*	*Females*	*Males*	*Females*
Heart	0·005	0·005	0·34	0·26	0·04	0·03
Liver	0·018	0·012	1·17	0·61	0·29	0·33
Gizzard (empty, lining removed)	0·060	0·071	3·96	3·70	0·81	1·06
Muscle	0·002	0·001	0·12	0·07	0·38[c]	0·30[c]
Kidney	0·023	0·026				
Proventriculus	0·020	0·016				
Lung	0·006	0·006				
Bone	0·003	0·011				
Skin and feathers	0·014	0·019				
Testes	0·010	—	0·66	—	0·02	—
Uterus	—	0·005	—	—	—	0·04
Oviduct minus uterus	—	0·009	—	—	—	0·18
Ovary minus three largest oocytes	—	0·189	—	—	—	0·67
Largest oocyte 1	—	0·633	—	—	—	7·80
Smaller oocyte 2	—	0·578	—	—	—	3·80
Smaller oocyte 3	—	0·587	—	—	—	1·40
Shelled egg	—	0·179	—	9·35	—	8·76

[a] $\text{day/kg} = \frac{\text{dpm/kg wet tissue}}{\text{dpm/daily air dry food intake}}$, $\text{day/tissue} = \frac{\text{dpm/total wet tissue}}{\text{dpm/daily air dry food intake}}$.

[b] Average daily dpm intake was 328 000 $\pm$ 28 500 and 417 100 $\pm$ 14 100 (mean $\pm$ standard deviation) for males and females, respectively.

[c] dpm for total muscle was calculated using 28·5% of final body weight. This value was obtained by excising and weighing the breast, leg, thigh, back and wing muscles (standard deviation was 2·1%).

Table IV

Tissue weight as a percentage of final body weight of Japanese quail

Tissue	*Males (n = 11)*			*Females (n = 3)*			*Literature*[a]	
	Mean	*Standard deviation*	*Coefficient of variation*	*Mean*	*Standard deviation*	*Coefficient of variation*	*Male*	*Female*
Heart	1·2	0·1	8	1·0	0·2	25	1·0	0·72
Liver	2·2	0·4	18	4·1	0·2	5	2·5	3·2
Gizzard (empty, lining removed)	1·8	0·3	14	2·1	0·1	5	3·1	2·9
Kidney	0·3	0·1	35	0·7	0·2	30	1·2	1·2
Proventriculus (empty)	0·4	0·1	12	0·7	0·1	18	—	—
Lung	0·9	0·2	20	0·7	0·1	18	—	—
Testes (both)	2·9	0·7	24	—	—	—	1·8	—
Oviduct	—	—	—	4·0	0·1	3	—	—
Ovary minus three largest oocytes	—	—	—	0·5	0·1	22	—	1·3
Largest oocyte 1	—	—	—	1·9	0·1	3	—	1·8
Smaller oocyte 2	—	—	—	1·0	0·01	2	—	1·5
Smaller oocyte 3	—	—	—	0·3	0·04	12	—	1·1
Shelled egg	—	—	—	7·1	0·5	7	—	7·9
Rest of carcass	80·0	—	—	62·3	—	—	—	—
Total	89·7[b]	1·1	1·2	87·4	4·5	5·2		

[a] Only one male and female were dissected. Body weights were 112 and 124 g respectively [13]. Our quail weighed 113·2 ± 11·4 and 131·5 ± 1·5 (mean ± standard deviation for 11 and 3 males and females, respectively).

[b] It appears that blood, food in gizzard and proventriculus and excrement accounted for about 10·3 to 12·6% of final body weight.

for females). Another estimate of the Tc percentage in female reproductive organs (22·6%) was obtained by summing the day/tissue transfer coefficients (ignoring the exponent) in Table III. Thus, our data show that the ^{95m}Tc content in males and females differs principally because Tc is preferentially deposited in oocytes and albumin.

The highest Tc concentrations in tissues of laying quail (dpm/g, Table II) were in oocytes, followed by the fully formed unlaid egg, and the ovaries (probably a reflection of oocytes in earlier maturation states). A similar pattern was evident for concentration ratios (Table III), except that the ovaries and shelled egg exhibited nearly identical values. The concentration ratio for shelled egg reflects the addition of water and albumin after oocyte maturation.

For edible tissues of females, the shelled egg contained five times the daily ingested dose (8·76%, Table III, day/tissue) found in muscle, liver, gizzard and heart combined (1·72%) and almost six times the daily ingested dose in male edible tissue. Transfer coefficients for males (day/tissue, Table III) indicated that gizzards contained about 53% of the total amount of Tc in edible tissues, followed by muscle, liver and heart (25, 19 and 3%, respectively). Both the concentration ratio and the weight-based transfer coefficient (day/kg) show that liver and gizzard present a greater potential for transfer by ingestion than do muscle or heart. If entire quail are consumed, the overall transfer coefficient for either sex is about 0·35 day/kg.

ACKNOWLEDGEMENTS

This work was performed under contract DE-ACO6-76RLO 1830 for the US Department of Energy. We thank M. I. Cochran, J. Downs-Berg, S. Fazzari and R. Webster for technical assistance.

REFERENCES

1. Hoffman, F. O., Huckabee, J. W., Lucas, D. M., Garten, C. T. Jr., Scott, T. G., Walker, R. L., Gouge, P. S. and Homes, C. V. (1980). Sampling of Technetium-99 in Vegetation and Soils in the Vicinity of Operating Gaseous Diffusion Facilities, ORNL/TN 7380. Oak Ridge, Tennessee, Oak Ridge National Laboratory.
2. Hoffman, F. O., Garten, C. T., Jr., Lucas, D. M. and Huckabee, J. W. (1982). Environmental behavior of technetium in soil and vegetation, implications for radiological assessments. *Envir. Sci. Technol.*, **16**, 214–17.

3. HOFFMAN, F. O., GARTEN, C. T. JR., HUCKABEE, J. W. and LUCAS, D. M. (1982). Interception and retention of technetium by vegetation and soil. *J. Envir. Qual.*, **11**, 134.
4. GARTEN, C. T. JR., HOFFMAN, F. O. and BONDIETTI, E. A. (1984). Field and greenhouse experiments on the fate of technetium in plants and soil. *Health Physics*, **46**, 647–56.
5. THOMAS, J. M., CADWELL, L. L., CATALDO, D. A., GARLAND, T. R. and WILDUNG, R. E. (1984). Concentration of orally administered and chronically fed 95m-Tc in Japanese quail eggs. *Health Physics*, **46**, 657–63.
6. COFFEY, J. L., HAYES, R. L., RAFTER, J. J., WATSON, E. E. and CARLTON, J. E. (1984). Radiation dosimetry and chemical toxicity considerations for 99-Tc. *Health Physics*, **46**, 418–22.
7. WILDUNG, R. E., GARLAND, T. R. and CATALDO, D. A. (1977). Accumulation of technetium by plants. *Health Physics*, **32**, 314–17.
8. WILDUNG, R. E., MCFADDEN, K. M. and GARLAND, T. R. (1979). Technetium sources and behavior in the environment. *J. Envir. Qual.*, **8**, 156–61.
9. TILL, J. E., HOFFMAN, F. O. and DUNNING, E. E. JR. (1979). A new look at ^{99}Tc releases to the atmosphere. *Health Physics*, **36**, 21–30.
10. ROBINSON, G. A. and TAM, K. H. (1974). Distribution of ^{95m}Tc and ^{131}I among tissues of the Japanese quail. *Poultry Sci.*, **53**, 2209–15.
11. ROCHE, J., MICHEL, R., VOLPERT, E. and SANZ, B. (1957). Sur la fixation du mangenese et du rhenium radioactifs par le corps thyroide et les oocytes de la poule pondeuse. *C. R. Soc. Biol.*, **151**, 1098–100.
12. CATALDO, D. A., WILDUNG, R. E. and GARLAND, T. R. (1978). Technetium accumulation, fate and behavior in plants. In: *Environmental Chemistry and Cycling Processes* (Ed. D. C. Adriano and I. L. Brisbin Jr.), DOE Symp. Series No. 45, Conf. 760429, Springfield, Virginia, National Technical Information Service, pp. 538–49.
13. WILSON, W. O., ABBOTT, U. K. and ABPLANALP, H. (1961). Evaluation of *Coturnix* (Japanese Quail) as a pilot animal for poultry. *Poultry Sci.*, **40**, 651–7.
14. WOODARD, A. E., ABPLANALP, H., WILSON, W. O. and VOHRA, P. (1973). Japanese Quail Husbandry in the Laboratory (*Coturnix coturnix japonica*). Department of Avian Sciences, University of California at Davis, Davis, California.
15. NG, Y. C., COLSHER, C. S. and THOMPSON, S. E. (1982). Transfer Coefficients for Assessing the Dose from Radionuclides in Meat and Eggs, NUREG/CR-2976, UCID-19464. Livermore, California, Lawrence Livermore National Laboratory.
16. WARD, G. M. and JOHNSON, J. J. (1984). Confirmation of Selected Milk and Meat Radionuclide Transfer Coefficients. Technical Report, Department of Animal Sciences, Colorado State University, Fort Collins, Colorado.
17. CATALDO, D. A., GARLAND, T. R. and WILDUNG, R. E. (1978). Nickel in plants I. Uptake kinetics using intact soybean seedlings. *Plant Physiol.*, **62**, 563–5.

31

The Significance of Environmental Exposure Pathways for Technetium

F. Owen Hoffman, Robert H. Gardner and Steven M. Bartell
Environmental Sciences Division, Oak Ridge National Laboratory, Oak Ridge, Tennessee 37831, USA

1. INTRODUCTION

Prior to the late 1970s, little attention was given to the investigation of the environmental behaviour and fate of ^{99}Tc. Environmental assessments of the radiological impact of technetium releases were based on the predictions of mathematical models that, in the absence of data on technetium, contained parameter values that were based on a series of assumptions relating the chemical behaviour of the pertechnetate anion (TcO_4^-) to that of iodide [1]. Radiological dose estimates calculated using these models for aquatic and atmospheric releases of Tc were well within acceptable limits [2, 3].

In 1974, data became available from potted plant experiments [4] that indicated that the Tc uptake from soil by vegetation may be substantially higher than was assumed in environmental assessment models. By 1977, other investigators had reported similar results [5–8]. To investigate the implications of these data for radiological impact assessment, Till *et al.* [9] substituted the assumed default value of 0·25 (pCi/kg, fresh wt tissue per pCi/kg, dry wt soil) with a value of 50 for the vegetation/soil concentration ratio ($CR_{v,s}$). Their calculations for an atmospheric release of 1 Ci/year from a hypothetical uranium enrichment facility produced dose estimates for adult members of critical population groups that infringed upon and/or exceeded radiation protection standards proposed by the US Environmental Protection Agency for the commercial uranium fuel cycle [9]. These results provided increased incentives for Tc research.

At present, the data that quantify the environmental behaviour of technetium are still far from complete, but the amount of information has increased substantially since 1979. Quantitative information is now available on the uptake of Tc from soil by vegetation under field conditions, the transfer of Tc from vegetation to animal food products, the mobility of Tc in surface soils, and the accumulation of Tc by aquatic organisms. A recent review by Till *et al.* [10] surveyed more than 130 references and recommended a probable range and an 'expected value' for many of the assessment model parameters used to predict the environmental transport and fate of Tc. The authors intended their estimates of these ranges to encompass the uncertainty associated with the quantification of these parameters.

In this report, we use this information to analyse the significance of the various environmental pathways of human exposure to Tc. The purpose of this new analysis is to reassess the results and implications of the 1979 estimates published by Till *et al.* [9] and to recommend research that will contribute to reducing the uncertainty associated with assessing the environmental impact of Tc.

2. METHODS

Expected values and estimates of the possible range of values for each of the parameters used in radiological assessment models were obtained from an analysis of the literature (Table I). These ranges represent

Table I
Parameter values for the prediction of the environmental transport and human dosimetry resulting from releases of ^{99}Tc

Input parameter	*Units*	*Minimum*	*Mode*	*Maximum*	*References*
Dose conversion factors					
Inhalation					
DCF_{inh}	mrem/pCi	—	1×10^{-6}	—	[17]
Ingestion					
DCF_{ing}	mrem/pCi	—	$1{\cdot}5 \times 10^{-6}$	—	[17]
Rates of intake					
I_{air}	m^3/y	—	$8{\cdot}4 \times 10^3$	—	[28]
I_{milk}	litre/y	—	$1{\cdot}1 \times 10^2$	—	[9]
I_{meat}	kg/y	—	$1{\cdot}1 \times 10^2$	—	[9]
I_{eggs}	eggs/y	$5{\cdot}2 \times 10^1$	$7{\cdot}3 \times 10^2$	$1{\cdot}1 \times 10^3$	assumed range
$I_{vegetables}$	kg/y	—	$6{\cdot}4 \times 10^1$	—	[9]
I_{water}	litre/y	—	8×10^2	—	[28]
$I_{aquatic\ organisms}$	kg/y	1×10^0	$1{\cdot}4 \times 10^1$	1×10^2	[29]

Table I—*continued*

Input parameter	*Units*	*Minimum*	*Mode*	*Maximum*	*References*
Concentration ratios[a]					
Milk/pasture vegetation					
$CR_{m,p}$	Ci/litre per Ci/kg	1×10^{-5}	1×10^{-3}	1×10^{-2}	[10]
Meat/pasture vegetation					
$CR_{f,p}$	Ci/kg per Ci/kg	8×10^{-6}	8×10^{-4}	8×10^{-3}	[10]
Eggs/pasture vegetation					
$CR_{eggs,p}$	Ci/egg per Ci/kg	9×10^{-5}	9×10^{-3}	9×10^{-2}	[10]
Vegetables/soil					
$CR_{v,s}$	Ci/kg per Ci/kg	1×10^{0}	2×10^{0}	2×10^{1}	[10]
Pasture vegetation/soil					
$CR_{p,s}$	Ci/kg per Ci/kg	1×10^{0}	2×10^{1}	2×10^{2}	[10]
Wet + dry deposition					
V_{total}	m/d	$1{\cdot}7 \times 10^{2}$	1×10^{3}	3×10^{3}	[30, 31]
Initial mass interception by vegetation					
Vegetables (fresh weight)					
r/Y_{veg}	m^2/kg	3×10^{-2}	$1{\cdot}2 \times 10^{-1}$	3×10^{-1}	[10]
Pasture vegetation (dry weight)					
$r/Y_{pasture}$	m^2/kg	7×10^{-1}	2×10^{0}	4×10^{0}	[10]
Effective half-life on vegetation					
T_w	days	5×10^{0}	$1{\cdot}3 \times 10^{1}$	3×10^{1}	[10]
Effective half-life in the root zone of soil[b]					
T_s	days	2×10^{2}	1×10^{3}	5×10^{3}	[10, 20, 32]
Effective density of the soil surface (to a depth of 15 cm)					
p	kg/m^2	1×10^{2}	$2{\cdot}1 \times 10^{2}$	3×10^{2}	[29]
Exposure time for pasture vegetation					
$t_{e,p}$	days	$1{\cdot}5 \times 10^{1}$	4×10^{1}	2×10^{2}	[29]
Exposure time for vegetables					
$t_{e,v}$	days	4×10^{1}	$7{\cdot}5 \times 10^{1}$	$1{\cdot}8 \times 10^{2}$	[29]
Exposure time for soil (30 y)					
t_s	days	—	$1{\cdot}1 \times 10^{4}$	—	
Aquatic food chain bioaccumulation					
Organism/water					
B_p	Ci/kg per Ci/litre	1×10^{0}	3×10^{1}	1×10^{3}	[10]

[a] Pasture and soil are given on a dry weight basis; all other values are given as fresh weight.
[b] Calculations based on the relationship between K_d and the downward movement of water in surface soil; total natural and artificial precipitation (irrigation) assumed to approximate 100 cm/y.

attempts to account quantitatively for the uncertainty associated with each parameter used to predict the environmental transport and food chain bioaccumulation of Tc. The sources of uncertainty encompassed by these ranges are (a) limitations in the quantity and scope of available data, (b) unexplained variability in the available data, and (c) differences between the conditions under which data were obtained and those for which model predictions are intended.

The uncertainty in model predictions is estimated by (a) assigning frequency distributions intended to encompass the uncertainty associated with each model parameter, (b) performing iterative model computations using parameter values selected randomly from the assigned distributions, and (c) statistically summarising the results [11]. All parameters, except I_{egg}, were varied over their ranges according to log-triangular distributions (I_{egg}, the rate of human consumption of eggs, was varied from a uniform distribution). The equations used in this analysis (see Appendix) are similar to those used in a variety of environmental radiological assessment models (for examples, see refs 12 and 13). Procedures to estimate the final uncertainties associated with model predictions can involve a variety of analytical and numerical methods [11, 14, 15].

The estimate of uncertainty in the model's prediction of dose is indicated in this analysis by two quantities: (a) an 'uncertainty factor' (UF) and (b) a probable range. The uncertainty factor is calculated as

$$\mathrm{UF} = (G_{sd})^2$$

where G_{sd} is the geometric standard deviation (the exponential of the standard deviation of log-transformed values).

The probable range is equivalent to a 95% probability interval for a lognormal distribution, whereby the maximum value of the range is

$$\max = G_m \times \mathrm{UF}$$

and the minimum value is

$$\min = G_m/\mathrm{UF}$$

where G_m is the geometric mean (the exponential of the mean of the log-transformed values) of the distribution of dose estimates.

An index of the relative importance of an exposure pathway or a specific model parameter is represented by the percentage contribution of their uncertainty to the uncertainty in the model's result. This index can be calculated in a number of ways. The ratio of the variance of model parameters and model predictions is often adequate for simple additive models or, for multiplicative models, the ratio of the variances of the log-transformed values [15]. The square of Spearman's correlation coefficient (the simple correlation coefficient of the rank-transformed values of model parameters and predictions) is useful when model predictions are monotonic functions of model parameters [15] and the parameters contribute at least 20% to the prediction uncertainty. For less sensitive parameters or parameters that are correlated with one another, the best estimates are based on multivariate regression techniques [14, 16].

The significance of potential exposure pathways is determined by the magnitude of the calculated dose and the percentage contributions of each exposure pathway to the uncertainty in the dose estimate. These results are then compared with the 1979 estimates published by Till *et al.* [9] using an assumed release of 1 Ci/year to the atmosphere and 100 Ci/year to a freshwater stream and substituting organ-specific dose conversion factors for dose conversion factors used to estimate effective dose equivalents (Table II). The estimates of effective and organ-specific dose equivalents made in this study are based on recently published internal dose conversion factors [17] adapted from ICRP Publication 30 dosimetric models and data bases [18]. An atmospheric release of 1 Ci/y is consistent

Table II
A comparison of dose conversion factors for internal exposure to ^{99}Tc

Organ	*Dose conversion factors (mrem/pCi)*			
	Inhalation		*Ingestion*	
	References			
	Till et al. *[9]*	*Killough and Eckerman [17]*	*Till* et al. *[9]*	*Killough and Eckerman [17]*
Effective	n.c.[a]	$1{\cdot}0 \times 10^{-6}$ D[b] $8{\cdot}3 \times 10^{-6}$ W[c]	n.c.	$1{\cdot}5 \times 10^{-6}$
Total body	$1{\cdot}7 \times 10^{-7}$	n.c.	$2{\cdot}1 \times 10^{-7}$	n.c.
Thyroid	$1{\cdot}1 \times 10^{-5}$	$4{\cdot}5 \times 10^{-5}$ D $4{\cdot}0 \times 10^{-5}$ W	$1{\cdot}4 \times 10^{-5}$	$6{\cdot}0 \times 10^{-6}$
GI tract	$5{\cdot}1 \times 10^{-7}$	$2{\cdot}1 \times 10^{-6}$ D $2{\cdot}3 \times 10^{-6}$ W	$3{\cdot}2 \times 10^{-6}$	$3{\cdot}8 \times 10^{-6}$

[a] n.c. = not calculated.
[b] D = clearance class for all compounds of Tc other than oxides, hydroxides, halides and nitrates.
[c] W = clearance class for Tc oxides, hydroxides, halides and nitrates.

with the value initially used by Till *et al.* in 1979 [9]. The assumption of a release of 100 Ci/y to freshwater streams is based on Till's recent estimates [10], indicating that aquatic releases of ^{99}Tc from uranium enrichment facilities may be as much as two orders of magnitude greater than releases to the atmosphere. In the United States, freshwater streams are the primary recipients of current environmental releases of ^{99}Tc, the principal source of which is liquid effluents from operating uranium enrichment facilities [10].

Using the same meteorological and release height assumptions employed by Till *et al.* in 1979 [9], the downwind air concentration at 1600 m resulting from a 1 Ci/year release is approximately 3×10^{-2} pCi/m^3. For the aquatic environment, a minimal rate of dilution of 40 m^3/s of water is assumed as an annual average, giving a ^{99}Tc water concentration of approximately 80 pCi/litre. This minimum rate of dilution is similar to assumptions made by other investigators, who have previously evaluated the potential impact of ^{99}Tc releases to the aquatic environment [3, 19].

The specific exposure pathways considered in this study are the transport of ^{99}Tc into (a) vegetables, (b) milk, (c) meat and (d) eggs from a unit air concentration for a 30-year continuous release of ^{99}Tc to the atmosphere, and the transport of ^{99}Tc into (e) aquatic biota from a unit concentration of ^{99}Tc in freshwater systems. The analysis is performed both with and without a correlation (r of $-0{\cdot}7$) between the uptake of ^{99}Tc from soil by vegetation ($\ln CR_{v,s}$ and $\ln CR_{p,s}$) and the effective half-life of ^{99}Tc in the root zone of soil ($\ln T_s$).

This correlation is based on the assumption that the migration of Tc out of the soil root zone and its uptake by vegetation should depend to a large extend on the mobility of Tc in soil. From their review of the literature, Baes and Sharp [20] report strong negative correlations ($r = -0{\cdot}83$) between the logarithms of soil solid phase/solution phase distribution coefficients (K_d) and plant/soil concentration ratios ($CR_{v,s}$ and $CR_{p,s}$) for 21 different elements. They estimated the same correlation from a single study [21] that reported on the relationship between the extractability of ^{95m}Tc from a Captina silt loam and its uptake by old field vegetation.

The inhalation and drinking water exposure pathways are also included in this study, but they are assumed to have negligible uncertainty. External exposure pathways to ^{99}Tc are unimportant because of the weak beta energy emitted by this isotope (0·292 MeV). The potential for uncertainty in the estimate of the conversion factors used to estimate dose equivalents from a unit intake of ^{99}Tc is also excluded from this analysis because this paper emphasises the evaluation of the relative importance of environmental pathways of exposure.

3. RESULTS

The calculated relationship between a concentration of ^{99}Tc in terrestrial and aquatic foods and a unit concentration of ^{99}Tc in air and water, respectively, is presented in Table III. The highest concentrations and smallest uncertainties are for the air–deposition–soil–vegetable pathway.

Table III

The calculated relationship between the concentration of ^{99}Tc in terrestrial and aquatic foods and the concentration in air and water, respectively

Food product	*Units*	*^{99}Tc concentrations*			
		Geometric mean	*Uncertainty factor*[a]	*Probable range*[b]	
				min	*max*
Terrestrial foods[c]					
Vegetables (pCi/kg, fresh wt per pCi/m^3, air)					
Covariance[d]		22 000	3·7	6 000	82 000
No covariance		22 000	6·9	3 200	150 000
Eggs (pCi/egg per pCi/m^3, air)					
Covariance[e]		340	23	15	7 900
No covariance		370	30	120	11 000
Meat (pCi/kg, fresh wt per pCi/m^3, air)					
Covariance[e]		31	24	1·3	760
No covariance		34	35	1·0	1 200
Milk (pCi/litre per pCi/m^3, air)					
Covariance[e]		46	22	2·1	1 000
No covariance		49	29	1·7	1 400
Freshwater aquatic organisms[f] (pCi/kg, fresh wt per pCi/litre, water)					
		31	17	1·8	530

[a] Uncertainty factor = (geometric standard deviation)^2.
[b] min = geometric mean/(geometric standard deviation)^2.
max = geometric mean × (geometric standard deviation)^2.
[c] Assumed release duration of 30 years.
[d] Assumed correlation between $CR_{v,s}$ and T_s ($r = -0{\cdot}7$).
[e] Assumed correlation between $CR_{p,s}$ and T_s ($r = -0{\cdot}7$).
[f] Accumulation in sediment and potential long-term remobilisation of Tc is not considered.

Table IV

The relative significance of environmental exposure pathways as indicated by predictions of the effective dose equivalent resulting from a ^{99}Tc concentration of 1 pCi/m^3 in above ground air (covariance is assumed between vegetation uptake and retention in soils with a correlation coefficient of −0·7)

Exposure pathway	*Effective dose equivalent (mrem/y)*			
	Geometric mean	*Uncertainty factor* [a]	*Probable range* [b]	
			min	*max*
All pathways	2·9	3·5	0·83	10
Inhalation	0·008 4	not computed		
Ingestion	2·9	3·5	0·83	10
Vegetables	2·1	3·7	0·57	7·6
Eggs	0·34	26	0·013	8·8
Meat	0·005 1	25	0·000 20	0·13
Milk	0·007 5	22	0·000 34	0·16

[a] Uncertainty factor = (geometric standard deviation)^2.
[b] Probable range is determined by
min = geometric mean/(geometric standard deviation)^2 and
max = geometric mean × (geometric standard deviation)^2,
approximating a 95% probability interval for a lognormal distribution.

For this pathway, probable ranges cover one to two orders of magnitude, depending on the assumption as to whether or not there is a correlation between the residence of ^{99}Tc in the root zone of soil and uptake by vegetation. For the other food types, probable ranges cover two to three orders of magnitude. The assumed relationship between plant uptake and soil residence exerts a more noticeable effect on the concentration in vegetables than for any other food type.

The calculated effective dose equivalents resulting from exposure to all terrestrial pathways are given in Tables IV and V. Table IV includes the correlation between plant uptake and ^{99}Tc residence in the soil root zone, whereas this effect is excluded from Table V. The dose due to exposure to all combined terrestrial pathways is dominated by the ingestion of vegetables. All other exposure pathways are negligible contributors to the total dose, with the exception of the consumption of ^{99}Tc in eggs which has the highest maximum value of the probable range when covariance is assumed between plant uptake and soil residence.

For aquatic pathways, effective dose equivalents from a unit concentration of ^{99}Tc in water are given in Table VI. Although the uncertainty

Table V

The relative significance of environmental exposure pathways as indicated by predictions of the effective dose equivalent resulting from a ^{99}Tc concentration of 1 pCi/m^3 in above ground air (no correlations assumed between vegetation uptake and retention in soil)

Exposure pathway	*Effective dose equivalent (mrem/y)*			
	Geometric mean	*Uncertainty factor*[a]	*Probable range*[b]	
			min	*max*
All pathways	3·4	7·2	0·47	24
Inhalation	0·008 4	not computed		
Ingestion	3·4	7·2	0·47	24
Vegetables	2·1	6·9	0·30	15
Eggs	0·37	33	0·011	12
Meat	0·005 6	35	0·000 16	0·20
Milk	0·008 2	29	0·000 28	0·24

[a] Uncertainty factor = (geometric standard deviation)^2.

[b] Probable range is determined by

min = geometric mean/(geometric standard deviation)^2 and

max = geometric mean × (geometric standard deviation)^2,

approximating a 95% probability interval for a lognormal distribution.

Table VI

The relative significance of environmental exposure pathways as indicated by predictions of the effective dose equivalent resulting from a ^{99}Tc concentration of 1 pCi/litre in fresh water

Exposure pathway	*Effective dose equivalent (mrem/y)*			
	Geometric mean	*Uncertainty factor*	*Probable range*	
			min	*max*
All pathways	0·002 2	4·0	0·001 2[a]	0·016[a]
Drinking water	0·001 2	not computed		
Ingestion of aquatic organisms	0·000 52	30	0·000 017	0·016

[a] Minimum value is determined by the drinking water pathway; maximum value is determined by the ingestion of aquatic organisms.

associated with exposure to combined pathways is totally controlled by the uncertainty in the water–aquatic organism–human pathway, the uncertainty factor for exposure to combined pathways is almost eight times less than that for the water–aquatic organism–human pathway alone. This is because the minimum value of the probable range of the dose from exposure to combined pathways is determined using the exposure to contaminated drinking water. This pathway is assumed to have no uncertainty.

For terrestrial and aquatic sources of ^{99}Tc the model parameters and

Table VII

The relative significance of environmental exposure pathways and model parameters as indicated by estimates of the percentage contribution to the uncertainty in the predicted dose

Dose estimate	*Exposure pathway*	*Parameter*	*Uncertainty contribution*[a]	
			Without covariance (%)	*With covariance (%)*
From all terrestrial pathways				
		V_{tot}	14	36
	Vegetables		80	67
		(T_s, $CR_{v,s}$)	60	34
	Eggs		20	32
		$CR_{egg,p}$	9	16
From ingestion of vegetables only				
	Uptake from soil		99	99
		($CR_{v,s}$, T_s)	75	49
		V_{tot}	16	34
	Direct deposition		<1	<1
From ingestion of eggs only				
		$CR_{egg,p}$	65	70
		($CR_{p,s}$, T_s)	28	19
		V_{tot}	5	6
From freshwater pathways				
	Ingestion of aquatic organisms	B_p	70	n.a.
		I_{ab}	30	n.a.

[a] Calculated using methods described by Hoffman and Gardner [15] and Gardner [14]; uncertainty contributions refer to uncertainty in dose estimates; analysis performed with and without assumed covariance between T_s and $CR_{v,s}$, and between T_s and $CR_{p,s}$ ($r = -0{\cdot}7$).
n.a. = not applicable.

exposure pathways that contribute most significantly to the uncertainty in the dose estimates are identified in Table VII. For terrestrial pathways, the amount of ^{99}Tc in both pasture vegetation and vegetables comes primarily from the soil. By contrast, direct interception of atmospheric deposits is negligible. The assumption of a correlation between soil retention of ^{99}Tc and vegetation uptake increases the importance of the air–deposition–soil–vegetation–chicken–egg pathway and decreases the importance of the uncertainties in the parameters $CR_{v,s}$, $CR_{p,s}$ and T_s. For the aquatic pathways, 70% of the total uncertainty is associated with the uncertainty in the estimate of the aquatic organism/water bioaccumulation factor (B_p).

The importance of these exposure pathways is put into better perspective by multiplying the normalised dose calculations in Tables IV and VI to air and water concentrations that are related to hypothetical releases of

Table VIII

Dose estimates for an assumed release of 1 Ci/y of ^{99}Tc to the atmosphere and 100 Ci/y to a freshwater stream

Exposure pathway	*Effective dose equivalent (mrem/y)*[a]		
	Geometric mean	*Probable range*	
		min	*max*
Combined terrestrial pathways[b]			
	0·09	0·03	0·3
Inhalation	0·000 3	assumed constant	
Ingestion	0·09	0·03	0·3
Vegetables	0·07	0·02	0·2
Eggs	0·01	0·000 4	0·3
Meat	0·000 2	6×10^{-6}	0·004
Milk	0·000 2	1×10^{-5}	0·005
Combined aquatic pathways[c]			
	0·2	0·1	1·0
Drinking water	0·1	assumed constant	
Ingestion of aquatic organisms	0·04	0·001	1·0

[a] The effective dose rate after an assumed release duration of 30 years; all values are rounded to one significant digit.

[b] The assumed downwind air concentration at 1600 m is 3×10^{-2} pCi/m^3; covariance is assumed between vegetation uptake and retention in soil ($r = -0{\cdot}7$).

[c] The assumed downstream water concentration is 80 pCi/litre; accumulation in sediment and potential long-term remobilisation of Tc is not considered.

1 Ci/year to the atmosphere and 100 Ci/year to a freshwater stream. For atmospheric releases, the maximum effective dose equivalent is 0·3 mrem/y (Table VIII). The minimum is three orders of magnitude lower. This minimum is dominated by direct inhalation of contaminated air. However, if the predominant chemical form of inhaled ^{99}Tc is a halide, hydroxide, nitrate or oxide, the inhalation dose would increase by nearly one order of magnitude. This is likely, given that the chemical form of the ^{99}Tc released to the atmosphere from uranium enrichment facilities is $^{99}TcF_6$ [10]. Considering the possibility that a freshwater stream receiving ^{99}Tc may not serve as a source of drinking water, estimates of effective dose equivalents range from 0·001 to 1 mrem/year for the aquatic exposure pathways.

Comparing our results to the 1979 estimates by Till *et al.* [9] shows a *marked* reduction in both the calculated doses and the contribution of milk and meat to the total dose from exposure to multiple terrestrial pathways (Table IX). Maximum dose estimates to the thyroid and gastrointestinal tract have been reduced by factors of approximately 60

Table IX
A comparison of dose estimates for atmospheric releases of ^{99}Tc

Exposure pathway	*Organ*	*Dose equivalent (mrem/y)*[a]		
		Till et al. *[9]*[b]	*This study*[c]	
			min	*max*
Combined terrestrial pathways	Thyroid	80	0·1	1·3
	GI tract	18	0·07	0·8
Vegetables	Thyroid	31	0·08	1·0
	GI tract	7	0·05	0·6
Meat	Thyroid	22	$2·5 \times 10^{-5}$	0·016
	GI tract	5	$1·6 \times 10^{-5}$	0·01
Milk	Thyroid	26	$4·3 \times 10^{-5}$	0·02
	GI tract	6	$2·7 \times 10^{-5}$	0·013
Eggs	Thyroid	n.c.	$4·1 \times 10^{-4}$	1·1
	GI tract	n.c.	$2·6 \times 10^{-4}$	0·8

[a] Assuming a 1 Ci/y release with a downwind air concentration at 1600 m of 3×10^{-2} pCi/m^3.
[b] The duration of the release is assumed at 15 y with no credit given for losses of Tc from the soil root zone.
[c] The duration of the release is assumed at 30 y; covariance is assumed between vegetation uptake and Tc retention in soil ($r = -0·7$); values given represent a probable range.

and 20, respectively. In the earlier estimates by Till *et al.* [9], milk and meat contributed almost as much to the total dose as did the vegetable pathway. In this study, dose estimates from the milk and meat pathways have been reduced by several orders of magnitude.

4. DISCUSSION AND CONCLUSIONS

The reasons for the large discrepancies between the results calculated in this analysis and those produced in 1979 are (a) data that show the uptake of ^{99}Tc from soil by field vegetation are from one to two orders of magnitude less than values reported from potted plant greenhouse experiments [21], (b) data indicating that the transfer of ^{99}Tc across the gastrointestinal tract of ruminants is several orders of magnitude less than it is for iodine [22–24], and (c) the fact that this analysis accounts for the removal over time of ^{99}Tc from the root zone of soil. The major difference between the calculational approach of this study and that of the earlier one is the application of numerical simulation (Monte Carlo) techniques, to propagate uncertainties of model parameters through mathematical models of similar structure.

In contrast to the 1979 results [9], none of the current dose estimates for releases of ^{99}Tc infringe on any known radiation protection standard. At present, the US Nuclear Regulatory Commission is considering the proposal of dose limits that would define levels below which the health risk to critical groups and average members of the public would not warrant additional measures to limit releases of radioactivity. These suggested *de minimis* levels are effective dose equivalents of 1 mrem/year to members of critical population subgroups and 0·1 mrem/year to members of the general public. In general, the effective dose equivalents, calculated in this study for hypothetical releases of 1 Ci/year to the atmosphere and 100 Ci/year to a freshwater stream, approach or exceed these *de minimis* levels at only the upper limits of their estimated range (Table VIII). Barring conditions that could produce substantially larger releases of ^{99}Tc than have been assumed here, further research to reduce the large uncertainties in the prediction of the environmental fate and behaviour of ^{99}Tc should lead to the conclusion that ^{99}Tc is not an environmentally important radionuclide.

In the terrestrial environment, priorities for further research should focus on reducing the uncertainty in predicting (a) the transfer of ^{99}Tc from soil to the edible portions of vegetation types of importance to the human diet; (b) the long-term residence of ^{99}Tc in the root zone of soil; (c) the relationship between soil residence and vegetation uptake of ^{99}Tc for specific soil properties, plant species and agricultural practices; (d) the transfer of ^{99}Tc into vegetation potentially consumed by poultry, with subsequent transfer into eggs; and (e) the processes of wet and dry deposition as they affect the transfer of ^{99}Tc from the atmosphere to the earth's surface.

In addition, research is needed to identify and investigate potentially important exposure pathways that have been excluded from this analysis. A specific example is ^{99}Tc cycling via transfer from contaminated groundwater into the tissues of deep-rooted vegetation and the subsequent deposition and accumulation of ^{99}Tc on the ground surface following tissue senescence and decomposition. Evidence for such cycling processes occurring in the vicinity of shallow land waste burial sites has been reported by Olsen *et al.* [25].

In the aquatic environment, research should focus on reducing the uncertainty associated with predicting the food chain bioaccumulation of ^{99}Tc in those species of aquatic organisms potentially consumed by humans. To date, few measurements on the uptake of ^{99}Tc by aquatic organisms have been made under field conditions [25, 26]. Because of the long half-life of ^{99}Tc, research should also determine the extent to which prolonged aquatic releases could result in large sediment concentrations and the conditions under which ^{99}Tc in sediment could be remobilised and incorporated into aquatic and terrestrial food products. The potential for sediment accumulation and long-term remobilisation of ^{99}Tc was not included in this analysis.

If the concept of *de minimis* levels of risk is not accepted by either the scientific community involved with radiological protection or the general public, then research on all aspects of the environmental behaviour of ^{99}Tc will be necessary to increase confidence in collective dose predictions. The prediction of collective dose will potentially involve extremely small exposures to individuals over time periods that extend for thousands of years into the future.

ACKNOWLEDGEMENTS

Research sponsored by the Office of Health and Environmental Research, US Department of Energy, under Contract No. DE-ACO5-840R21400 with Martin-Marietta Energy Systems, Inc.

REFERENCES

1. NG, Y. C., BURTON, C. A., THOMPSON, S. E., TANDY, R. K., KRETNER, H. K. and PRATT, M. W. (1968). Prediction of the maximum dosage to man from the fallout of nuclear devices. In: *Handbook for Estimating the Maximum Internal Dose from Radionuclides Released to the Biosphere*, UCRL-50163, Part IV. Livermore, California, Lawrence Radiation Laboratory.
2. HOFFMAN, F. O. and KAYE, S. V. (1976). Terrestrial exposure pathways: Potential exposures of man from the environmental transport of waste nuclides. In: *Management of Wastes from the LWR Fuel Cycle* (Proc. Symp.). Energy Research and Development Administration, CONF-76-0701, pp. 524–38.
3. BLAYLOCK, B. G. and WITHERSPOON, J. P. (1978). Evaluation of radionuclides released from the light water reactor nuclear fuel cycle to the aquatic environment. In: *Environmental Chemistry and Cycling Processes* (Ed. D. C. Adriano and I. L. Brisbin), US Dept. of Energy Symp. Series No. 45, CONF-760429, pp. 851–65.
4. WILDUNG, R. E., GARLAND, T. R. and CATALDO, D. A. (1974). Preliminary studies on the uptake of technetium by soybeans. In: *Pacific Northwest Laboratory Annual Report for 1974 to the USAEC Division of Biomedical and Environmental Research, Part 2, Ecological Sciences*, Batelle-Northwest Laboratory, BNWL-1950PT2.
5. GAST, R. G. (1975). *The Behavior of Technetium-99 in Soils and Plants*. Progress Report for the Period: April 1, 1974–March 31, 1975, USAEC Contract No. AT(11-1)-2447, COO-2447-1.
6. GAST, R. G., LANDA, E. R. and THORVIG, L. J. (1976). *The Behavior of Technetium-99 in Soils and Plants*. Progress Report for the Period: April 1, 1974–March 31, 1977, USAEC Contract No. Ey-76-5-20-2447, COO-2447-5.
7. CATALDO, D. A., WILDUNG, R. E. and GARLAND, T. R. (1978). Technetium accumulation, fate, and behavior in plants. In: *Environmental Chemistry and Cycling Processes* (Ed. D. C. Adriano and I. L. Brisbin), US Dept. of Energy Symp. Series No. 45, CONF-760429, pp. 538–49.
8. WILDUNG, R. E., GARLAND, T. R. and CATALDO, D. A. (1977). Accumulation of technetium by plants. *Health Physics*, **32**, 314–17.
9. TILL, J. E., HOFFMAN, F. O. and DUNNING, D. E. JR (1979) A new look at ^{99}Tc releases to the atmosphere. *Health Physics*, **36**, 21–30.
10. TILL, J. E., SHOR, R. E. and HOFFMAN, F. O. Environmental Effects of the Uranium Fuel Cycle—A Review of Data for Technetium. NUREG/CR-3738, Oak Ridge, Tennessee, Oak Ridge National Laboratory (in press).
11. GARDNER, R. H., ROJDER, B. and BERGSTRÖM, U. (1983). PRISM: A systematic method for determining the effect of parameter uncertainties on model predictions. Studsvik Energiteknik AB report/NW-83/55, Nyköping, Sweden.
12. US NUCLEAR REGULATORY COMMISSION REGULATORY GUIDE 1.109, Rev. 1 (1977). *Calculation of Annual Doses to Man from Routine Releases of Reactor Effluents for the Purpose of Evaluating Compliance with 10 CFR Part 50, Appendix I*. US Nuclear Regulatory Commission, Washington, DC.
13. INTERNATIONAL ATOMIC ENERGY AGENCY (1982). *Generic Models and Parameters for Assessing the Environmental Transfer of Radionuclides from Routine Releases*. Safety Series No. 57, Vienna, IAEA.

14. GARDNER, R. H. (1984). A unified approach to sensitivity and uncertainty analysis. In: *Applied Simulation and Modelling* (Ed. M. H. Hamza), Proc. ISTED International Symposium, San Francisco, USA, June 4–6, 1984. Anaheim, CA, ACTA Press, pp. 155–7.
15. HOFFMAN, F. O. and GARDNER, R. H. (1983). Evaluation of uncertainties in radiological assessment models. In: *Radiological Assessments*, Ch. 11 (Ed. J. E. Till and H. R. Meyer), US Nuclear Regulatory Commission, Washington, DC. NUREG/CR-3332, ORNL-5968.
16. DOWNING, D. J., GARDNER, R. H. and BARTELL, S. M. Univariate and multivariate methods for analysis of sets of predictions from Monte Carlo simulations (submitted to *Communications in Statistics*).
17. KILLOUGH, G. G. and ECKERMAN, K. F. (1983). Internal dosimetry. In: *Radiological Assessments*, Ch. 7 (Ed. J. E. Till and H. R. Meyer), US Nuclear Regulatory Commission, Washington, DC. NUREG/CR-3332, ORNL-5968.
18. INTERNATIONAL COMMISSION ON RADIOLOGICAL PROTECTION (1979). *Limits for Intakes of Radionuclides by Workers*, Annals of the ICRP, Publication 30, Supplement to Part 1, Vol. 3, Nos 1–4, New York, Pergamon Press.
19. BLAYLOCK, B. G. and WITHERSPOON, J. P. (1975). Dose estimation and prediction of radiation effects on aquatic biota resulting from radioactive releases from the nuclear fuel cycle. In: *Impacts of Nuclear Releases into the Aquatic Environment* (Sym. Proc.), Vienna, IAEA, STI/PUB/406, pp. 377–93.
20. BAES, C. F. III (1982). Prediction of radionuclide K_d values from soil–plant concentration ratios. *Trans. Am. Nucl. Soc.*, **41**, 53–4.
21. GARTEN, C. T. JR, HOFFMAN, F. O. and BONDIETTI, E. A. (1984). Field and greenhouse experiments on the fate of technetium in plants and soil. *Health Physics*, **46**, 647–56.
22. JONES, BERNT E. V. (1979). *Metabolism of Technetium-95m Pertechnetate in Goats and Pigs*, Dept. of Clinical Chemistry, College of Veterinary Medicine, Swedish University of Agricultural Sciences, S-750 07, Uppsala, Sweden.
23. WIECHEN, A., HEINE, K. and HAGEMEISTER, H. (1983). A contribution to the question of the transfer of technetium into milk (in German). *Wissenschaft und Umwelt*, **11**, 41–6.
24. BONDIETTI, E. A. and GARTEN, C. T. JR. Transfer of ^{131}I and ^{95m}Tc from pasture to goat milk (this volume, pp. 339–47).
25. OLSEN, C. R., LOWRY, P. D., LEE, S. Y., LARSEN, I. L. and CUTSHALL, N. H. (1983). *Chemical, Geological, and Hydrological Factors Governing Radionuclide Migration from a Formerly Used Seepage Trench: A Field Study*. ORNL-TM-8839/NFW-83/15, Oak Ridge, Tennessee, Oak Ridge National Laboratory, pp. 38–9.
26. BLAYLOCK, B. G., FRANK, M. L., HOFFMAN, F. O. and DEANGELIS, D. L. Behavior of technetium in freshwater environments (this volume, pp. 79–89).
27. BEASLEY, T. M. and LORZ, H. V. A review of the biological and geochemical behaviour of technetium in the marine environment (this volume, pp. 197–216).
28. INTERNATIONAL COMMISSION ON RADIOLOGICAL PROTECTION (1975). *Report of the Task Group on Reference Man*. ICRP Publication 23, Oxford, Pergamon Press.

29. HOFFMAN, F. O., GARDNER, R. H. and ECKERMAN, K. F. (1982). *Variability in Dose Estimates Associated with the Food Chain Transport and Ingestion of Selected Radionuclides*, Oak Ridge, Tennessee, Oak Ridge National Laboratory, NUREG/CR-2612, ORNL/TM-8099.
30. BONDIETTI, E. A., HOFFMAN, F. O. and LARSEN, I. L. (1984). Air-to-vegetation transfer rates of natural submicron aerosols. *J. Envir. Radioact.*, **1**, 5–27.
31. HOFFMAN, F. O., MILLER, C. W. and NG, Y. C. Uncertainties in radioecological assessment models. In: *Seminar on the Environmental Transfer to Man of Radionuclides Released from Nuclear Installations*, Brussels, Belgium, 17–21 October 1983. International Atomic Energy Agency and the Commission of the European Communities, IAEA-SR-85/H (in press).
32. BAES, C. F. III and SHARP, R. D. (1983). A proposal for estimation of soil leaching and leaching constants for use in assessment models. *J. Envir. Qual.*, **12**, 17–28.

APPENDIX

Equations for the calculation of the environmental transport and dose resulting from a unit concentration of ^{99}Tc in air and water

Terrestrial pathways

Dose from combined pathways

$$D_{\text{total}} = D_{\text{ing}} + D_{\text{inh}} \tag{1}$$

Dose from inhalation

$$D_{\text{inh}} = \text{DCF}_{\text{inh}} * I_{\text{air}} * C_{\text{air}} \tag{2}$$

Dose from combined ingestion pathways

$$D_{\text{ing}} = \text{DCF}_{\text{ing}} * \{(I_{\text{milk}} * C_{\text{milk}}) + (I_{\text{meat}} * C_{\text{meat}}) + (I_{\text{eggs}} * C_{\text{eggs}}) + (I_{\text{vegetables}} * C_{\text{vegetables}})\} \tag{3}$$

Concentration in milk

$$C_{\text{milk}} = C_{\text{pasture}} * \text{CR}_{\text{m,p}} \tag{4}$$

Concentration in meat

$$C_{\text{meat}} = C_{\text{pasture}} * \text{CR}_{\text{f,p}} \tag{5}$$

Concentration in eggs

$$C_{\text{eggs}} = C_{\text{pasture}} * \text{CR}_{\text{eggs,p}} \tag{6}$$

Concentration in vegetables

$$C_{\text{vegetables}} = C_{\text{(v) direct deposition}} + C_{\text{(v) uptake from soil}} \tag{7}$$

Concentration via direct deposition from the atmosphere

$$C_{\text{(v) direct deposition}} = C_{\text{air}} * V_{\text{total}} * r/Y_{\text{vegetables}} * T_{\text{w}}/\ln 2 * \{1 - \exp[-(\ln 2/T_{\text{w}} * t_{\text{e,v}})]\} \tag{8}$$

Concentration via uptake from soil

$$C_{\text{(v) soil uptake}} = C_{\text{air}} * V_{\text{total}} * \text{CR}_{\text{v,s}} * T_{\text{s}}/\ln 2 * 1/p * \{1 - \exp[-(\ln 2/T_{\text{s}} * t_{\text{s}})]\} \tag{9}$$

Concentration in pasture

$$C_{\text{pasture}} = C_{\text{(p) direct deposition}} + C_{\text{(p) uptake from soil}} \quad (10)$$

Concentration via direct deposition

$$C_{\text{(p) direct deposition}} = C_{\text{air}} * V_{\text{total}} * r/Y_{\text{pasture}} * T_{\text{w}}/\ln 2 * \{1 - \exp[-(\ln 2/T_{\text{w}} * t_{\text{e}})]\} \quad (11)$$

Concentration via uptake from soil

$$C_{\text{(p) uptake from soil}} = C_{\text{air}} * V_{\text{total}} * \text{CR}_{\text{p,s}} * T_{\text{s}}/\ln 2 * 1/p * \{1 - \exp[-(\ln 2/T_{\text{s}} * t_{\text{s}})]\} \quad (12)$$

Aquatic pathways

Dose from combined water pathways

$$D_{\text{water}} = \text{DCF}_{\text{ing}} * \{(I_{\text{water}} * C_{\text{water}}) + (I_{\text{aquatic biota}} * C_{\text{aquatic biota}})\} \quad (13)$$

Concentration in aquatic biota

$$C_{\text{aquatic biota}} = C_{\text{water}} * B_{\text{p}} \quad (14)$$

32

Estimations des Risques Radiologiques Liés à un Rejet Concerté de Technétium dans l'Environnement

J. Quinault et A. Grauby
CEA/IPSN/DERS/SERE, CEN Cadarache,
BP No. 1, 13115 Saint Paul-lez-Durance, France

Le calcul de l'estimation des risques radiologiques liés à un rejet concerté de technétium dans l'environnement est soumis à la variabilité de la nature et à la méconnaissance de certains paramètres.

Afin d'obtenir un ordre de grandeur et pouvoir comparer les impacts du technétium et d'autres radioéléments, Iode, Strontium, Cobalt, Cesium, nous avons effectué des calculs en un lieu fictif de la vallée du Rhône en utilisant les formules, et les valeurs des paramètres correspondants, du 'Manuel de Radioécologie'.

Le manuel de radioécologie a été conçu comme une synthèse des informations disponibles, que l'on a souhaité cohérente mais simplificatrice et globalisante. Dans un premier temps, nous nous sommes limités volontairement au cas des rejets continus. Aussi les aliments de l'homme, d'origine végétale ou animale, seront-ils toujours supposés obtenus à maturité, avec des concentrations des mileux vecteurs constantes pendant sa croissance.

Pour permettre la comparaison entre rejets dans un fleuve ou dans l'atmosphère nous avons simulé les deux types de transfert à partir d'une émission de 1 curie par an dans chaque cas. Les conditions retenues et les paramètres utilisés sont présentés en annexe, ou sont ceux du manuel.

Ce travail nous conduit à l'estimation de l'ingestion annuelle de radioélément, en moyenne durant la vie de l'installation nucléaire considéré.

Pour un rejet de 1 Ci/an dans un fleuve utilisé pour l'irrigation des cultures, l'ingestion annuelle de technétium est supérieure, d'un facteur 10 pour le cobalt à 2·5 pour les Strontium et Césium, à celle des autres radioéléments. Ce 'score' est dû principalement à la forte valeur de transfert après une contamination en surface, avec du technétium, juste avant la plantation.

Pour un rejet de 1 Ci/an dans l'atmosphére, l'ingestion annuelle de technétium est sept fois plus faible que celle d'iode élémentaire et deux fois plus forte que pour les autres radioéléments. Ce résultat est du à la forte valeur du coefficient de transfert apparent à l'équilibre obtenu dans la synthèse Quinault-Angeletti à partir des mesures effectuées à Jülich et à Idaho-falls.

En ce qui concerne les risques liés aux ingestions annuelles précédentes rappelons que selon la CIPR No. 26 il convient de différentier les risques non stochastiques n'apparaissant jamais en deça d'un certain seuil, pouvant être spécifique d'un organe, et les risques stochastiques (aléatoires) pouvant apparaître plusieurs années après l'exposition sur un nombre variable de sujets exposés. La gravité de l'effet ne dépend pas de la 'dose' reçue, seul le nombre de sujets atteints en dépend.

Par suite pour qu'un individu présente un risque acceptable, au sens de la CIPR, il faut que deux limites soient respectées simultanément:

— La première, destinée à éliminer toute possibilité d'apparition d'effets non stochastiques (certains): *équivalent de dose engagé par organe et par an* $\leq$ *0·05 Sv an*$^{-1}$, (0·03 Sv an^{-1} pour les cristallins).
— La seconde garantit, pour les effets stochastiques (aléatoires), un niveau de risque éventuel acceptable pour l'ensemble de la vie de l'individu: *équivalent de dose effectif total engagé par an* $\leq$ *0·005 Sv an*$^{-1}$ (DE).

Compte tenu de la définition de l'équivalent de dose effectif total, les résultats des estimations seront analysés de la façon suivante:

(a) $DE \leq 1{\cdot}5 \times 10^{-3}$ (Sv an^{-1})
  — l'ingestion n'entraînera aucun effet non stochastique, le niveau de risque stochastique lié à cette ingestion annuelle est acceptable.

(b) $1{\cdot}5 \times 10^{-3} < DE \leq 5 \times 10^{-3}$ (Sv an^{-1})
  — Il convient d'effectuer le calcul des équivalents de dose engagés annuellement, organe par organe, pour s'assurer qu'aucun ne dépasse les limites recommandées par la CIPR.
  — Le niveau de risque stochastique lié à l'ingestion annuelle est acceptable.

(c) $5 \times 10^{-3} < \mathrm{DE}\ (\mathrm{Sv\ an}^{-1})$

— Le niveau de risque stochastique lié à l'ingestion est supérieur aux valeurs recommandées par la CIPR.

Compte tenu des faibles niveaux de radioactivité rencontrés généralement dans l'environnement, sauf éventuellement en cas d'accident, le cas (a) est le plus fréquent, par suite le risque stochastique devient le risque majeur du fait de l'absence de seuil et de la linéarité, supposé aux faibles doses, de la loi dose-effet.

Nous avons calculé le débit annuel d'équivalent de dose effectif total, engagé sur 10 ans, à partir des valeurs publiées dans la méthodologie NRPB-CEA. Le coefficient de dose étant beaucoup plus faible pour le technétium, d'un facteur 400 à 28, la 'dose' finale est plus faible malgré l'ingestion annuelle plus importante.

Pour les rejets dans un fleuve la dose technétium est plus faible d'un facteur 4 à 77, et pour les rejets atmosphériques d'un facteur 13 à 2800. Mais ces résultats sont obtenus avec des approximations parfois grossières et pour minimiser les incertitudes liées à la connaissance des transferts il conviendrait sans doute d'étudier encore:

— le transfert du technétium consécutif à une irrigation par aspersion ou à la raie,
— le transfert par dépôt sec et par la pluie,
— le lessivage dans le sol au cours du temps en function d'un cycle irrigation-pluies,
— l'évolution de la biodisponibilité au cours du temps,
— la synergie avec d'autres éléments stables ou radioactifs.

Il faut noter qu'un effort est également à faire dans l'évaluation prévisionnelle du transfert des Iode, Strontium, Cobalt, Cesium, pour ne prendre que les éléments de mon exemple, qui d'ores et déjà conduisent à des coefficients de dose généralement élevés, notament pour l'Iode, dont nous pouvons remarquer le manque de données concernant les transferts sol $\rightarrow$ plante.

Tableau I
Transferts par irrigation

Rejet 1 Ci an^{-1} ⇒	Conc. de l'eau ≃ 0·8 Bq m^{-3}
débit moyen	1500 $m^3 s^{-1}$
irrigation	400 litres m^{-2} an^{-1}

Consommation des animaux

bovins végétaux	57 kg j^{-1}
eau	18 litres j^{-1}
volailles végétaux	0·06 kg j^{-1}
eau	0·2 litres j^{-1}

Transfert	^{99}Tc	^{129}I	^{90}Sr	^{60}Co	^{134}Cs	^{136}Cs
F_s^a	$0{\cdot}12^c$	10^{-7}	3×10^{-3}	10^{-2}	3×10^{-3}	3×10^{-3}
F_h^b	2	? (1)	0·23	4×10^{-2}	10^{-2}	10^{-2}

[a] $m^2 kg^{-1}$ frais.
[b] F_h: Facteur de transfert sol-plante pour une répartition homogène avant le semi (kg de sol sec k^{-1} frais).

[c] $F_s \simeq 40 \dfrac{\text{kg sol sec}}{\text{kg frais}} \Rightarrow \dfrac{40}{300} = 0{\cdot}12 \dfrac{m^2}{\text{kg frais}}.$

F_s: Facteur de transfert sol-plante pour un apport chronique en surface de la végétation (m^2 de sol kg^{-1} frais).

Tableau II
Taux de pertes (φ)

	Tc	*I*	*Sr*	*Co*	*Cs*
Taux de perte par lixiviation (φ_l)					
Kd_{sol}	5	25	21	400	10 000
φ_l	0·5	0·1	0·12	6×10^{-3}	0
Taux de perte par exportation des récoltes (φ_r)					
φ_r	0·14	7×10^{-3}	5×10^{-3}	2×10^{-3}	5×10^{-4}
Taux de perte total (φ_t) *et accumulation sol* (A_s)					
φ_t	0·6	0·1	0·1	8×10^{-3}	5×10^{-4}
$T_{1/2}$(an)	$2{\cdot}1 \times 10^5$	$1{\cdot}6 \times 10^7$	2·8	5·2	^{134}Cs 2·2 ^{137}Cs 30
A_s	1·7	6	2·7	5·4	^{134}Cs 3 ^{137}Cs 11·4

Tableau III
Depots atmospheriques

Rejet 1 Ci an^{-1} $\Rightarrow$ C air $\simeq 1{\cdot}2 \times 10^{-3}$ Bq m^{-3} pour CTA $= 1 \times 10^{-6}$
Vitesse de dépôt sec 10^{-2} m s^{-1} iode (I_2)
10^{-3} m s^{-1} aérosols
Facteur de transfert apparent à l'équilibre
$1{\cdot}5 \times 10^4$ ^{129}I (Iz)
2×10^3 ^{137}Cs (aérosols) et autres aérosols (Tc)

Pluies 800 litres m^{-2} an^{-1}

$$Cv = Fp.Rp.\overline{Cp}$$
$$Fp = 0{\cdot}6\ I_2 \text{ (sur prairie)}$$
$$= 1 \text{ pour les autres (majorant)}$$
$$Rp = 10 \text{ pour tous}$$
$$\overline{Cp} = Fa.\sqrt{2\pi}.\bar{\sigma}_3.\overline{Ca}$$
$$Fa \simeq 0{\cdot}1\ (m^2\,C^{-1})$$
$$\sqrt{2\pi}.\bar{\sigma}_3 \simeq 300$$
$$\overline{Cp} \simeq 3{\cdot}6 \times 10^{-2} \text{ Bq litre}^{-1}$$

F_p: facteur pluies vers le végétal pour une seule pluie (1 kg^{-1} frais)
R_p: coéfficient de rétention moyenne du dépôt pour la pluie
$\overline{Cp}$: concentration de l'eau de pluie
Fa: facteur de transfert atmosphérique eau de pluie (m^2 . l^{-1})
$\overline{Ca}$: concentration moyenne annuelle dans l'atmosphère.

Tableau IV
Ingestion annuelle (Bq an^{-1}) pour un rejet de 1 Ci an^{-1} dans un fleuve utilisé pour l'irrigation des cultures

Aliments	^{99}*Tc*	^{129}*I*	^{90}*Sr*	^{60}*Co*	^{134}*Cs*	^{137}*Cs*
Végétaux	3 800	630	180	360	90	90
Lait	3 400	250	25	100	84	84
Boeuf	1 400	170	5	20	72	72
Volaille	3	2	0	?	6	6
Oeuf	80	23	0	?	0	0
Eau de boisson	800	800	800	800	800	800
Poisson	290	58	2 900	29	2 900	2 900
Total	9 800	1 900	3 900	1 300	4 000	4 000
Comparaison	1	0·2	0·40	0·1	0·4	0·4

Tableau V

Ingestion annuelle (Bq an^{-1}) pour un rejet de 1 Ci an^{-1} dans l'atmosphère

Aliments	^{99}Tc	^{129}I	^{90}Sr	^{60}Co	^{134}Cs	^{137}Cs
Végétaux	320	3 000	260	250	250	250
Lait	280	1 200	40	70	240	240
Boeuf	120	840	6	12	20	20
Volaille	0	8	0	?	16	16
Oeuf	5	110	0	?	0	0
Eau de boisson	0	0	0	0	0	0
Poisson	0	0	0	0	0	0
Total	730	5 200	310	330	530	530
Comparaison	1	7	0·4	0·5	0·7	0·7

Tableau VI

Equivalent de dose effectif engagé pour un rejet de 1 Ci an^{-1} soit dans un fleuve, soit dans l'air

	^{99}Tc	^{129}I	^{90}Sr	^{60}Co	^{134}Cs	^{137}Cs
Dose effective engagée NRPB-CEA						
Fac. dose Sv Bq	$2{\cdot}5 \times 10^{-10}$	$9{\cdot}9 \times 10^{-8}$	$3{\cdot}3 \times 10^{-8}$	$7{\cdot}6 \times 10^{-9}$	$2{\cdot}0 \times 10^{-8}$	$1{\cdot}4 \times 10^{-8}$
Indice de comparaison	1	400	130	28	80	56
Rejets liquides						
Débit dose Sv an^{-1}	$2{\cdot}5 \times 10^{-6}$	$1{\cdot}9 \times 10^{-4}$	$1{\cdot}3 \times 10^{-4}$	$9{\cdot}2 \times 10^{-6}$	$8{\cdot}0 \times 10^{-5}$	$5{\cdot}6 \times 10^{-5}$
Indice de comparaison	1	77	53	4	33	23
Rejets atmosphériques						
Débit dose Sv an^{-1}	$1{\cdot}8 \times 10^{-7}$	$5{\cdot}1 \times 10^{-4}$	$1{\cdot}0 \times 10^{-5}$	$2{\cdot}3 \times 10^{-6}$	$1{\cdot}1 \times 10^{-5}$	$7{\cdot}4 \times 10^{-6}$
Indice de comparaison	1	2 800	56	13	58	41

Tableau VII

Matrice des quanta de 'dose'-liquide (QDL)

QDL	^{99}Tc	^{129}I	^{90}Sr	^{60}Co	^{134}Cs	^{137}Cs
Végétaux	$1{\cdot}3 \times 10^{-8}$	$8{\cdot}7 \times 10^{-7}$	$8{\cdot}3 \times 10^{-8}$	$3{\cdot}6 \times 10^{-8}$	$2{\cdot}5 \times 10^{-8}$	$1{\cdot}8 \times 10^{-8}$
Lait	$7{\cdot}5 \times 10^{-9}$	$2{\cdot}3 \times 10^{-7}$	$7{\cdot}4 \times 10^{-9}$	$6{\cdot}2 \times 10^{-9}$	$1{\cdot}5 \times 10^{-8}$	$1{\cdot}1 \times 10^{-8}$
Boeuf	$7{\cdot}5 \times 10^{-9}$	$3{\cdot}7 \times 10^{-7}$	$3{\cdot}7 \times 10^{-9}$	$2{\cdot}7 \times 10^{-9}$	$2{\cdot}7 \times 10^{-8}$	$2{\cdot}1 \times 10^{-8}$
Volailles	$5{\cdot}0 \times 10^{-11}$	$9{\cdot}9 \times 10^{-9}$	$1{\cdot}7 \times 10^{-10}$	?	$7{\cdot}5 \times 10^{-9}$	$5{\cdot}3 \times 10^{-9}$
Oeuf	$1{\cdot}6 \times 10^{-9}$	$1{\cdot}2 \times 10^{-7}$	$1{\cdot}7 \times 10^{-9}$	?	0	0
Eau de boisson	$2{\cdot}5 \times 10^{-10}$	$9{\cdot}9 \times 10^{-8}$	$3{\cdot}3 \times 10^{-9}$	$7{\cdot}1 \times 10^{-9}$	$2{\cdot}0 \times 10^{-8}$	$1{\cdot}4 \times 10^{-8}$
Poisson	$7{\cdot}6 \times 10^{-9}$	$6{\cdot}0 \times 10^{-7}$	$1{\cdot}0 \times 10^{-5}$	$2{\cdot}1 \times 10^{-8}$	$6{\cdot}0 \times 10^{-6}$	$4{\cdot}2 \times 10^{-6}$

Dose = ration alimentaire annuelle * QDL * C eau (= équivalent de dose effectif engagé) (* = produit matriciel).

Tableau VIII
Matrice des quanta de 'dose'-air

QDA	^{99}Tc	^{129}I	^{90}Sr	^{60}Co	^{134}Cs	^{137}Cs
Végétaux	$8{\cdot}0 \times 10^{-7}$	$3{\cdot}0 \times 10^{-3}$	$8{\cdot}7 \times 10^{-5}$	$1{\cdot}8 \times 10^{-5}$	$5{\cdot}1 \times 10^{-5}$	$3{\cdot}6 \times 10^{-5}$
Lait	$4{\cdot}5 \times 10^{-7}$	$7{\cdot}7 \times 10^{-4}$	$9{\cdot}0 \times 10^{-6}$	$3{\cdot}2 \times 10^{-6}$	$3{\cdot}1 \times 10^{-5}$	$2{\cdot}2 \times 10^{-5}$
Boeuf	$4{\cdot}5 \times 10^{-7}$	$1{\cdot}3 \times 10^{-3}$	$3{\cdot}0 \times 10^{-6}$	$1{\cdot}3 \times 10^{-6}$	$6{\cdot}2 \times 10^{-5}$	$4{\cdot}3 \times 10^{-5}$
Volailles	$3{\cdot}0 \times 10^{-9}$	$3{\cdot}6 \times 10^{-5}$	$1{\cdot}8 \times 10^{-7}$	?	$1{\cdot}5 \times 10^{-5}$	$1{\cdot}0 \times 10^{-5}$
Oeuf	$9{\cdot}1 \times 10^{-6}$	$7{\cdot}7 \times 10^{-4}$	$1{\cdot}8 \times 10^{-6}$	?	0	0
Eau de boisson	0	0	0	0	0	0
Poisson	0	0	0	0	0	0

Dose = ration *QDL*C air (*: produit matriciel) (équivalent de dose effectif engagé).

33

Interaction entre Technétium et Molybdène dans l'Inhibition de la Nitrogénase d'*Azotobacter chroococcum*

C. M. Vandecasteele,[a] J. Henrot,[a] J. M. Pieri,[b] C. Myttenaere,[c] M. Cogneau[d] et C. Van Hove[a]

[a] *Laboratoire de Physiologie Végétale, Université Catholique de Louvain, Place Croix du Sud, 4, B-1348 Louvain-la-Neuve, Belgique*

[b] *Laboratoire de Biochimie, Université de Nantes, Rue de la Houssinière, 2, F-44072 Nantes Cedex, France*

[c] *Programme Biologie-Radioprotection, Commission des Communautes Européennes, DG XII, Rue de la Loi, 200, B-1049 Bruxelles, Belgique*

[d] *Laboratoire de Chimie Nucléaire, Université Catholique de Louvain, Chemin du Cyclotron, 2, B-1348 Louvain-la-Neuve, Belgique*

1. INTRODUCTION

Les études portant sur la toxicité du technétium chez les microorganismes les décrivent pour la plupart comme peu sensibles et n'accumulant que peu cet élément [1–4]. Ces considérations ne s'appliquent toutefois pas aux organismes diazotrophes lorsque ceux-ci sont cultivés en milieu dépourvu d'azote combiné [5]. Dans ces conditions, les effets du technétium se manifestent à des concentrations relativement faibles, de l'ordre de 1 μM litre^{-1}, par une inhibition de la croissance qui survient d'autant plus rapidement que le niveau de contamination du milieu est élevé; en présence de nitrates dans la solution nutritive, le seuil de toxicité est déplacé vers des concentrations en technétium plus élevées, aux environs de 1 mM litre^{-1}. D'autre part, il a également été observé que, lorsque l'azote atmosphérique constitue pour ces procaryotes la seule source azotée, l'accumulation de ce radiopolluant est nettement plus importante qu'en présence de nitrates. En outre, il a été démontré par autoradiographie que cet élément s'accumule préférentiellement dans les cellules ou tissus qui sont le siège de la diazotrophie [5].

Ces résultats suggèrent une interaction spécifique entre le technétium et le métabolisme de la fixation du diazote, confirmée par le fait que la nitrogénase, enzyme responsable de la réduction du N_2 en NH_4^+ est rapidement inhibée par la présence de technétium et que son inhibition précède l'apparition de symptômes de toxicité sur la croissance globale de la culture mesurée par colorimétrie [6].

Plusieurs hypothèses peuvent être avancées pour justifier cette interaction spécifique entre technétium et nitrogénase, parmi lesquelles la substitution par ce polluant radioactif, du molybdène présent au niveau du site actif d'un des composés du complexe nitrogénasique qu'est la protéine MoFe. Cette hypothèse se fonde sur la similitude des propriétés chimiques de ces deux éléments et sur le fait que ce genre de substitution a déjà été rapportée pour un autre métal, le tungstène, donnant lieu à la formation d'une nitrogénase inactive [7–9]. Nous avons voulu la vérifier en étudiant l'influence du technétium sur le métabolisme du molybdène et la distribution de ces deux éléments au niveau intracellulaire. La nitrogénase a été isolée de bactéries contaminées ou non en technétium et analysée pour sa teneur en ces deux éléments potentiellement antagonistes.

2. MATERIEL ET METHODES

Azotobacter chroococcum (souche K 5/1), un diazotrophe aérobe libre a été cultivé en milieu liquide décrit par Burk (in Newton *et al.* [10]), dont la teneur en fer a été réduite à 0·5 mg $litre^{-1}$ afin d'éviter l'apparition de précipités de phosphates ferro-calciques. Dans le but d'accentuer les interactions entre molybdène et technétium, le dixième de la quantité de molybdate préconisée a été effectivement apporté au milieu (soit 0·1 mg $litre^{-1}$). Les cultures ont été réalisées en flacons de 20 litres contenant 15 litres de milieu dont l'aération et l'agitation étaient obtenues par bullage d'air comprimé (6·5 litres min^{-1}); la température était de 28°C.

Le molybdène stable a été tracé par du ^{99}Mo (Na_2MoO_4) fourni par l'Ire. Le technétium-99 a été apporté sous forme de pertechnétate de sodium (NEN) à raison de 50 μg $litre^{-1}$ et tracé à l'aide de ^{95m}Tc ($KTcO_4$) produit au cyclotron de Louvain-la-Neuve.

Après croissance des cultures en présence ou en absence de technétium, les bactéries, toujours en phase linéaire de croissance, sont récoltées par centrifugation. Les culots de cellules sont lavés deux fois à l'aide d'un milieu de composition identique au milieu de culture mais exempt de traceurs radioactifs et remis en suspension dans un tampon tris-HCl 20 mM $litre^{-1}$ de pH 7·2. Les cellules sont ensuite broyées à l'aide d'une presse de French (Ribi-cell fractionator, Sorvall) travaillant sous une pression de 16 000 à 20 000 psi. Les débris de parois et les cellules encore intactes sont alors séparés de la fraction soluble par centrifugation (10 min × 20 000 g). L'extraction de la nitrogénase présente dans le surnageant en tant qu'enzyme cytosoluble est réalisée selon la méthode décrite par Burns et Hardy [11]. Après une centrifugation de 16 heures à 20 000 *g*, l'extrait cytosoluble laisse apparaître trois fractions: un culot et un surnageant formé d'une couche supérieure faiblement colorée représentant 5 à 10% du volume qui est éliminée et d'une fraction de coloration brun foncé qui est récupérée et définie comme l''extrait brut'. Cette dernière fraction est traitée au sulfate de protamine pour en précipiter les acides nucléiques qui sont éliminés par centrifugation. Le surnageant (PS I) est rapidement porté à une température de 60 °C et incubé à cette température durant 7 min; les protéines dénaturées par la chaleur sont ensuite éliminées par centrifugation. Les composés de la nitrogénase présente dans le surnageant (Δ60) sont précipités par un second traitement au sulfate de protamine, isolés par centrifugation et resolubilisés par addition de cellulose phosphate (Cellex-P, Bio-Rad) fournissant la fraction PS II. Cette fraction est chargée sur une colonne de DEAE-cellulose (DE 52 Whatman).

Après lavage de la colonne à l'aide d'un tampon tris-HCl 20 mM $litre^{-1}$ de pH 7·2 contenant 0·15M $litre^{-1}$ de NaCl, la protéine MoFe est éluée par le même tampon contenant cette fois 0·25M $litre^{-1}$ de NaCl (Ac 1).

La teneur en protéine des différentes fractions a été estimée par la méthode de Ehresmann *et al.* [12]. Les activités en ^{99}Mo et ^{95m}Tc ont été mesurées par scintillation solide (Autogamma 5220, Packard).

3. RESULTATS ET DISCUSSION

Dans nos conditions expérimentales et pour une concentration en molybdène dans le milieu égale à 0·1 mg $litre^{-1}$ (soit $\simeq 1\ \mu M\ litre^{-1}$), la présence de technétium (0·5 $\mu M\ litre^{-1}$) provoque une inhibition de 15 à 20% de l'activité nitrogénasique mesurée *in vivo* par le test à l'acétylène et réduit l'accumulation du molybdène chez la bactérie par rapport à celle observée dans le témoin: les facteurs de transfert pour le Mo calculés sur base du poids frais des cellules ($\mu g\ g^{-1}$ poids frais des cellules/$\mu g\ ml^{-1}$ solution nutritive) sont réduits d'environ 30% en présence de technétium dans le milieu et passent de 70·5 à 47·7. Ces résultats témoignent d'un antagonisme entre ces deux métaux au niveau de leur accumulation par les micro-organismes. L'accumulation du technétium est supérieure à celle du molybdène: son facteur de transfert estimé sur base du poids frais des organismes s'élève à 162·7.

Au niveau intracellulaire, ces deux éléments se retrouvent essentiellement dans la fraction cytosoluble: après broyage des cellules et centrifugation, plus de 90% des quantités accumulées de ces deux métaux sont présents dans le surnageant tandis que 5 à 6% du Mo et 8% du Tc sont associés au culot formé de débris de parois et de cellules ayant échappé au broyage.

Tableau I

Teneurs spécifiques en Mo et en Tc (en ng mg^{-1} de protéine) dans les différents extraits provenant de bactéries contaminées ou non en technétium

Traitement	*− Tc*	*+ Tc*	
	Mo	*Mo*	*Tc*
Extrait brut	62·0	42·6	60·1
PS I	70·4	49·7	64·5
Δ60	166·4	113·7	70·5
PS II	211·7	158·0	42·1
Ac 1	201·3	194·2	28·0

Le Tableau I ci-dessous présente, pour les extraits bactériens obtenus en présence ou en absence de technétium, les teneurs en ces deux éléments rapportées à la quantité de protéine dans les différentes fractions successives de la purification de la nitrogénase.

Qu'il s'agisse de cellules contaminées ou non en technétium, on observe au fur et à mesure des étapes de la purification de la nitrogénase, une augmentation de la teneur spécifique en molybdène dans les différentes fractions. De plus, partant de teneurs spécifiques différent de près de 30% dans l'extrait brut, selon le traitement, reflétant en cela les différences constatées pour l'accumulation, les teneurs spécifiques en Mo des protéines MoFe isolées de bactéries contaminées ou non en technétium sont sensiblement identiques. Une certaine augmentation de la teneur spécifique en technétium s'observe également au cours des premières étapes, quoique nettement inférieure à celle obtenue pour le molybdène; par contre, pour ce qui est des fractions provenant des dernières étapes, au sein desquelles la nitrogénase représente une part de plus en plus importante de la quantité des protéines, la teneur spécifique en technétium se voit fortement réduite. La nitrogénase, et la protéine MoFe en particulier, ne semble donc témoigner que d'une faible affinité pour la fixation de cet élément, à moins que la liaison du technétium à cette protéine soit plus labile que la liaison du molybdène et qu'il se décroche de ses sites de fixation durant la purification.

4. CONCLUSIONS

Si le technétium provoque une inhibition de l'activité de la nitrogénase et de l'accumulation du molybdène dans les cellules, il n'affecte pas la teneur en cofacteur métallique Mo de la protéine MoFe de la nitrogénase. De plus, son affinité pour la protéine MoFe semble être inférieure à celle qu'il présente pour les composés protéiques en général. Ceci nous amène à conclure qu'il est peu probable que l'inhibition de la nitrogénase par le technétium résulte d'une substitution du molybdène par ce radiopolluant.

RÉFÉRENCES

1. Gearing, P. J., Van Baalen, C. and Parker, P. L. (1971). In: *Radionuclides in Ecosystems* (Ed. D. J. Nelson), Proc. 3rd Natl Symp. Radioecology, Vol. 2, Oak Ridge, pp. 857–65.
2. Gearing, P. J., Van Baalen, C. and Parker, P. L. (1975). *Plant Physiol.*, **55**, 240–6.
3. Fisher, N. S. (1982). *Envir. Sci. Technol.*, **16**, 579–81.
4. Dartevelle-Moureau, Z., Jakubowski-Mania, B., Oleksy-Gebski, M.-J., Bossus, A., Vandecasteele, C. M., Cogneau, M., Bonotto, S. and Van der Ben, D. Workshop on the *Role of microflora on long-term availability of radionuclides in marine ecosystems*, Santa Teresa (I), April 27–29, 1983 (in press).
5. Vandecasteele, C. M., Delmotte, A., Roucoux, P. and Van Hove, C. (1982). In: *Environmental Migration of Long-lived Radionuclides*, Proc. Int. Symp. on the Migration in the terrestrial environment of long-lived radionuclides from the nuclear fuel cycle, IAEA-SM-257/76, pp. 275–86.
6. Vandecasteele, C. M., Delmotte, A., Henrot, J., Van Hove, C. and Cogneau, M. (1984). In: *Role of Microorganisms on the Behaviour of Radionuclides in Aquatic and Terrestrial Systems and their Transfer to Man*, Proc. of the workshop (Ed. E. Bonnijns-Van Gelder and R. Kirchmann), pp. 158–72.

7. Benemann, J. R., Smith, G. M., Kostel, P. J. and McKenna, C. E. (1973). *FEBS Lett.*, **29**(3), 219–21.
8. Bishop, P. E., Jarlenski, D. M. L. and Hetherington, D. R. (1980). *Proc. Natl Acad. Sci. USA*, **77**(12), 7342–6.
9. Nagatani, H. H. and Brill, W. J. (1974). *Biochem. Biophys. Acta*, **362**, 160–6.
10. Newton, J. W., Wilson, P. W. and Burris, R. H. (1953). *J. Biol. Chem.*, **204**, 445–51.
11. Burns, R. C. and Hardy, R. W. F. (1972). In: *Methods in Enzymology* (Ed. A. San Pietro), Vol. XXIVb. London, Academic Press, pp. 480–96.
12. Ehresmann, B., Imbault, P. and Weil, J. A. (1973). *Anal. Biochem.*, **54**, 454–63.

34

Toxicity of Long-term Application of Dietary Technetium to Rats and their Offspring

R. VAN BRUWAENE,* M. HEGELA, G. B. GERBER, R. KIRCHMANN and J. R. MAISIN

Department of Radiobiology, Centre d'Etude de l'Energie Nucléaire, B-2400 Mol, Belgium

1. INTRODUCTION

The low specific activity of technetium-99 and its potential accumulation in the environment make it possible that technetium could exert toxic effects by means of its radiation as well as by means of its chemical properties. Indeed, the toxic properties of technetium to plants are well documented [1] and could result from the interference of technetium with enzymes requiring molybdenum as trace element. Little information concerning the toxicity of technetium to mammals is however available. Most likely such toxic effects would appear in organs in which technetium is concentrated, as in the thyroid [2] and/or might affect in utero or postnatal development. We have therefore studied the effects of different doses of technetium given in food to rats during pregnancy and thereafter to the mothers and their offspring. Thyroid function was tested on the basis of uptake of iodine, T3, T4 and TSH serum concentrations and thyroid histology. The weights of the total body and of a variety of organs of the young rats were also measured, blood cells were counted and organ uptake of technetium was determined. Fertility of rats exposed for long periods of time to technetium was assayed in a preliminary way.

2. MATERIALS AND METHODS

Adult 3 months old female Wistar rats received pertechnetate Tc-99 in their food at various doses ranging from 0·5 μg to 50 μg per g food.

* Deceased 27 August 1984.

Ammonium pertechnetate-99 obtained from NEN was mixed to finely ground rat food (Hope Farms N.V., iodine content 0·52 μg/g), water was added to obtain a paste which was then dried and cut into portions. The doses of technetium and times of applications are indicated in the tables. Technetium addition had no measurable influence on food consumption which was of the order of 17 g/day for young adult males and of 13 g/day for females. Technetium feeding was initiated after successful mating as indicated by vaginal smears. Feeding was continued after birth and the offspring were sacrificed 7 or 17 weeks after starting technetium feeding, i.e. at an age of 4 or 14 weeks.

Iodine uptake by the thyroid was measured 24 h following injection of a tracer dose of I-131 and expressed as percentage of iodine injected taken up by the thyroid. The thyroid hormones T3 and T4 were determined in rat serum by radioimmuno-assay using a commercially available kit from the IRE (Institut National des Radioelements, Fleurus, Belgium). The data are expressed in ng of hormone per ml serum. TSH was assayed by using a specific rat anti-TSH antiserum in the laboratory of Professor J. E. Dumont, Brussels. These experiments are not yet completed and will form part of a later publication.

At sacrifice, the weight of the body and of organs was recorded. Blood cells were counted by conventional methods. Technetium-99 concentration was measured by liquid scintillation spectrometry. Thyroid samples were fixed in Carnoy's solution, embedded in paraffin, sectioned and stained with haematoxyline-eosine. A few thyroid samples obtained from rats given the highest Tc doses were also studied by autoradiography.

3. RESULTS

Technetium appears not to be very toxic to adult and developing rats. Body and organ weights are not significantly altered at an age of 4 or 24 weeks even at doses as high as 15 μg of technetium-99/g food. Moreover, no changes in the blood formula could be discerned. Technetium feeding affects, however, thyroid function to a certain degree. Analysis of serum hormones indicates no changes in T3 and T4 even at doses of 50 μg Tc/g food in the adult mothers treated for 26 weeks (Table I). A significant reduction in T3 is noted only at an age of 17 weeks in the female offspring given 10 μg/g. T4 decreases significantly at an age of 17 weeks, especially after high Tc doses. Preliminary experiments indicate that serum TSH was significantly increased in mothers as well as in the offspring. Twenty-

Table I
T3 and T4 concentrations in serum (ng/ml) after feeding different amounts of technetium-99 for various lengths of time

	Age (weeks)	*T3*			*T4*		
		Control	*10 μg/g*	*50 μg/g*	*Control*	*10 μg/g*	*50 μg/g*
Mothers	20	1·17 ± 0·18	1·87 ± 0·38	1·11 ± 0·17	44·0 ± 9·0	54·0 ± 9·0	36·0 ± 1·0
	29	0·90 ± 0·09	1·00 ± 0·06	1·28 ± 0·24	36·0 ± 4·0	41·0 ± 1·6	33·7 ± 1·7
Offspring	12	0·94 ± 0·12	1·38 ± 0·06		78·0 ± 12	64·0 ± 8·0	
(females)	17	2·28 ± 0·25	1·34 ± 0·18		60·3 ± 3·7	32·7 ± 1·2	
	20	1·87 ± 0·32	1·88 ± 0·30	1·38 ± 0·15	50·4 ± 5·7	46·4 ± 7·2	34·0 ± 4·0
Offspring	12	1·38 ± 0·05	1·29 ± 0·19	1·15	59·0 ± 1·0	72·0 ± 12·0	31·7
(males)	17	2·11 ± 0·16	1·80 ± 0·03	1·47 ± 0·50	70·7 ± 5·7	46·0 ± 0·6	28·7 ± 3·3
	20	1·93 ± 0·58	2·22 ± 0·32	1·52 ± 0·40	67·5 ± 2·5	50·0 ± 4·0	42·0 ± 3·1

four-hour uptake of iodine by adult controls is about 18% of the dose in females and somewhat lower in males. Treatment with 10 or 50 μg/g of technetium reduces significantly iodine uptake in the mothers receiving 10 and 50 μg/g Tc and in the male offspring receiving 50 μg/g Tc (Table II).

No characteristic histological changes were seen in the thyroid of the treated rats. The acini had almost normal size and appearance, and no necrotic areas were found. The number of pyknotic nuclei was slightly increased, and the cells showed a greater variability in size.

After feeding technetium, the rat thyroids accumulated large amounts of Tc-99 (about 50 μg/g after a dose of 15 μg/g food) (Table III), i.e. in the order of 10% of their iodine content. In agreement with the data on sheep and earlier data on rats [4, 5], large amounts of technetium were also found in kidney followed by the other parenchymal organs. Muscle and brain contain only little technetium. Comparison of the data after 10 and 20 weeks of feeding indicates that already after 10 weeks, equilibrium concentrations have been attained in most organs.

Table II
Twenty-four-hour I-131 uptake in percentage of dose injected after feeding different amounts of technetium-99

	Age (weeks)	*Control*	*10 μg/g*	*50 μg/g*
Mothers	29	18·0 ± 1·5	13·1 ± 0·5	4·2 ± 1·0
Offspring (females)	17	18·2 ± 1·3	20·1 ± 1·3	
Offspring (males)	17	11·6 ± 1·3	7·6 ± 1·3	1·9 ± 0·1

Table III
Technetium content (pCi/g) in organs of offspring after feeding different amounts of technetium-99

Organ	*Age (weeks)*	*0·5 μg/g*	*5 μg/g*	*10 μg/g*	*15 μg/g*	*50 μg/g*
Kidney	7	2 744 ± 14	14 500 ± 1 080	24 847 ± 1 600	19 496 ± 2 756	50 131 ± 6 321
	17	2 199 ± 23	13 970 ± 187	122 892 ± 596	31 041 ± 920	52 150 ± 919
Liver	7	345 ± 80	3 552 ± 400	7 827 ± 293	11 251 ± 128	16 905 ± 1 027
	17	292 ± 28	4 048 ± 383	8 147 ± 212	12 000 ± 1 664	19 150 ± 650
Pancreas	7	62 ± 50	942 ± 125	3 092 ± 724	2 994 ± 390	11 228 ± 47
	17	N.D.	1 710 ± 350	2 473 ± 433	4 171 ± 286	20 212 ± 4 311
Spleen	7	225 ± 108	1 799 ± 257	4 515 ± 695	5 017 ± 268	11 979 ± 1 462
	17	N.D.	2 782 ± 46	4 737 ± 80	6 168 ± 95	24 333 ± 722
Thyroid	7	161 453	—	515 611 ± 12 300	355 054	—
	17	—	—	—	826 940	—
Brain	7	N.D.	125 ± 7	343 ± 72	518 ± 86	2 383 ± 523
	17	N.D.	279 ± 121	416 ± 77	624 ± 16	3 200 ± 701
Muscle	7	N.D.	800 ± 12	742 ± 101	1 377 ± 201	4 146 ± 418
	17	N.D.	934 ± 51	1 463 ± 415	1 746 ± 10	3 767 ± 567
Lung	7	200 ± 50	—	2 933 ± 103	5 794 ± 602	11 160 ± 367
	17	151 ± 19	2 423 ± 163	4 987 ± 737	8 023 ± 859	16 530 ± 1 500

N.D.: Non-detectable.

Table IV
Fertility of female rats fed technetium-99 for various lengths of time

Br.	*F.*	*Control*			*10 μg/g*			*50 μg/g*		
		R	*F*	*M*	*R*	*F*	*M*	*R*	*F*	*M*
1st	3	6/6	11	9	4/6	3	8	5/6	4	9
2nd	14	5/6	10	12	3/6	11	8	2/6	0	1
3rd	17	5/6	20	11	5/6	20	7	5/6	0	5

R = ratio of females pregnant to impregnated; M = male offspring; F = female offspring; Br. = breeding; F = feeding weeks.

Fertility was assessed in a preliminary experiment after long term feeding of technetium. A significant reduction in litter size and/or number of pregnant females was seen only for the second and third breeding after a dose of 50 μg/g technetium (Table IV). Preliminary experiments on male rats also suggest that technetium feeding (50 μg/g) affects fertility in males.

4. DISCUSSION AND CONCLUSIONS

All in all, the data indicate that technetium is not very toxic to mammals. Changes in thyroid function are noted only at technetium concentrations which are in the order of several percent of the iodine content. It should however be pointed out that the food given to the rat contained relatively large amounts of iodine. Toxicity thresholds may be lower under conditions of iodine deficiency. A reduced fertility is also observed only after very high technetium doses and may indeed be mediated via an effect on the thyroid. Studies in rats relating technetium toxicity to iodine deficiency are now under way and are to be completed by corresponding studies on Tc metabolism in sheep under different dietary regimens [4]. Nevertheless, the data presented make it unlikely that technetium concentrations in the environment could ever attain levels at which they might become toxic to mammals or to man. The present data do not allow a decision whether the effects observed are due to chemical or the radiological toxicity of technetium. The dose rate in the thyroid for the highest dose is about 0·12 Gy/day at equilibrium, i.e. slightly more than 10 Gy total dose during the 14 weeks of postnatal treatment and somewhat more for the mothers. This seems to be too small to account for effects to the thyroid but this aspect requires more investigation if possible by using the longer lived technetium-98 of still lower specific activity.

ACKNOWLEDGEMENTS

This work was supported by contract BIO-D-378-81-B of the Radiation Protection Programme of the Commission of the European Communities, publication No. 2118. We are grateful to Mme M. Lambiet-Collier for preparing the histological sections.

REFERENCES

1. Wildung, R. E., Garland, T. R. and Cataldo, D. A. (1977). Accumulation of technetium by plants. *Health Physics*, **32**, 314–17.
2. Gast, R. G., Thorvig, L. H., Landa, E. R. and Gallaher, K. J. (1978). Technetium toxicity to plants and sorption by soils. In: *Environmental Chemistry and Cycling Processes* (Ed. D. C. Adriano and I. L. Brisbin Jr), DOE Sym. Series 45, CONF 760429, pp. 550–67.
3. Carmody, R. and Highman, J. H. (1975). Uptake of pertechnetate by mammary tissue and thyroid after pregnancy. *Brit. J. Radiol.*, **48**, 63–70.
4. Wolff, J. and Maurey, J. R. (1962). Thyroidal iodide transport. III. Comparison of iodide with anions of periodic group VIIA. *Biochim. Biophys. Acta*, **57**, 422–6.

5. Van Bruwaene, R., Gerber, G. B., Kirchmann, R., Vankerkom, J., Colard, J. and Cogneau, M. (1983). Technetium metabolism in sheep. *Ann. Assoc. Belge Radioprot.*, **8**, 244–8.
6. Van Bruwaene, R., Gerber, G. B., Kirchmann, R. and Vankerkom, J. (1983). Technetium (Tc-95m) metabolism in sheep after an intravenous injection. Proc. 7th ICRR, Review and summaries, pp. 603–11.
7. Van Bruwaene, R., Gerber, G. B. and Kirchmann, R. (1980). Technetium transfer through the food chain: an analysis of present knowledge and of perspectives for future research. *Ann. Assoc. Belge Radioprot.*, **5**, 25–33.
8. Beasley, T. M., Palmer, H. E. and Nelp, W. B. (1966). Distribution and excretion of technetium in humans. *Health Physics.*, **12**, 1425–35.
9. Hays, M. T. (1973). Pertechnetate transport in man: Absorption after subcutaneous and oral administration; secretion into saliva and gastric juice. *J. Nucl. Med.*, **133**, 331–5.

Summary of the Oral Presentations

Among the different steps of a nuclear fuel cycle both uranium enrichment plants and reprocessing plants have been calculated to be the main potential sources of release. The largest anticipated release source of ^{99}Tc is via the liquid effluent at the U enrichment facilities. However, not enough data are available in this case to calculate the ratio of the amount released to the amount of ^{99}Tc coming into the nuclear enrichment plant (confinement factor).

Smaller confinement factors have been calculated for reprocessing plants giving smaller releases. In the European situation small ^{99}Tc releases are mostly from reprocessing plants. Use of the bioindicators *Patella* (a mussel) and *Fucus* (a brown alga) showed the presence of ^{99}Tc in samples taken in the English Channel and the Arctic waters.

In the session dealing with the chemical behaviour of technetium in different ecosystems and its relation to transfer and transfer coefficients, Tc sorption in many soils has been shown to depend on various parameters such as organic carbon content, total nitrogen, extractable iron, extractable aluminium, particle size distribution and pH. In aerobic conditions this sorption is low in general; it is considerably enhanced when anoxic conditions occur. Such anoxic conditions may even exist in regularly aerated soils in so-called microsites. Micro-organisms have been repeatedly identified as playing a crucial role, although the precise mechanism was not elucidated. After reduction of TcO_4^- to Tc(IV) a fairly good stabilisation in this chemical state can be brought about by humic acids, preferably of low molecular weight. However, at sediment–water interfaces of organic-rich seawater sediments, and particularly in anoxic basins, some Tc removal to underlying sediments could occur. On the other hand, in oxygenated seawater binding of pertechnetate with inorganic particulate matter is slight.

In the session on the impact of the binding of technetium to macromolecules on their transfer in the food chain, much evidence was presented demonstrating the large accumulation of Tc by brown algae compared to green and red algae. This phenomenon makes brown algae attractive for use as bioindicators for the distribution of Tc in the oceanic environment except at high latitudes. Also, TcO_4^- absorbed and translocated in plant tissues was shown to be reduced, especially in chlorophyllous tissues. Many different ligands are apparently able to bind the reduced Tc, as was shown by chromatographic analysis. Although many bioorganic molecules have been suggested as candidates for binding the reduced Tc, hardly any qualitative description of their chemical identity has been given. Comparisons between Tc and elements like S and Se have been tentatively made. There was no substantial evidence, however, that their behaviour is comparable in plants. The original idea that in animals Tc may act like radioiodine has been seriously challenged. There were conflicting results on whether the uptake in the bloodstream after gastrointestinal ingestion is influenced by the chemical form of technetium or not.

From a point of view of uncertainties in the predictive models, better information on the long-term behaviour of ^{99}Tc and its transfer to edible plant parts should be established. Since in a first approach eggs seemed to be a quantitatively important step in the food chain, the study of the transfer of ^{99}Tc into vegetation potentially consumed by poultry, with subsequent transfer into eggs, should be given a certain priority. In the actual models moreover, some, perhaps less obvious, exposure pathways are not included, prompting a future effort to account for some pathways such as ^{99}Tc cycling via transfer from contaminated groundwater into the tissues of deep-rooted vegetation and the subsequent deposition and accumulation of ^{99}Tc on the soil surface by plant litter decomposition.

Summaries of the Discussions of the Working Groups

WORKING GROUP ON PROBLEMS OF SOURCES AND DISTRIBUTION OF MAN-MADE TECHNETIUM AND ITS RADIOLOGICAL CONSEQUENCES

Chairman
A. AARKROG

The group identified the following topics which need developing:

1. There exists no or very little information
   (a) on liquid and gaseous releases from waste treatment or from vitrification plants in earlier years;
   (b) on possible releases in the Eastern countries.
2. The models which have been used do not consider the binding of Tc to organic matter or, for the aquatic environment, the binding to the sediments.
3. The ranges given in the health assessments are based on judgement rather than on the real variation normally occurring in the environment.
4. Little is known about the behaviour of Tc in freshwater systems.

The group identified the studies which are needed:

1. Realistic long-term uptake experiments in plants, e.g. using the fallout ^{99}Tc.
2. Uptake experiments of Tc in vegetables and fruits from irrigation.

3. Collection of concentration factors for fish in the freshwater and marine environment.
4. The chemical form of the Tc discharged from various installations needs investigation.
5. For doses to man: Individual and collective doses from present discharges ought to be known.

The group drew the conclusions that:

1. The main uncertainty lies in the source term (magnitude of discharges and chemical form).
2. More field studies are needed.
3. Cheap analytical procedures for low-level ^{99}Tc analysis should be applied.
4. The reliability of the analysis should be enhanced by intercalibration exercises.
5. Vegetables and very probably eggs are important pathways to man.
6. Technetium-99 is an excellent long-distance tracer in the marine environment.

WORKING GROUP ON THE CHEMICAL BEHAVIOUR

Chairman
M. SHEPPARD

The basic facts revealed to be predominant were:

1. The reactions between technetium and the soil sediments depend on pH, E_h, organic matter, presence of oxides, sulphides and microbes in the soil and its previous history.
2. The reduction in soil is slow.
3. The re-oxidation of reduced complexes is probably a mixture of first-order reactions.
4. Technetium in soil will probably become less available with time due to a shift to less readily oxidisable forms and eventually become bound or occluded into organic, oxide or sulphide forms.

The group identified the further studies which are needed on:

1. The aging of soil, cycling and recycling of technetium, both vertically and laterally.

2. The rate of re-oxidation under realistic conditions, including sinks due to local reduction in, for example, micropores.
3. The reaction products of Tc reduction and speciation of reduced Tc.
4. The rate of conversion of Tc to complexed forms.

The group drew some main conclusions:

1. The basic soil chemistry for Tc is not well known. From a methodological point of view attention should be given to
   (a) standard methods for soil extraction and the preparation of the plant and soil for analysis;
   (b) the interference of other ions like molybdenum in ^{95m}Tc analysis and ruthenium in ^{99}Tc analysis;
   (c) the need for both a consistent expression and a test for the REDOX system.

   In connection with these points, the group recommended investigation of whether it would be possible to reproduce the aging of a natural soil system in the laboratory and determine the long-term available pool of Tc.
2. For the Working Group, it became obvious that one needs to know more about 'soil aging' of Tc, cycling and recycling, and speciation with time to reliably predict the long-term fate and behaviour of Tc.

WORKING GROUP ON THE BIOCHEMICAL REACTIONS AND BINDING WITH MACROMOLECULES IN VARIOUS LIFE SYSTEMS

Chairman
J. PIERI

The group identified further studies which are needed on:

1. The distribution of Tc in the marine ecosystem arising from different sources of contamination (fallout, nuclear power and reprocessing plant releases). Observations 'in the field' should be parallelled by laboratory research.
2. The different chemical forms (TcO_4^- and its reduction to Tc(IV) or TcO_2) and their behaviour in the environment.
3. The role of micro-organisms at the water–sediment interface.

4. The role of the estuaries and coastal waters, where sediments may impinge on technetium transfer.
5. Whether toxicity of Tc could possibly be included in health assessment models.

The group stressed the necessity to improve the analytical methods for the measurement of Tc in the different compartments of the environment.

GENERAL CONCLUSIONS

Chairman
A. CIGNA

Four main points have been raised concerning:

1. Sources: Information on releases from enrichment plants and reprocessing plants became available only recently. For the future, data should be collected on possible releases from waste repositories.
2. Levels of technetium in the environment: It was suggested that simple standardised methods for Tc measurements should be utilised. A fallout sample file should be set up especially for the terrestrial environment.
3. Setting up of models of the Tc cycle: Efforts to validate them should be initiated.
4. From a point of view of radiation protection it became fully obvious to the seminar's participants that for technetium one should consider

   (a) effective dose equivalents for critical groups;
   (b) collective dose commitments for the general population;
   (c) dosimetric evaluation.

5. The discussions revealed that research outlined as follows will be needed in the future:

   (a) Research of those ecological compartments highlighted by sensitivity analysis.
   (b) Research of the mechanisms responsible for the high concentration factors to explain the differences between field and laboratory results.
   (c) Some attenuation should be paid to the influence of climate on the Tc accumulation in the biosphere.
   (d) Possible important technetium reservoirs should be located.

List of Participants

The seminar was attended by the following scientists:

AARKROG, A., Risø National Laboratory, Roskilde; APROSI, G., EDF, Chatou; BALLESTRA, S., IAEA, Oceanographic Museum, Monaco; BEASLEY, T., Oregon State University, Newport, Oregon; BENCO, C. (Miss), ENEA-CREA, S. Teresa; BLAYLOCK, B. G., Oak Ridge National Laboratory, Oak Ridge; BOHN-BLUM, B. E., KFZ, Karlsruhe; BOJANOWSKI, R., IAEA, Oceanographic Museum, Monaco; BONOTTO, S., CEN-SCK, Mol; CAMUS, H., CEA-CEN, Cadarache; CANNARSA, S., ENEA, S. Teresa, La Spezia; CARLSEN, L., Risø National Laboratory, Roskilde; CHIAPPINI, R., CEA-CEN, Bruyere-le-Chatel; CIGNA, A., ENEA, S. Teresa; CREMERS, A., KUL, Leuven; DECOBERT, V., SGN, Filiale COGEMA, St Quentin Yvelines; DESMET, G., CEC, Brussels; FOULQUIER, L., CEA-CEN, Cadarache; GARCÍA-LEÓN, M., Universidad de Sevilla; GARTEN, C. T. JR, Oak Ridge National Laboratory; GRAUBY, A., CEA-CEN, Cadarache; GREEN, N., NRPB, Chilton; GUETAT, P., CEA-CEN, Fontenay-aux-Roses; HANKE, C., Technische Hochschule, Darmstadt; HEGELA MOSTAPHA, K., SCK/CEN, Mol; HOFFMAN, O., Oak Ridge National Laboratory; HOLM, G. E. G., IAEA, Musée Océanographique, Monaco; HUGON, L., CEA-CEN, Cadarache; ITSCHERT, A., UCL, Louvain-la-Neuve; KIRCHMANN, R., CEN-SCK, Mol; LAUDELOUT, H., UCL, Louvain-la-Neuve; LEMBRECHTS, J. F. M. M., Stichting ITAL, Wageningen; LIESER, K. H., Technische Hochschule, Darmstadt; LUYKX, F., CEC, Luxembourg; MAES, A., KUL, Leuven; MASSON, M., CEA-CEN, Cherbourg; MAZZI, U., ICTR/CNR, Padova; MYTTENAERE, C., CCE, Bruxelles; PAAKKOLA, O.,

Finnish Centre for Radiation and Nuclear Safety, Helsinki; PATTI, F., CEA-CEN, Fontenay-aux-Roses; PIERI, J., Université de Nantes; PROEHL, G., GSF, Neuherberg; QUINAULT, J. M., CEA-CEN, Cadarache; RANCON, D., CEA-CEN, Cadarache; REBOUL, J., CEA-CEN, Saclay; REFOSCO, F., CNR-ICTR, Padova; RIOSECO, J., Lunds Universitet; ROUCOUX, P., CEA-CEN, Cadarache; SAAS, A., CEA-CEN, Cadarache; SCHULTE, E. H., ENEA-CREA, S. Teresa; SCOPPA, P., ENEA-CREA, S. Teresa; SHEPPARD, M. I., AECL, Whiteshell Nuclear Research Est.; SHEPPARD, S., Whiteshell Nuclear Research Est.; SJOBLOM, K. L., Finnish Centre for Radiation and Nuclear Safety, Helsinki; SMITH, H., NRPB, Chilton; STALMANS, M., KUL, Leuven; STEFANOVIC, D., Inst. des Sciences Nucléaires B. Kidric, Belgrade; THOMAS, J., Battelle Pacific Northwest Lab., Richland; TILL, J., Radiological Assessments Corp., Neeses; TISATO, F., CNR, Padova; VANDECASTEELE, C. M., UCL, Louvain-la-Neuve; VAN DER BEN, D., IRSNB, Bruxelles; VAN LOON, L., ITAL, Wageningen; VERGNAUD, G., COGEMA Spr., Cherbourg; VERTHÉ, C., Université de Liège; WATTERS, R., US Dept. of Energy, Washington; WILDUNG, R., Battelle Pacific Northwest Lab., Richland; ZATTERA, A., ENEA-CRE, S. Teresa.

Index

GPSR Compliance

The European Union's (EU) General Product Safety Regulation (GPSR) is a set of rules that requires consumer products to be safe and our obligations to ensure this.

If you have any concerns about our products, you can contact us on ProductSafety@springernature.com

In case Publisher is established outside the EU, the EU authorized representative is:

Springer Nature Customer Service Center GmbH
Europaplatz 3
69115 Heidelberg, Germany

Batch number: 10607212

Printed by Printforce, the Netherlands